Lecture Notes in Computer Science   10609

Commenced Publication in 1973
Founding and Former Series Editors:
Gerhard Goos, Juris Hartmanis, and Jan van Leeuwen

More information about this series at http://www.springer.com/series/7409

Christian Beecks · Felix Borutta
Peer Kröger · Thomas Seidl (Eds.)

Similarity Search and Applications

10th International Conference, SISAP 2017
Munich, Germany, October 4–6, 2017
Proceedings

 Springer

Editors
Christian Beecks
Fraunhofer Institute
  for Applied Information Technology
Sankt Augustin
Germany

Felix Borutta
Ludwig-Maximilians-Universität München
Munich
Germany

Peer Kröger
Ludwig-Maximilians-Universität München
Munich
Germany

Thomas Seidl
Ludwig-Maximilians-Universität München
Munich
Germany

ISSN 0302-9743   ISSN 1611-3349 (electronic)
Lecture Notes in Computer Science
ISBN 978-3-319-68473-4   ISBN 978-3-319-68474-1 (eBook)
DOI 10.1007/978-3-319-68474-1

Library of Congress Control Number: 2017954912

LNCS Sublibrary: SL3 – Information Systems and Applications, incl. Internet/Web, and HCI

Printed on acid-free paper

This Springer imprint is published by Springer Nature
The registered company is Springer International Publishing AG
The registered company address is: Gewerbestrasse 11, 6330 Cham, Switzerland

Preface

This volume contains the papers presented at the 10th International Conference on Similarity Search and Applications (SISAP 2017) held in Munich, Germany, during October 4–6, 2017. SISAP is an annual forum for researchers and application developers in the area of similarity data management. It focuses on the technological problems shared by numerous application domains, such as data mining, information retrieval, multimedia, computer vision, pattern recognition, computational biology, geography, biometrics, machine learning, and many others that make use of similarity search as a necessary supporting service.

From its roots as a regional workshop in metric indexing, SISAP has expanded to become the only international conference entirely devoted to the issues surrounding the theory, design, analysis, practice, and application of content-based and feature-based similarity search. The SISAP initiative has also created a repository (http://www.sisap. org/) serving the similarity search community, for the exchange of examples of real-world applications, source code for similarity indexes, and experimental test beds and benchmark data sets.

The call for papers welcomed full papers, short papers, as well as demonstration papers, with all manuscripts presenting previously unpublished research contributions. At SISAP 2017, all contributions were presented both orally and in a poster session, which facilitated fruitful exchanges between the participants.

We received 53 submissions from authors based in 17 different countries. The Program Committee (PC) was composed of 43 international members. Reviews were thoroughly discussed by the chairs and PC members: Each submission received at least three reviews. Based on these reviews as well as the subsequent discussions among PC members, the PC chairs accepted 23 full papers to be included in the conference program and the proceedings.

The proceedings of SISAP are published by Springer as a volume in the *Lecture Notes in Computer Science* (LNCS) series. For SISAP 2017, as in previous years, extended versions of five selected excellent papers were invited for publication in a special issue of the journal *Information Systems*. The conference also conferred a Best Paper Award, as judged by the PC co-chairs and Steering Committee.

Beside the presentations of the accepted papers, the conference program featured three keynote presentations from exceptionally skilled scientists: Christian S. Jensen, from Aalborg University, Denmark, Reinhard Förtsch, from the Deutsches Archäologisches Institut, Germany, and Cyrus Shahabi, from University of Southern California, USA.

We would like to thank all the authors who submitted papers to SISAP 2017. We would also like to thank all members of the PC and the external reviewers for their effort and contribution to the conference. We want to express our gratitude to the members of the Organizing Committee for the enormous amount of work they did.

We also thank our sponsors and supporters for their generosity. All the submission, reviewing, and proceedings generation processes were carried out through the EasyChair platform.

August 2017

Christian Beecks
Felix Borutta
Peer Kröger
Thomas Seidl

Organization

General Chairs

Peer Kröger   Ludwig-Maximilians-Universität München, Germany
Thomas Seidl   Ludwig-Maximilians-Universität München, Germany

Program Committee Chairs

Christian Beecks   Fraunhofer Institute for Applied Information
                                Technology FIT, Germany
Peer Kröger   Ludwig-Maximilians-Universität München, Germany

Program Committee

Giuseppe Amato	ISTI-CNR, Italy
Laurent Amsaleg	CNRS-IRISA, France
Hiroki Arimura	Hokkaido University, Japan
Benjamin Bustos	University of Chile, Chile
Selçuk Candan	Arizona State University, USA
Lei Chen	Hong Kong University of Science and Technology, Hong Kong, SAR China
Edgar Chávez	CICESE, Mexico
Paolo Ciaccia	University of Bologna, Italy
Richard Connor	University of Strathclyde, UK
Michel Crucianu	CNAM, France
Bin Cui	Peking University, China
Andrea Esuli	ISTI-CNR, Italy
Fabrizio Falchi	ISTI-CNR, Italy
Claudio Gennaro	ISTI-CNR, Italy
Michael Gertz	Heidelberg University, Germany
Magnus Hetland	NTNU, Norway
Michael E. Houle	National Institute of Informatics, Japan
Yoshiharu Ishikawa	Nagoya University, Japan
Björn Þór Jónsson	IT University of Copenhagen, Denmark
Guoliang Li	Tsinghua University, China
Jakub Lokoč	Charles University in Prague, Czech Republic
Rui Mao	Shenzhen University, China
Stéphane Marchand-Maillet	Viper Group – University of Geneva, Switzerland
Henning Müller	HES-SO, Switzerland
Beng Chin Ooi	National University of Singapore, Singapore
Vincent Oria	New Jersey Institute of Technology, USA
Deepak P.	Queen's University Belfast, Ireland

Rasmus Pagh	IT University of Copenhagen, Denmark
Marco Patella	DISI – University of Bologna, Italy
Oscar Pedreira	Universidade da Coruña, Spain
Miloš Radovanović	University of Novi Sad, Serbia
Kunihika Sadakane	The University of Tokyo, Japan
Shin'Ichi Satoh	National Institute of Informatics, Japan
Erich Schubert	Heidelberg University, Germany
Yasin Silva	Arizona State University, USA
Matthew Skala	North Coast Synthesis Ltd., Canada
Nenad Tomašev	Google DeepMind, UK
Agma Traina	University of São Paulo at São Carlos, Brazil
Takashi Washio	ISIR, Osaka University, Japan
Pavel Zezula	Masaryk University, Czech Republic
Zhi-Hua Zhou	Nanjing University, China
Arthur Zimek	University of Southern Denmark, Denmark
Andreas Züfle	George Mason University, USA

Contents

Approximate Similarity Search

The Power of Distance Distributions: Cost Models and Scheduling Policies for Quality-Controlled Similarity Queries

Paolo Ciaccia and Marco Patella[✉]

DISI, University of Bologna, Bologna, Italy
{paolo.ciaccia,marco.patella}@unibo.it

Abstract. Approximate similarity queries are a practical way to obtain good, yet suboptimal, results from large data sets without having to pay high execution costs. In this paper we analyze the problem of understanding how the strategy for searching through an index tree, also called *scheduling policy*, can influence costs. We consider quality-controlled similarity queries, in which the user sets a quality (distance) threshold θ and the system halts as soon as it finds k objects in the data set at distance $\leq \theta$ from the query object. After providing experimental evidence that the scheduling policy might indeed have a high impact on paid costs, we characterize the policies' behavior through an analytical cost model, in which a major role is played by parameterized local distance distributions. Such distributions are also the key to derive new scheduling policies, which we show to be optimal in a simplified, yet relevant, scenario.

1 Introduction

The challenge of any approximate query processing (AQP) technique is to yield the best possible result given a certain amount of resources (CPU time, disk accesses, etc.) to be spent or, conversely, minimize the number of used resources for a given quality level.

A fundamental observation which is relevant for the purpose of this paper is that good (possibly optimal) strategies for delivering exact results are not necessarily good (optimal) when coming to consider AQP. For instance, classical join algorithms (e.g., nested loops) used in relational databases perform poorly for approximate aggregate queries, for which ad-hoc algorithms, such as ripple joins [HH99], have been developed.

Based on above observation, in this paper we study the problem of search strategies for approximate similarity queries. As also acknowledged in [PC09, CP10], most of the existing approaches to approximating the resolution of similarity queries have concentrated on aggressive pruning strategies, underestimating the problem of early delivering good results. Notice that, for the case of index-based query processing, this translates to consider the order under which index nodes are visited (which we call *scheduling policy*) as a first-class citizen.

© Springer International Publishing AG 2017
C. Beecks et al. (Eds.): SISAP 2017, LNCS 10609, pp. 3–16, 2017.
DOI: 10.1007/978-3-319-68474-1_1

Indeed, albeit the so-called MINDIST scheduling policy is known to be optimal for k-NN queries [BB+97, HS03], its performance for approximate queries has been proven inferior to other strategies [BN04, PC09]. Previous studies on alternative scheduling policies, however, have only provided empirical results, and a thorough analysis of the behavior of such policies for approximate similarity search is still lacking.

With the aim of filling this gap, in this paper, after confirming the different behaviors observed with different scheduling policies, we validate the observed results using information-theoretic arguments (Sect. 2). Then, in Sects. 3 and 4 we introduce two cost models for approximate similarity queries, both based on the concept of distance distribution [CPZ98]. Clearly, being able to estimate the cost of an approximate query is a fundamental step for understanding what influences processing costs. Finally, in Sect. 5 we introduce *optimal* scheduling policies, which again exploit distance distributions.

1.1 Preliminaries

The problem we consider in this paper can be precisely defined as follows: Given a metric space $\mathcal{M} = (\Omega, d)$, where Ω is a domain, also called the *object space*, and $d : \Omega \times \Omega \to \Re_0^+$ is a non-negative and symmetric binary function that also satisfies the triangle inequality, and a data set of objects $X \subseteq \Omega$, retrieve the object(s) in X which are closest, according to d, to a user-specified query object $q \in \Omega$ [CN+01, ZA+06]. The queries we consider here are *quality-controlled k-NN* queries, i.e., k nearest neighbor queries where the user specifies a quality (distance) threshold θ and the search stops as soon as k objects are found at a distance not higher than θ from the query object q. The value of θ can be obtained from either knowledge from the domain (e.g., "give me the three closest gas stations within 10 Km") or by recurring to probabilistic considerations, like those proposed by the PAC technique [CP00]. Indeed, although in [CP00] only 1-NN queries were considered, the definition of $r_{\delta,\epsilon}^q$ given therein (which coincides with the notion of threshold θ used in this paper) can be appropriately extended to the case $k > 1$ (see Appendix A).

The scenario we consider includes an index tree $\mathcal{T}$ built over the data set $X \subseteq \Omega$, where each node N_i corresponds to a *data region*, $\Omega_i \subseteq \Omega$. Node N_i stores a set of *entries*: entries in internal nodes consist of (at least) a pointer to a child node N_c and a description of Ω_c, with $\Omega_c \subseteq \Omega_i$; entries in leaf nodes are (pointers to) indexed objects. The set X_i of all the objects reachable from (a path starting from) node N_i is guaranteed to be contained in the data region Ω_i, i.e., $X_i \subseteq \Omega_i$. Index trees satisfying the above definition provide a recursive decomposition of the space [HNP95], so that objects reachable from a node are obtained as the union of objects reachable from its children nodes. For the case of M-tree [CPZ97], which we consider here, regions are *balls*, i.e.:

$$\Omega_i = \{p \in \Omega \mid d(c_i, p) \le r_i\} \tag{1}$$

where c_i is the routing object of node N_i and r_i its covering radius.

2   Performance of Schedules

The goal of quality-controlled queries is to minimize the cost (measured as the number of accessed nodes) to return the result; as also argued in [BN04], the strategy used to decide which index node to visit first, which we call *scheduling policy*, can influence costs, because different scheduling policies can lead to find earlier objects close to the query point q.

Implementing a specific scheduling policy might require that, besides the geometric description of the region of a node N_i, also available is some information on the actual data indexed by the node itself. For instance, the DBIN method [BFG99] organizes objects into clusters and assumes that they are samples of a mixture of Gaussian distributions (one per cluster), whose parameters (means, variances, and weights) are off-line estimated and stored.

In the following, we collectively refer to all the information maintained about a node N_i as the *statistics* of N_i, denoted $stats(N_i)$. By exploiting statistics on nodes, one can derive from them several *indicators*, Ψ_A, Ψ_B, etc., each assigning to node N_i a value, $\Psi_A(q, stats(N_i)), \Psi_B(q, stats(N_i)), \ldots$, that depends on the specific query q. For simplicity, in the following we write, with a slight abuse of notation, $\Psi(q, N_i)$ in place of $\Psi(q, stats(N_i))$. We also assume that nodes with a lower value of the chosen indicator will be accessed first.

The indicators we consider in this paper are:

- MINDIST: the minimum distance between q and Ω_i, MINDIST $(q, N_i) = \max\{0, d(q, c_i) - r_i\}$;
- MAXDIST: The maximum distance between q and Ω_i, MAXDIST $(q, N_i) = d(q, c_i) + r_i$;
- MINMAXDIST: The distance between q and the center of Ω_i, MINMAXDIST $(q, N_i) = d(q, c_i)$;

Note that MINDIST $(q, N_i) \leq d(p, q) \leq$ MAXDIST (q, N_i) $\forall p \in X_i$, whereas $d(p, q) \leq$ MINMAXDIST (q, N_i) $\forall p \in X_i$ only when no deletions occur (which justifies the name for this indicator).

As a first step we show experimental results conducted on the *real* data sets described in Table 1.

Table 1. Description of the data sets used in the experiments: d is the data set dimensionality, N the cardinality of the data set and n the number of leaf nodes for an M-tree built on the data set.

Name	Feature description	d	N	n
Corel	Color histogram	32	68,040	4,389
Airphoto	texture (Gabor filter)	60	274,424	18,386
EEG	EEG electrode reads	64	2,824,483	174,429

Results included in Figs. 1, 2 and 3 show that the considered indicators perform consistently over all the used data sets and values of $k \in \{1, 10, 50\}$.

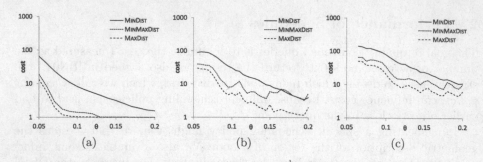

Fig. 1. Average query cost as a function of the quality threshold for different indicators on the `Corel` data set: $k = 1$ (a), $k = 10$ (b), and $k = 50$ (c).

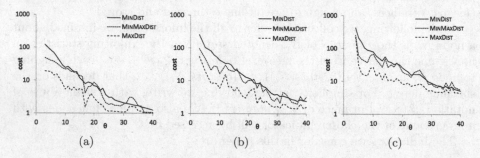

Fig. 2. Average query cost as a function of the quality threshold for different indicators on the `Airphoto` data set: $k = 1$ (a), $k = 10$ (b), and $k = 50$ (c).

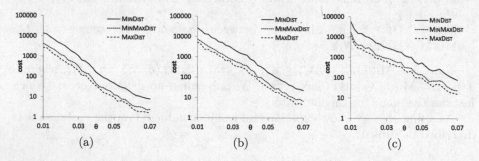

Fig. 3. Average query cost as a function of the quality threshold for different indicators on the `EEG` data set: $k = 1$ (a), $k = 10$ (b), and $k = 50$ (c).

The MINDIST policy, which is optimal for the case of exact k-NN queries [BB+97], attains the worst performance, while MAXDIST leads to the lowest costs.

2.1   Understanding the Behavior of Scheduling Policies

The natural question that arises when looking at the results in Figs. 1, 2 and 3 is the following: Why is MINDIST such a bad indicator with respect to MAXDIST and MINMAXDIST? The approach we propose to understand the behavior of scheduling policies based on indicator's values is to shift the analysis on the relationship existing between indicators' values and probability of finding close-enough objects. In doing this, however, we do not consider classical correlation-based measures, since they would require us to hypothesize in advance the specific form of correlation to be sought (e.g., linear, quadratic). For the sake of simplicity, we limit the analysis to the case of 1-NN search, although for general k-NN queries a similar characterization can be provided using the results in Appendix A.

In order to present our model, a few preliminary definitions are useful. Let $F()$ be the distance distribution corresponding to distance d, $F(x) = \Pr\{d(\mathbf{q}, \mathbf{p}) \leq x\}$, where $\mathbf{p}$ and $\mathbf{q}$ are random points in Ω. We define $d_{X_i}(q)$ to be the distance of the closest point to query q in N_i, i.e., the distance of the NN *local* to N_i. We also denote with $G_i(\theta)$ the *distance distribution* of $d_{X_i}(\mathbf{q})$, $G_i(\theta) = \Pr\{d_{X_i}(\mathbf{q}) \leq \theta\}$, i.e., the probability that a random query $\mathbf{q}$ would find in N_i a point whose distance from $\mathbf{q}$ is not higher than θ.

By adopting an information-theoretic approach, we consider how much information an indicator Ψ can provide us about the probability of finding a point at distance $\leq \theta$ from q. Then, one would choose, among all the available indicators, the one that maximizes such information.

Consider now the probabilistic event of finding a result with distance not higher than θ, $found_i(\theta)$, in node N_i, whose entropy is:

$$H(found_i(\theta)) = -G_i(\theta)\log_2 G_i(\theta) - (1 - G_i^1(\theta))\log_2(1 - G_i(\theta)) = \mathcal{H}(G_i(\theta))$$

where $\mathcal{H}(s)$ is the entropy of a Bernoulli variable with probability of success s. $H(found_i(\theta))$ represents the uncertainty of the event "the termination threshold is reached in node N_i" for a given value of θ. The use of an indicator Ψ should allow us to reduce such uncertainty. If Ψ can assume m distinct values, $\psi^1, \ldots, \psi^m$, one can compute the *conditional entropy* of success assuming Ψ, i.e.:[1]

$$H(found_i(\theta)|\Psi) = \sum_{j=1}^m \Pr(\psi^j) H(found_i(\theta)|\psi^j) = \sum_{j=1}^m \Pr(\psi^j) \mathcal{H}(G_i(\theta|\psi^j))$$
(2)

where $G_i(\theta|\psi^j) = \Pr\{d_{X_i}(\mathbf{q}) \leq \theta|\Psi(\mathbf{q}, N_i) = \psi^j\}$ is the conditional distribution of NN in node N_i, given that for N_i we observed the indicator value ψ^j.

By considering a randomly chosen node (since scheduling policies are not an issue at this point), we can finally define $H(found(\theta)) = \sum_{i=1}^n H(found_i(\theta))/n$ and $H(found(\theta)|\Psi) = \sum_{i=1}^n H(found_i(\theta)|\Psi)/n$, where n is the number of index nodes.

The *mutual information* $I(found_i(\theta), \Psi)$ represents the information about the termination event contained in the Ψ indicator, and is defined as:

$$I(found(\theta), \Psi) = H(found(\theta)) - H(found(\theta)|\Psi)$$
(3)

---

[1] Here we assume, for simplicity, that indicator Ψ is of discrete type. For continuous types, our results still apply when density functions are used in place of probabilities.

while the *normalized mutual information* is the ratio between $I\left(found\left(\theta\right),\Psi\right)$ and $H\left(found\left(\theta\right)\right)$:

$$NI\left(found\left(\theta\right),\Psi\right) = \frac{I\left(found\left(\theta\right),\Psi\right)}{H\left(found\left(\theta\right)\right)} = 1 - \frac{H\left(found\left(\theta\right)|\Psi\right)}{H\left(found\left(\theta\right)\right)} \qquad (4)$$

Since it is $H\left(found\left(\theta\right)|\Psi\right) \le H\left(found\left(\theta\right)\right)$, it is $I\left(found\left(\theta\right),\Psi\right) \ge 0$ and $NI\left(found\left(\theta\right),\Psi\right) \in [0,1]$.

The higher the value of I (and of NI) for a given indicator, the better the estimation of the success probability given by the Ψ indicator, and the sooner we are expected to reach the termination threshold for a query if we schedule nodes according to values of Ψ. This suggests that the indicator that maximizes I (NI) for a given data set is the one that provides the best schedule.

Figure 4 shows that MINDIST is indeed a poor indicator for the probability of stopping the query, since the information we gain is the worst (among considered indicators) for almost all considered values of threshold θ (in the remainder of the paper, for the sake of brevity we show results of our experimentation on the Corel data set only, since results for other data sets are similar). This would suggest that, even if MINDIST is optimal for exact queries, its performance would rapidly deteriorate when considering approximate queries. On the other hand, the MAXDIST indicator provides the maximum information and, consequently, attains the best performance over all the considered range of θ values.

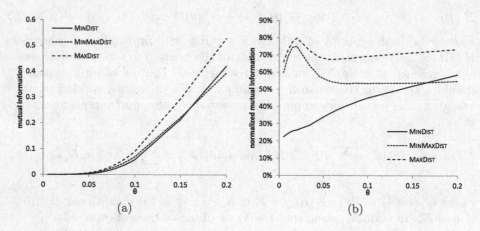

(a)   (b)

Fig. 4. Mutual information I (a), and normalized mutual information NI (b) for different indicators on the Corel data set.

3   A Basic Cost Model for Quality-Controlled Queries

In this section, we introduce a first cost model based on distance distributions for predicting the expected number of nodes to be fetched to find an object having a distance from q not higher than θ.

The model is valid for the case of 1-NN search and 2-levels trees, since this is the scenario more amenable to be formally characterized. Then, the index tree consists of a root and n leaf nodes, and any scheduling policy can be viewed as a way to permute the set $\{1, \ldots, n\}$ obtaining a schedule $\Pi = (\Pi_1, \Pi_2, \ldots, \Pi_i, \ldots, \Pi_n)$, where N_{Π_i} is the leaf that schedule Π will fetch at step i.

To estimate the number of nodes read before the algorithm stops, in [PC09] we considered the probability, $p_{stop}(c, \theta; \Pi)$, that the search algorithm, using schedule Π, will find in no more than c steps $(1 \leq c \leq n)$ a point whose distance from the query is not higher than θ:

$$p_{stop}(c, \theta; \Pi) = \Pr\left\{\min_{i \leq c}\left\{d_{X_{\Pi_i}}(\mathbf{q})\right\} \leq \theta\right\} = 1 - \prod_{i=1}^{c} \Pr\left\{d_{X_{\Pi_i}}(\mathbf{q}) > \theta\right\}$$

$$= 1 - \prod_{i=1}^{c}\left(1 - \Pr\left\{d_{X_{\Pi_i}}(\mathbf{q}) \leq \theta\right\}\right) = 1 - \prod_{i=1}^{c}\left(1 - G_{\Pi_i}(\theta)\right) \tag{5}$$

Note that, since the search stops after fetching *all* nodes, it is $p_{stop}(n, \theta; \Pi) = 1$. The expected cost for processing a random query $\mathbf{q}$ with threshold θ using schedule Π, $E[Cost(\mathbf{q}; \Pi, \theta)]$, can then be derived by observing that the probability of finding a result in *exactly* c steps is $p_{stop}(c, \theta; \Pi) - p_{stop}(c-1, \theta; \Pi)$:

$$E[Cost(\mathbf{q}; \Pi, \theta)] = \sum_{c=1}^{n} c \cdot \left(p_{stop}(c, \theta; \Pi) - p_{stop}(c-1, \theta; \Pi)\right)$$

$$= \sum_{c=1}^{n} c \cdot p_{stop}(c, \theta; \Pi) - \sum_{c=0}^{n-1}(c+1) \cdot p_{stop}(c, \theta; \Pi)$$

$$= \sum_{c=1}^{n} c \cdot p_{stop}(c, \theta; \Pi) - \sum_{c=1}^{n-1} c \cdot p_{stop}(c, \theta; \Pi) - \sum_{c=1}^{n-1} p_{stop}(c, \theta; \Pi)$$

$$= n \cdot p_{stop}(n, \theta; \Pi) - \sum_{c=1}^{n-1} p_{stop}(c, \theta; \Pi) = n - \sum_{c=1}^{n-1} p_{stop}(c, \theta; \Pi)$$

$$= n - \sum_{c=1}^{n-1}\left(1 - \prod_{i=1}^{c}\left(1 - G_{\Pi_i}(\theta)\right)\right) = 1 + \sum_{c=1}^{n-1}\prod_{i=1}^{c}\left(1 - G_{\Pi_i}(\theta)\right) \tag{6}$$

Above cost model will largely overestimate query costs, since it does not consider at all *geometric pruning* (GP), which for the case of M-tree arises whenever $\textsc{MinDist}(q, N_i) > \theta$. In order to take GP into account, in the above cost model it is sufficient to discard all the c values for which N_{Π_c} would be pruned, i.e.:

$$E\left[Cost^{GP}(\mathbf{q}; \Pi, \theta)\right] = 1 + \sum_{\substack{c=1 \\ N_{\Pi_c} \text{ not pruned}}}^{n-1} \prod_{i=1}^{c}\left(1 - G_{\Pi_i}(\theta)\right) \tag{7}$$

The problem with the GP cost model is that its estimates are (almost) independent of the particular query at hand. Indeed, the local NN distance

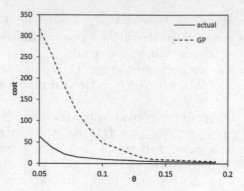

Fig. 5. Performance of the GP cost model for the MINDIST scheduling policy on the `Corel` data set.

distributions $G_i(\theta)$ do not depend on the specific q, thus the model always provides the same estimates for any value of θ; the relative positioning of the query with respect to nodes is only taken into account when geometric pruning is considered by Eq. 7. Also observe that for all queries q that lead to the same schedule, for example according to the MINDIST indicator, Eq. 6 yields the same prediction regardless of the actually observed MINDIST values. Thus, even when considering geometric pruning, costs will be largely overestimated, as demonstrated by Fig. 5, in which actual and predicted costs are shown.

4   The Query-Sensitive Cost Model

To improve the estimates with respect to the GP cost model, we again put *indicators* into play. As we discussed in Sect. 2, indicators are values derived from statistics of nodes and are used to schedule nodes during the search process. Therefore, they represent what the query "knows" about an index node before accessing it. The key idea to develop a *query-sensitive* cost model, i.e., a model able to adapt its estimates to the specific query, is to make the local NN distance distribution dependent on the values of indicators observed by the query at hand.

By looking at indicator's values one can then estimate $p_{stop}(c, \theta; \Pi)$ as:

$$p_{stop}(c, \theta; \Pi) = 1 - \prod_{i=1}^{c} \left(1 - G_{\Pi_i}\left(\theta|\psi^j\right)\right) \tag{8}$$

where now we explicit the use of conditional probabilities $G_i(\theta|\Psi(\mathbf{q}, N_{\Pi_i}) = \psi^j)$. By substituting Eq. 8 into the GP cost model of Eq. 7, we obtain the new query-sensitive (QS) cost model:

$$E\left[Cost^{QS}(\mathbf{q}; \Pi, \theta)\right] = 1 + \sum_{c=1}^{n-1} \prod_{i=1}^{c} \left(1 - G_{\Pi_i}\left(\theta|\psi^j\right)\right) \tag{9}$$

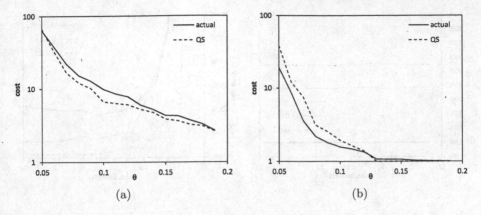

Fig. 6. Performance of the QS cost model based on conditional probabilities on the Corel data set for the MINDIST (a) and the MINMAXDIST (b) schedules.

Figure 6 shows the estimates of the QS cost model when the MINDIST and the MINMAXDIST indicators are used. We immediately see that predicted costs are indeed much better than those of the GP cost model for all considered values of θ.

4.1 Storing the NN Distance Distributions

In order to obtain accurate predictions from the QS cost model, we should appropriately store the conditional probabilities in order to compute, for each node N_i, the value of $G_i\left(\theta|\psi^j\right)$ given the actual value of θ and the observed value of indicator Ψ. It is clear that, the more precisely such distributions are stored, the more accurate the predictions of the cost model will be. We store the $G_i\left(\theta|\psi^j\right)$ probabilities as 2-dimensional histograms: to reduce the complexity of building and maintaining such histograms, for each value of θ the conditional probability is stored as an uni-dimensional equi-depth histogram. It follows that, for a given value of θ, the accuracy of predictions only depends on the number of buckets allocated for storing the ψ^j values. Figure 7 shows the relative error of the QS cost model, i.e., the relative difference between actual and estimated costs, for different values of θ and of the number of buckets for storing ψ^j.

Graphs for the MINMAXDIST schedule confirm our intuition: increasing the number of buckets leads to better estimations, until a limit is reached when further increasing the number of buckets does not improve model predictions. The optimal number of buckets is around $15 \div 20$. Behavior of the QS cost model for the MINDIST scheduling policy is however somehow surprising, since the quality of model estimates does not depend on the size of the histograms. The reason for this performance is explained by Fig. 8(a), where we plot values of $\Pr\left\{\text{MINDIST}\left(\mathbf{q}, N_i\right) = 0 | d_{X_i}\left(\mathbf{q}\right) \leq \theta\right\}$, i.e., the probability that a node solving the query has MINDIST $= 0$. For all the considered range of θ values, there is a high probability of finding the result in nodes having MINDIST $= 0$,

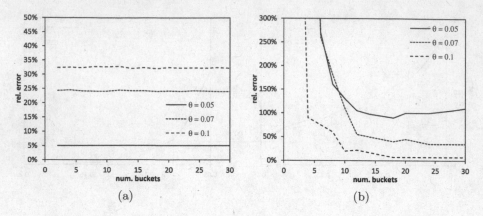

Fig. 7. Relative error of the QS cost model on the `Corel` data set with varying storage size for the $G_i\left(\theta|\psi^i\right)$ distributions: (a) MINDIST and (b) MINMAXDIST schedule.

thus nodes for which MINDIST > 0 are almost never accessed. Note also that, on average, 70% of the nodes have MINDIST $= 0$, i.e., their region includes the query. The net result is that, no matter how many buckets we use to store $G_i\left(\theta|\text{MINDIST}\left(\mathbf{q}, N_i\right) = \psi^j\right)$, only the first bucket (i.e., the one for which MINDIST $\left(\mathbf{q}, N_i\right) = 0$) will be accessed in most of the cases, thus 2 buckets are enough for the MINDIST indicator: one for nodes having MINDIST $= 0$ and one for those with MINDIST > 0. We finally note that, as shown in Fig. 8(b), performances of the cost model for the MINDIST schedule actually *degrade* for high values of θ, i.e., when the query gets easier (this is not the case, as shown in Fig. 7(b), for the MINMAXDIST indicator). This, again, is due to the fact that the $G_i\left(\theta|\text{MINDIST}\left(\mathbf{q}, N_i\right) = 0\right)$ probability is decreasing with respect to values of θ, thus it gets harder, for the model, to predict accurately when the query will be stopped.

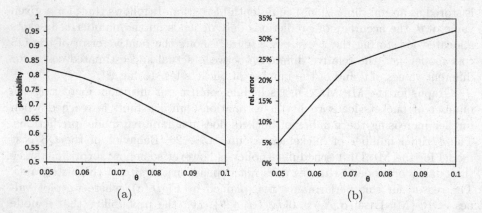

Fig. 8. Probability that nodes solving the query have MINDIST $= 0$ as a function of θ (a) and relative error of the cost model for the MINDIST schedule (b).

5   Optimal Schedules

The QS cost model introduced in Sect. 4 allows us to accurately predict the cost of an approximate similarity query, for any given scheduling policy. It is now natural to consider, given a certain distance threshold, which is the scheduling policy leading to the lowest cost.

We can partially answer this question as follows. From the examination of Eq. 9, it is easy to derive that, *for any fixed indicator Ψ*, the optimal choice is to choose, at each step of the schedule for answering query q, the node i for which $G_i\left(\theta|\psi^j\right)$ is the maximum among yet-to-be-fetched nodes. To see why this is the case, let $x_i = 1 - G_i\left(\theta|\psi^j\right)$. Then, the right-hand side of Eq. 9 equals $1 + \sum_{c=1}^{n-1} \prod_{i=1}^{c} x_i$, that is:

$$1 + x_1 + x_1 \cdot x_2 + \ldots + x_1 \cdot x_2 \cdot \ldots \cdot x_{n-1} \tag{10}$$

which is minimized when $x_1 \leq x_2 \leq \ldots \leq x_{n-1}$.

Each Ψ-optimal schedule should be contrasted to the corresponding Ψ-based schedule: while the latter orders nodes by just looking at the observed values of the indicator Ψ, the former uses such values to select from each node N_i the local NN distance distribution to be used for the query at hand, and then order nodes based on the value of such distribution for the specific distance threshold θ.

Figure 9 shows costs for Ψ-optimal and Ψ-based schedules for the MINDIST and MINMAXDIST indicators.

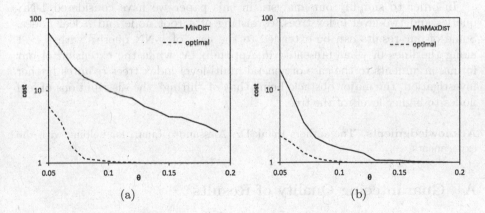

(a)   (b)

Fig. 9. Comparison between different schedules based on the (a) MINDIST and (b) MINMAXDIST indicators on the `Corel` data set.

The schedule based on the MINDIST (respectively, MINMAXDIST) indicator alone lead to costs that are about 10 (resp., 2) times higher than the corresponding optimal schedule. In particular, for values of θ higher than 0.12 (resp., 0.1) the optimal schedule is able to stop the search *by visiting a single leaf node*!

By comparing the optimal schedules in Figs. 9(a) and (b), we see that MIN-MAXDIST is again a better indicator than MINDIST, since it leads to lower costs with an average 50% difference in the interval [0.05,0.1]. This is consistent with the analysis performed in Sect. 2.1; however, in the general case, one cannot claim that the optimal schedule for indicator Ψ_A would lead to lower costs than the optimal schedule for indicator Ψ_B for all possible data sets.

From a practical point of view, the choice of the indicator to be used for a specific data set should be based on the mutual information analysis performed in Sect. 2.1.

6   Conclusions

In this paper we have investigated the effect that different scheduling policies can have on the costs of solving quality-controlled approximate similarity queries, i.e., similarity queries that are stopped as soon as a "good-enough" result is found, with index trees. In particular, any scheduling policy uses one indicator to determine in which order the tree nodes have to be accessed. We have shown how, for any given indicator Ψ, one can derive a corresponding Ψ-optimal schedule. This has exploited a novel query-sensitive cost model for which each node of an index tree needs to store statistical information about past queries.[2] We remark that maintaining statistics for predicting query processing costs and for guaranteeing the quality of approximate queries is a common trend, as also exemplified by the actual interest in database systems [CDK17].

In order to simplify our analysis, in this paper we have considered 1-NN queries and two-level index trees, consisting of a root node and n leaf nodes. Some of our results can be extended to the case of k-NN queries with $k > 1$ along the lines of what presented in Appendix A, while the extension of our formal arguments to the case of general multi-level index trees requires further investigation, the major obstacle being that of "lifting" the distributions of leaf nodes to higher levels of the tree.

Acknowledgments. The authors thank Dr. Alessandro Linari for helping with the experiments.

A   Guaranteeing Quality of Results

In this Appendix, we show how the threshold θ of a quality-controlled query can be chosen so as to provide probabilistic guarantees on the quality of results, extending the results presented in [CP00] to the case $k > 1$. The approximate result $\widetilde{\mathcal{R}}$ of a query is a list of k objects that may, however, not be the k closest ones to the query q. A possible way to define the quality of $\widetilde{\mathcal{R}}$ is in terms of the relative error Err wrt the exact result $\mathcal{R}$. Let us denote the i-th NN of a query

---

[2] In [PC09] we introduced only the query-independent cost model for approximate queries.

q in a set of objects X as $p_X^i(q)$ and with $\widetilde{p}_X^i(q)$ the i-th NN of q in $\widetilde{\mathcal{R}}$. In the (simplest) case presented in [CP00], when $k = 1$, the error is computed as:

$$Err = \frac{d\left(q, \widetilde{p}_X^1(q)\right)}{d\left(q, p_X^1(q)\right)} - 1 \tag{11}$$

This can be extended to the case $k > 1$ by using the error on the k-th nearest neighbor (see also [AM+98, ZS+98]):

$$\boxed{Err \stackrel{\text{def}}{=} \frac{d\left(q, \widetilde{p}_X^k(q)\right)}{d\left(q, p_X^k(q)\right)} - 1} \tag{12}$$

which reduces to Eq. 11 when $k = 1$.

The type of guarantee provided on the quality of results in [CP00] has the form: *"with probability at least $1 - \delta$ the error does not exceed ϵ"*, that is $\Pr\{\mathbf{Err} \le \epsilon\} \ge 1 - \delta$, where $\epsilon \ge 0$ is an accuracy parameter, $\delta \in [0, 1)$ is a confidence parameter, and $\mathbf{Err}$ is the random variable obtained from Eq. 12 applied to a random query $\mathbf{q}$. In [CP00], this is computed by using $G(x)$, i.e., the distance distribution of the 1-NN, which can be obtained from the distance distribution $F(\cdot)$ as:

$$G(x) \stackrel{\text{def}}{=} \Pr\left\{d\left(\mathbf{q}, p_X^1(\mathbf{q})\right) \le x\right\} = 1 - (1 - F(x))^N \tag{13}$$

For $k \ge 1$, this generalizes to $G^k(x)$, i.e., the probability to find at least k objects at a distance $\le x$, which can be computed as:

$$G^k(x) \stackrel{\text{def}}{=} \Pr\left\{d\left(\mathbf{q}, p_X^k(\mathbf{q})\right) \le x\right\} = 1 - \sum_{j=0}^{k-1} \binom{N}{j} \cdot F(x)^j \cdot (1 - F(x))^{N-j} \tag{14}$$

The probability that the result of an approximate k-NN query θ is correct, i.e., that the approximate k-th NN, $\widetilde{p}_X^k(q)$, is indeed the correct one, is given by $G^k\left(d\left(q, \widetilde{p}_X^k(q)\right)\right)$. Since we want to bound the error with confidence $1 - \delta$, we obtain the following guarantee on the error:

$$Err \le \epsilon = \frac{d\left(q, \widetilde{p}_X^k(q)\right)}{\sup\left\{x | G^k(x) \le \delta\right\}} - 1$$

because the probability that $d\left(q, p_X^k(q)\right) < \sup\left\{x | G^k(x) \le \delta\right\}$ is less than δ. The (one-sided) confidence interval corresponding to $1 - \delta$ is therefore $[0, \epsilon]$. Thus, the threshold θ for a quality-controlled query can be obtained as $(1 + \epsilon)$ $\sup\left\{x | G^k(x) \le \delta\right\}$. Clearly, if $G^k()$ is invertible, it is: $\sup\left\{x | G^k(x) \le \delta\right\} = (G^k)^{-1}(\delta)$.

References

[AM+98] Arya, S., Mount, D.M., et al.: An optimal algorithm for approximate nearest neighbor searching. JACM **45**(6), 891–923 (1998)

[BB+97] Berchtold, S., Böhm, C., et al.: A cost model for nearest neighbor search in high-dimensional data space. In: Proceedings of PODS 1997, Tucson, AZ, pp. 78–86 (1997)

[BFG99] Bennett, K.P., Fayyad, U.M., Geiger, D.: Density-based indexing for approximate nearest-neighbor queries. In: Proceedings of KDD 1999, San Diego, CA, pp. 233–243 (1999)

[BN04] Bustos, B., Navarro, G.: Probabilistic proximity searching algorithms based on compact partitions. JDA **2**(1), 115–134 (2004)

[CDK17] Chaudhuri, S., Ding, B., Kandula, S.: Approximate query processing: no silver bullet. In: Proceedings of SIGMOD 2017, Chicago, IL (2017, to appear)

[CN+01] Chávez, E., Navarro, G., et al.: Proximity searching in metric spaces. ACM Comp. Sur. **33**(3), 273–321 (2001)

[CP00] Ciaccia, P., Patella, M.: PAC nearest neighbor queries: approximate and controlled search in high-dimensional and metric spaces. In: Proceedings of ICDE 2000, San Diego, CA, pp. 244–255 (2000)

[CP10] Ciaccia, P., Patella, M.: Approximate and probabilistic methods. SIGSPATIAL Spec. **2**(2), 16–19 (2010)

[CPZ97] Ciaccia, P., Patella, M., Zezula, P.: M-tree: an efficient access method for similarity search in metric spaces. In: Proceedings of VLDB 1997, Athens, Greece, pp. 426–435 (1997)

[CPZ98] Ciaccia, P., Patella, M., Zezula, P.: A cost model for similarity queries in metric spaces. In: Proceedings of PODS 1998, Seattle, WA, pp. 59–68 (1998)

[HH99] Haas, P.J., Hellerstein, J.M.: Ripple joins for online aggregation. In: Proceedings of SIGMOD 1999, New York, NY, pp. 287–298 (1999)

[HNP95] Hellerstein, J.M., Naughton, J.F., Pfeffer, A.: Generalized search trees for database systems. In: Proceedings of VLDB 1995, Zurich, Switzerland, pp. 562–573 (1995)

[HS03] Hjaltason, G.R., Samet, H.: Index-driven similarity search in metric spaces. ACM TODS **28**(4), 517–580 (2003)

[PC09] Patella, M., Ciaccia, P.: Approximate similarity search: a multi-faceted problem. JDA **7**(1), 36–48 (2009)

[ZA+06] Zezula, P., Amato, G., et al.: Similarity Search: The Metric Space Approach. Springer, Heidelberg (2006)

[ZS+98] Zezula, P., Savino, P., et al.: Approximate similarity retrieval with M-trees. VLDBJ **7**(4), 275–293 (1998)

Cache and Priority Queue Based Approximation Technique for a Stream of Similarity Search Queries

Filip Nalepa$^{(\boxtimes)}$, Michal Batko, and Pavel Zezula

Faculty of Informatics, Masaryk University, Brno, Czech Republic
f.nalepa@gmail.com

Abstract. Content-based similarity search techniques have been employed in a variety of today applications. In our work, we aim at the scenario when the similarity search is applied in the context of stream processing. In particular, there is a stream of query objects which need to be evaluated. Our goal is to be able to cope with the rate of incoming query objects (i.e., to reach sufficient throughput) and, at the same time, to preserve the quality of the obtained results at high levels. We propose an approximation technique for the similarity search which combines the probability of an indexed object to be a part of a query result and the time needed to examine the object. We are able to achieve better trade-off between the efficiency (processing time) and the quality (precision) of the similarity search compared to traditional priority queue based approximation techniques.

1 Introduction

Large quantities of unstructured data are being produced these days due to the digital media explosion. One of common subtasks while processing such data is searching in the data. Traditional approaches based on exact match of data attributes are not usually appropriate for these data types. Instead, content-based similarity search techniques are a valid option. Often *k-nearest-neighbors queries* (kNN) are applied, which retrieve the k objects that are the most similar to a given query object.

One of contemporary paradigms for processing large amounts of data is stream processing when there is a potentially infinite sequence of data items which are continuously being created and have to be continuously processed. For example, consider a text search-engine which continuously receives images from external sources and needs to continuously annotate them by textual descriptions according to the image content so that they are subsequently findable by the textual keywords. As another example, a spam filter receives incoming emails and compares them to some learned spam knowledge base so that spam messages can be detected. A news notification system needs to compare the newly published articles to the profiles of all the subscribed users to find out who should be notified.

© Springer International Publishing AG 2017
C. Beecks et al. (Eds.): SISAP 2017, LNCS 10609, pp. 17–33, 2017.
DOI: 10.1007/978-3-319-68474-1_2

All these applications have to deal with some form of content-based searching while processing the streamed data. An important characteristic is that the data do not need to be processed immediately as in interactive applications, but some delay is acceptable. The performance of these applications is mostly determined by the number of processed data items in a given time interval, i.e., the throughput is the most important metric. The individual query search times can be improved by applying some similarity indexing technique, for which there are efficient algorithms based on the metric model of similarity [16]. On the other hand, one should care also about the effectiveness of the search results. These two metrics (throughput and effectiveness) usually go against each other; improving the effectiveness worsens the throughput and vice versa.

I/O costs typically have a significant effect on the performance of similarity search techniques. In our previous work [10], we exploit the fact that some orderings of the processed queries can result into considerably lower I/O costs and overall processing times than a random ordering. This is based on the observation that two similar queries need to access similar data of the search index. By obtaining an appropriate ordering of queries, the accessed data can be cached in the main memory and reused for evaluation of similar queries lowering down the I/O costs. We proposed a technique to dynamically reorder the incoming queries which, according to our experiments, allows to significantly improve the throughput.

In a typical stream-processing scenario, the input data come from an external source and the application cannot control the speed of the entering data. Therefore it is important that the application can adapt to the changes in the input frequency so that it is possible to process all the data. One way of reacting is to alter the quality of the results: when the incoming stream is too fast, the search quality is lowered so that the throughput is increased and vice versa.

A common way to control the level of result quality in a similarity search engine is to employ approximation where only the most promising indexed objects are examined to form an approximate answer. In such an indexing engine, the order of examining the indexed objects is often determined by a *priority queue*, and the examination of the queue proceeds until a stop condition is hit. A good approximation technique should place the most promising indexed objects (i.e., the ones with the highest probability of being in the precise result) at the beginning of the queue, thus ensuring that they are examined before the stop condition is met.

In this work, we propose an approximation technique which combines the priority queue based approach with the caching system we presented in our previous work. We make use of the fact that the cached objects can be examined in significantly lower time than the uncached ones.

The next section summarizes works which use some sort of caching in the similarity search. The formal definition of the problem which we solve is provided in Sect. 3. The paper proceeds with a detailed description of the approximation technique in Sect. 4. A score function is used to determine which indexed objects should be examined; its properties are studied in Sect. 5. Conducted experiments are presented in Sect. 6.

2   Related Work

The usage of a caching mechanism in similarity search has been proposed in several papers to reduce the amount of I/O operations. In [8], the authors propose caching of similarity search results and reusing them to produce approximate results of similar queries. The concept of caching similarity search results is used also in [12]. The paper focuses on caching policies which incrementally reorganize the cache to ensure that the cached items cover the similarity space efficiently. The Static/Dynamic cache presented in [14] consists of a static part to store queries (along with their results) that remain popular over time and a dynamic part to keep queries that are popular for a short period of time. Authors of [5] present a caching system to obtain quick approximate answers. If the cache cannot provide the answer, the distances computed up to that moment are used to query the index so that the computations are not wasted. Caching of data partitions is presented in [4] for simultaneous solving of multiple queries so that each data partition is read at most once. The data partition caching is used also in [7] complemented with caching previous answers which serve to set initial search radius for similar kNN queries. The authors of the paper [13] target the situations when the distance computation itself is an expensive operation. They propose D-cache which stores distances computed during previous queries to avoid some distance computations of the subsequent queries. A cache of distances is used also by the Snake Table [2] which is designed for processing streams of queries with snake distribution (i.e., consecutive query objects are similar). In [1], an inverted cache index maintains "usefulness" statistics which are explored to reorder a priority queue to increase the effectiveness of accessing data partitions. A different approach is described in [15] to process a stream of queries which is based on parallel locality sensitive hashing.

In our previous work [10], we proposed a technique for enhancing the throughput of processing a stream of similarity query objects by reordering the query objects combined with caching previously accessed data partitions. In this paper, we extend our previous approach in order to improve the trade-off between the throughput and the quality of the similarity search results. The combination of the caching mechanism with the priority queue based query evaluation differentiates us from the other approaches.

3   Problem Definition

In this section, we formally define the problem which we solve in this work. Similarity can be universally modeled using metric space (D, d) abstraction [16], where D is a domain of arbitrary objects and d is a total distance function $d : D \times D \to R$. The distance between any two objects from D corresponds to the level of their dissimilarity ($d(p, p) = 0$, $d(o, p) \geq 0$). Let $X \subseteq D$ be a database of objects.

Let $s = ((q_1, t_1), (q_2, t_2), \ldots)$ be a stream of pairs (q_i, t_i) where q_i is a query object and t_i is its creation time (e.g. when it has entered the application), where

$t_i \leq t_{i+1}$ for each $i \geq 1$. For each query object q_i in s, a k-nearest-neighbors query $NN(q_i, k)$ is executed which returns k nearest objects from the database X to the query object.

In practice, approximate techniques are usually applied to accelerate the query execution. In such a case, the results may contain objects which are not among the actual precise k nearest objects. The quality of the approximate results is often expressed by the *precision* metric. Let A be the set of the k nearest objects to the query object and B be the set of objects returned by an approximate technique. The *precision* is computed as $prec = \frac{|A \cap B|}{k}$ where $|A| = |B| = k$.

Our goal is to obtain results of high precision while still being able to keep up with the rate of incoming query objects. Formally, let T be a time limit, BL be a backlog limit constraining the maximum number of arrived unprocessed query objects at each time $t \leq T$ (i.e., any such unprocessed query objects q_i for which $t_i \leq t$). The task is to maximize the average precision of the executed kNN queries until the time T while obeying the backlog limit (i.e., achieving sufficient throughput).

4   Approximation Technique

In our approach we consider a generic metric index which uses data partitioning $R = \{r_1, \ldots, r_n\}$ where $r_i \subseteq X$. When evaluating a query, a subset of the partitions $Q \subseteq R$ needs to be accessed, typically from a disk storage [16]. A usual bottleneck of similarity search techniques is the reading the partitions from the disk during query evaluations. In our previous work [10], we proposed a technique to speedup the processing by caching the partitions (and the contained objects) in the main memory in order to reduce I/O costs of subsequent queries. The number of cached objects is restricted by a limit $maxCacheSize$, and the *least recently used* policy is used to replace the content of the cache whenever it is full. Using this caching system allows to achieve better processing times since some partitions can be quickly obtained from the cache instead of reading them slowly from the disk.

The aim of indexing methods is to generate partitions in such a way that any two objects o_i, o_j in a given partition are similar to each other, i.e., $d(o_i, o_j)$ is small. A priority queue based strategy [16] for evaluating a query $NN(q_i, k)$ works as follows. The data partitions are ordered according to their likeliness of containing any of the k nearest objects to the query object. By expanding the individual partitions, eventually a *priority queue* $(o_1, o_2, \ldots, o_h)$ of *indexed objects* of the database X is generated where $h \leq |X|$[1]. The indexed objects o_i are examined (i.e., $d(o_i, q_i)$ is computed) in this order during the query evaluation. The goal of search techniques is to achieve such an ordering so that the probability of an indexed object being a part of the answer decreases with an

---

[1] Some data partitions and the contained objects can be skipped during the query evaluation due to search space pruning rules employed by the index structure. For the purposes of this paper, we omit such objects from the priority queue.

increasing position in the priority queue. Note that the priority queue is often generated incrementally during the query evaluation.

To employ an approximation technique, the indexed objects are examined according to the priority queue, and once a stop condition is fulfilled, the processing is stopped and k closest objects to the query object, which have been examined so far, form the answer set. Frequently, the stop condition is specified as the maximum number of objects examined during a single query.

We propose an approximation technique in which we exploit the caching system presented in our previous work. The approximation technique works not only with the position of an object in the priority queue (like traditional approximation techniques) but also with the information whether the object is or is not cached. Based on these data, a score of the object is computed which determines whether the object should be examined. More specifically, the position of the object is used to derive the probability that it is a part of the precise answer set (the probability should decrease with an increasing position). The information about the caching state is used to estimate the time to examine the object. The score should increase with an increasing probability of an object being a part of the answer set and decrease with an increasing examination time. The score determines the worthiness of the object to be examined by balancing the contribution of the object to the quality of the final query result and the time spent by examining the object.

Let us provide formal definitions of the used terms.

$$answerProbability(pos) = prob$$

is the probability that an object at the pos^{th} position in the priority queue is a part of the precise answer set (i.e., it is one of the real k nearest neighbors). This should be a nonincreasing function since the intention of similarity search techniques which use the priority queue is to examine more promising objects before the less promising ones, i.e., the probability of an object being a part of the answer set should not increase with an increasing position in the priority queue:

$$answerProbability(pos_1) \geq answerProbability(pos_2) \text{ where } pos_1 < pos_2$$

In practice, the function is meant to be specified empirically based on the properties of the used indexing and search technique, and on the properties of the indexed data and the expected queries.

For the examination time, let there be a function

$$time(isCached) = t \text{ where } isCached \in \{true, false\}$$

providing the time to examine a single object of the priority queue depending on whether the object is or is not cached. The following should hold:

$$time(true) \leq time(false)$$

That is, the cached objects are processed much faster than the others; the validity of this is studied in our previous work [10].

Whether an object is worth examining is determined by computing $score(pos, isCached)$ which has the following properties based on the characteristics of $answerProbability$ and $time$ functions:

$$score(pos, true) \geq score(pos, false)$$

$$score(pos_1, isCached) \geq score(pos_2, isCached) \text{ where } pos_1 < pos_2$$

Before processing a query, $minScore$ parameter is set. Only objects which have at least as high score as $minScore$ are examined during the query evaluation. As a consequence of the $score$ function properties, two position limits PL_1 and PL_2 can be computed. PL_1 is the maximum position for which $score(PL_1, false) \geq minScore$; PL_2 is the maximum position for which $score(PL_2, true) \geq minScore$ (note that $PL_1 \leq PL_2$). It means that an object is examined during processing of a query if and only if it is not cached and its position in the priority queue is at maximum PL_1, or it is cached and its position in the priority queue is at maximum PL_2. If PL_1 equals PL_2, we get the traditional stop condition based solely on the maximum number of examined objects.

It is worth noting how $answerProbability$ (and transitively the $score$ function) is related to the precision of query results (our optimization criteria). First, realize that

$$\sum_{pos=1}^{|X|} answerProbability(pos) = k$$

Let $P = \{pos | 1 \leq pos \leq h \wedge score(pos, isCached_{pos}) \geq minScore\}$ be the set of positions whose objects are examined during the given kNN query where h is the length of the priority queue and $isCached_{pos}$ is true iff the object at the position pos is cached at the time of the evaluation. Then the expected precision is computed as follows:

$$\frac{\sum_{pos \in P} answerProbability(pos)}{k} = precision \qquad (1)$$

That is the sum of the $answerProbability$ values of the examined objects determines the expected precision of the query result according to the assumed probabilities.

5   Score Function

In the previous section, we presented the properties which the $score$ function has to hold; this section is devoted to the actual specification of the score function.

Let us simplify the original problem definition a little bit. Let $u = (q_{i_1}, q_{i_2}, \ldots, q_{i_n})$ be a finite sequence of query objects. The task is to process all the query objects of u in the given order within the time T while achieving the maximum average precision. The items of the sequence u are selected and reordered out of the query objects of the original (infinite) sequence s. The way to generate u is outlined later.

The simplified problem can be also defined as an instance of a *0/1 knapsack problem*. Let there be a set of items $I = \{x_1, \ldots, x_m\}$ where each item x_i has its weight w_i and value v_i. The task is to select a subset of I so that the sum of the weights of the selected items does not exceed a given weight limit W and the sum of the values of the selected items is maximized.

Let $(o_{j1}, o_{j2}, \ldots, o_{ju_j})$ be the priority queue for the query object q_j. The set of items I comprises of all the objects of the priority queues of each query object in u:

$$I = \{o_{i_11}, o_{i_12}, \ldots, o_{i_1u_{i_1}}, o_{i_21}, \ldots, o_{i_nu_{i_n}}\}$$

The weight of each o_{ij} equals the time needed for examination of the given object:

$$w_{ij} = time(isCached_{ij})$$

where $isCached_{ij}$ is true iff the object o_{ij} is cached at the moment of the query evaluation.

The value of each o_{ij} equals its probability of being contained in the answer set:

$$v_{ij} = answerProbability(j)$$

where j is the position of o_{ij} in the corresponding priority queue.

The weight limit W equals the time until which the processing of the whole sequence u of the query objects has to be finished: $W = T$.

In other words, the task is to examine such objects o_{ij} so that the overall time does not exceed the time limit T and the sum of their probabilities being in the answer sets is maximized, hence the maximum average precision is achieved according to Formula 1.

There are a number of approaches solving the *0/1 knapsack problem* in general. However, the problem of our scenario is that the values (the probability of an object to be a part of the answer) and the weights (derived from the actual content of the cache) are not known in advance since the priority queue is determined incrementally while a query is being processed. One of the approaches to solving the *0/1 knapsack problem* is to compute the density of each item x_i: $d(x_i) = \frac{v_i}{w_i}$. The items are sorted by their densities in a descending order, and they are put into the knapsack greedily in this order until the weight limit W is reached. This approach provides the optimal solution if it is allowed to put a fraction of an item into the knapsack. In such a case, the last item placed into the knapsack may be fractioned so that the weight limit is not exceeded.

Since we work with inexact data *answerProbability* and *time* to examine an object, in practice we do not require the theoretical solution of the knapsack problem to strictly obey the weight limit. In other words, we can tolerate a little excess of the time limit T, thus we do not need to fraction the last item put into the knapsack. If the density of the last item placed into the knapsack following the described algorithm was known ahead, we would be able to solve our problem by examining only the objects having at least such a density.

This means we would be able to achieve the optimal solution even without knowing all the values and the weights ahead.

Consequently, the density is an appropriate *score* function:

$$score(pos, isCached) = \frac{answerProbability(pos)}{time(isCached)}$$

It remains to determine the minimal density limit (the density of the last item placed into the knapsack), which is, in our terms, the *minScore.* How to overcome the limitation of not knowing the values and the weights ahead is discussed in Sect. 5.2.

5.1   Knapsack Problem Analysis

In this section, we briefly analyze the input and the solution of the knapsack problem. Let us have an instance of the knapsack problem with an optimal solution having the sum of the values V. A higher overall value V (and hence a higher average precision) can be achieved by reducing weights of the items (i.e., by enlarging the maximal number of cached objects $maxCacheSize$). Another way to improve the average precision is to have the cached objects (light items) at the beginning of the priority queues (valuable items), i.e., to have light valuable items. This is influenced by the ordering of the query objects in the sequence u. If two consecutive query objects q_1, q_2 are similar to each other, also the sets of objects at the beginning of the corresponding priority queues are similar (i.e., there is a significant intersection). This results in a high concentration of cached objects at the beginning of the priority queue when processing q_2.

The problem of the query ordering is dealt with in our previous work [10] which is aimed at improving the throughput of processing of a stream of query objects. We denote the set of arrived query objects which are waiting for their processing as a *buffer*. Whenever a new query object arrives at the application, it is placed to the buffer. The query object is removed from the buffer right before its processing. The proposed algorithm dynamically reorders the query objects within the buffer so that sequences of similar query objects are obtained. Combined with keeping recently examined objects in the main memory cache, this results in high reuse of the cached objects and thus improved throughput.

In fact, the result of the query ordering creates the bridge between the original problem defined in Sect. 3 and the problem simplification in Sect. 5 since it enables us to obtain the ordered subsequence of query objects u out of the original sequence of query objects s. The subsequence u is generated incrementally as query objects are pulled from the buffer, and it stops when the time limit T is reached.

5.2   Minimal Density

Up to now, we have specified the *score* function as the density of an object referring to the knapsack problem; we have also shown how the average precision of evaluated queries can be improved by enlarging the cache size and/or by reordering the query objects. In this section, we describe a method to set the *minScore* limit (a.k.a. the minimal density) in order to decide which objects are worth examining during a query evaluation.

Since the densities of the objects are not known in advance, we modify the minimal density limit dynamically. The strategy is to maintain the number of arrived unprocessed query objects (the query objects in the buffer) at the maximum backlog limit BL (defined in Sect. 3). According to our previous work [10], by maintaining a larger buffer, more effective query ordering can be generated (i.e., having more light valuable items in the knapsack as described in Sect. 5.1). If the current size of the buffer is over the backlog limit BL, the minimal density limit is raised causing the buffer size to be lowered. Analogically, if the current buffer size is below BL, the minimal density limit is lowered. A particular way of how the density limits can be modified is provided in the section with experiments.

Using the presented strategy, the backlog limit BL constraint can be temporarily violated since we consider it as a soft limit, i.e., it can be exceeded, but appropriate actions should be taken to push the current backlog down. If this is unacceptable in real application, the soft limit can be set to a lower value so that the violations of the hard limit BL are avoided.

6 Experiments

In this section, we provide experimental evaluation of the proposed approach.

6.1 Setup

Let us start with describing the setup of the experiments.

We use the M-Index [11] structure to index the metric-space data. It employs practically all known principles of metric space partitioning, pruning, and filtering, thus reaching high search performance. The actual data are partitioned into buckets which are stored as separate files on a disk and read into the main memory during query evaluations.

For the experiments, we use the Profimedia dataset of images [6]. We created two different subsets of the images and extracted their visual-feature descriptors. The generated datasets are: 1 million Caffe descriptors [9] (4096 dimensional vectors) and 10 million MPEG-7 descriptors [3].

Separately, we created two sequences of 10,000 images represented by corresponding descriptors which are used as *query objects* in the experiments. The sequences are already ordered so that consecutive images are similar to each other. During experiments, images from the respective collection are continuously streamed and processed by approximate 30-NN queries.

We also created two more sequences of 10,000 images which serve as the *training* data to specify the *answerProbability* function (details in Sect. 6.2).

The least recently used policy is used when inserting to the full cache. In particular, the data partitions with the oldest last access time are discarded and replaced with the new partitions of the current query so that the maximum size of the cache is obeyed.

Similarity search techniques (including M-Index) usually make the decisions whether objects of the priority queue should be examined on per partition basis, i.e., it is decided for the whole data partition whether its objects should be examined. However, we use an *object* as the unit of the data access for the analytical purposes of this paper since the occupation of individual data partitions typically differs across the dataset. Theoretically we decide for each object whether it should be examined based on its score; practically we make the decision for the whole data partition by considering the score of the first object in the given partition. The difference between the theoretical *per object* processing and practical *per partition* processing is that, in practice, the examined objects with a score lower than *minScore* occur if they reside in the data partition whose first object has a score at least as high as the *minScore* limit. This situation can happen for at most two data partitions per a query since there are two position limits PL_1, PL_2; meaning the difference is negligible.

6.2   Answer Probability

In this section, we verify the assumption of good indexing and search techniques when the priority queue is employed. In particular, we verify whether the function $answerProbability(pos)$ is nonincreasing for M-Index and the experimental data.

We evaluated 10,000 30-NN *training* queries for each dataset and computed the probability that an object is a part of the answer set given its position in the priority queue. The probability was determined as $answerProbability(pos) = \frac{resultCount}{queryCount}$ where *resultCount* is the number of times an object at the position *pos* was a part of the precise answer set; *queryCount* is the number of queries (i.e., 10,000). The results can be seen in Fig. 1; they are aggregated by 500 positions, i.e., the first point in the graph shows the overall probability for the positions 1–500; the next one for the positions 501–1,000 and so on. The graphs show the results only for a part of the priority queue, but there is no raise in the probability at later positions. It can be observed that our assumption is valid for M-Index and the tested datasets. The difference in the probability values between the two graphs is caused mainly by the difference in the dataset sizes. These results serve as the specification of the *answerProbability* function which is needed for experiments in Sects. 6.5 and 6.6.

6.3   Position Limits

The experiments in this section are carried out to analyze the impact of the position limits PL_1, PL_2 on the effectiveness and the efficiency of the processing. For the MPEG-7 dataset, the cache size was limited by 150,000 objects. For the set of PL_1 limits {5,000; 10,000; 20,000; 30,000; 40,000}, a series of

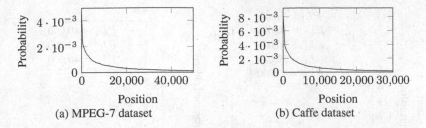

Fig. 1. Answer probability according to the object position in the priority queue

experiments was conducted for various PL_2 starting from $PL_2 = PL_1$ and ending at $PL_2 = 300{,}000$. During each experiment, 10,000 queries were evaluated. Figure 2a depicts the results. There is a single curve for each PL_1 showing the processing time and the achieved average precision for individual PL_2 limits. The first (bottom left) point of each curve expresses the result when $PL_2 = PL_1$, i.e., the traditional approximation technique when the stop condition is defined as the maximum number of examined objects. The graph presents the improvement of the proposed approach compared to the traditional approximation technique. For instance, using the traditional technique, the precision of 62% can be achieved in 52 min ($PL_1 = PL_2 = 20{,}000$) whereas the precision of 64% can be reached in just 31 min using the new approach ($PL_1 = 10{,}000$; $PL_2 = 40{,}000$), which is 60% of the original time.

Figure 2b shows the average amount of examined objects per a query. For example, for $PL_1 = PL_2 = 10{,}000$, there are 10,000 objects examined per each query. As PL_2 rises, there are more objects examined; the raise is not linear since only cached objects are examined at the positions above PL_1. For instance, if $PL_1 = 10{,}000$; $PL_2 = 300{,}000$, there are 94,000 examined objects on average per each query.

6.4 Cache Size

As presented in Sect. 5.1, the size of the cache plays the decisive role in the ratio of heavy and light items of the knapsack problem. In this section, we show how the cache size practically influences the efficiency and the effectiveness of the processing. The experiments were conducted for the cache size ranging from 80,000 to 700,000 objects for the MPEG-7 dataset. We used the following position limits: $PL_1 = 14{,}500$; $PL_2 = 101{,}500$. (These are one of the optimal position limits used in the experiments in Sect. 6.5). In Fig. 3a, we can observe the processing times. When the cache size of only 80,000 is used, the processing time is very high since a lot of objects which are within PL_1 in the priority queue (and therefore have to be processed regardless of whether they are cached or not) are loaded from the disk. As the cache size approaches 300,000 objects, the time quickly drops since most of the objects which are within PL_1 can be retrieved from the cache. The reason of the subsequent raise of the processing time is depicted in Fig. 3b capturing the numbers of examined objects for individual cache sizes.

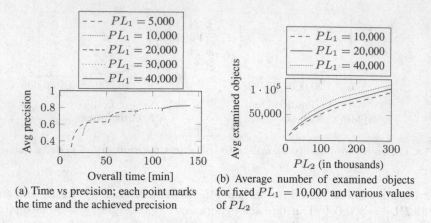

(a) Time vs precision; each point marks the time and the achieved precision

(b) Average number of examined objects for fixed $PL_1 = 10,000$ and various values of PL_2

Fig. 2. Results for experiments with various PL_1 and PL_2; 10,000 30-NN queries; MPEG-7 dataset

As the cache enlarges, so does the amount of objects which appear within PL_2 in the priority queue and which are cached at the same time (and therefore are examined). As the number of examined objects rise so does the precision (Fig. 3c) going from the precision of 0.67 to 0.80 for the cache sizes 80,000 and 700,000 respectively.

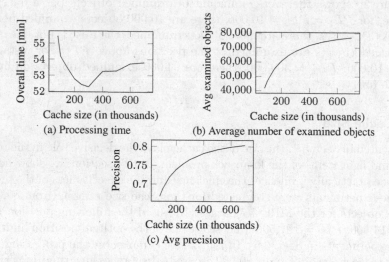

(a) Processing time

(b) Average number of examined objects

(c) Avg precision

Fig. 3. Cache size influence on the efficiency and the effectiveness; 10,000 30-NN queries; MPEG-7 dataset; $PL_1 = 14,500$; $PL_2 = 101,500$

6.5   Optimal Position Limits

In Sect. 6.3, we showed how the precision and the processing time change when PL_1 is fixed and PL_2 is altered. In this section, we operate with the density as a definition of the *score* function to set the optimal position limits (details in Sect. 5). Specifically, given *minScore* (alias minimal density), PL_1 and PL_2 are computed to be the maximum values satisfying $score(PL_1, false) \geq minScore$ and $score(PL_2, true) \geq minScore$ (see Sect. 4 for further details).

Let us start with the MPEG-7 dataset and the cache size limit of 150,000 objects. Figure 4a compares the traditional approach (i.e., $PL_1 = PL_2$) and the proposed approach when the position limits are set according to the minimal density limit. The new approach defeats the traditional one according to the results, i.e., a better precision is achieved for a given processing time, or a given precision can be achieved faster. Figure 4b compares the optimal setting of the position limits to other position limit settings as presented in Sect. 6.3. The optimal setting (solid line) does not drop below the other position limit settings (dashed lines), thus the optimality of the position limits is experimentally verified.

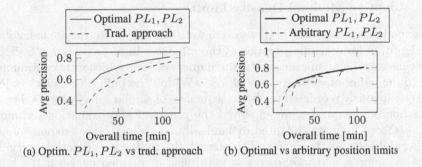

(a) Optim. PL_1, PL_2 vs trad. approach   (b) Optimal vs arbitrary position limits

Fig. 4. Optimal PL_1, PL_2 setting for MPEG-7 dataset; 10,000 30-NN queries

We also carried out experiments with high precision search using the MPEG-7 dataset. The cache size limit was set to 500,000 (i.e., 5% of the dataset). In Fig. 5, we can see the proposed approach outperforms the traditional technique even for the precisions higher than 0.9. For instance, to reach the precision of 0.94, it is possible to save more than 15% of the processing time.

Concerning the Caffe dataset, Fig. 6 shows the comparison of optimal position limit settings and the traditional approach with the cache size limit of 150,000 objects. The improvement for the Caffe dataset is not as large as for the MPEG-7 dataset, but still we are, for example, able to achieve the precision of 0.85 with more than 15% time savings.

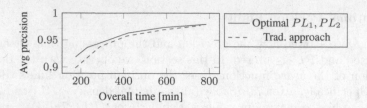

Fig. 5. Optimal PL_1, PL_2 vs trad. approach; high precision search; MPEG-7 dataset

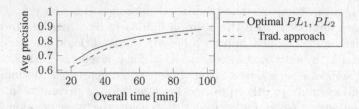

Fig. 6. Optimal PL_1, PL_2 vs traditional approach; Caffe dataset

6.6 Adaptive Minimal Density Limit

The experiments presented in this section were carried out to analyze behavior of the algorithm for dynamic adaption of the *minScore* limit presented in Sect. 5.2. This time we took an unordered stream of query objects which were continuously sent to the buffer of waiting query objects. Within the buffer, the query objects were continuously reordered to obtain sequences of similar query objects. For the evaluation of the queries, we employed the proposed approximation technique. The *minScore* limit was adapted dynamically according to the current number of query objects in the buffer. The size of the buffer was checked every 100 processed queries and *minScore* was multiplied or divided by 1.5 if the buffer size was too high or too low, respectively. The initial *minScore* was set so that $PL_1 = 5,000$.

We used the MPEG-7 dataset; the backlog limit BL was set to 50,000 query objects. The maximal cache size was 150,000 objects. The experiment was run for 5 h in total, and we changed the input rates of incoming query objects. For the first 2 h, a new query object was added to the buffer every 60 ms, every 600 ms for another 1 h, every 70 ms for the next hour and every 300 ms for the final hour.

Figure 7a captures the evolution of the number of query objects waiting for processing in the buffer throughout the time. The moments when the input frequency is changed are depicted by the vertical lines. At first, the buffer size grows far over the backlog limit before it settles on the 50,000 line. This is because of the nature of the query reordering method which needs some "warm up" time before it can produce effective ordering of the queries. After that we can see that the size of the buffer is stabilized at the backlog limit even though the rate of incoming query objects changes.

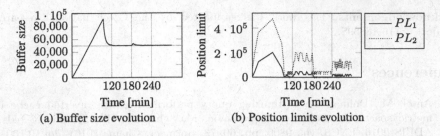

Fig. 7. Adaptive minimal density; input frequencies: 60 ms, 600 ms, 70 ms, 300 ms

Figure 7b tracks the applied position limits during the experiment. Except for the "warm up" phase, the position limits correlate with the input frequency as expected enabling to maintain the stable buffer size while keeping high precision of the results.

In Table 1, we summarize the parameters used by the presented approach.

Table 1. System parameters

$answerProbability(pos)$	Set automatically during the training phase
$time(isCached)$	Set automatically during the training phase
PL_1, PL_2	Adjusted automatically according to the backlog
$maxCacheSize$	Defined by a user
BL	The backlog limit defined by a user
adaptive min. dens. factor	Defined by a user; influences the speed of adaption of PL_1, PL_2

7 Conclusion

We have presented an approximation technique designed for processing a stream of kNN similarity search queries. A score is computed for individual objects of the priority queue by a combination of the probability of the object to be a part of the answer set and the time needed to examine the object. Based on the score, it is determined whether the object is worth examining during the query evaluation.

Compared to the baseline of a traditional approximation technique using the number of examined objects as the stop condition, the proposed approach allows to achieve better trade-off between the processing time and the precision of the similarity search. We have also shown a practical use of the proposed approach by showing its ability to react to the changes in the frequency of incoming streamed query objects.

Acknowledgements. This work was supported by the Czech national research project GA16-18889S.

References

1. Antol, M., Dohnal, V.: Optimizing query performance with inverted cache in metric spaces. In: Pokorný, J., Ivanović, M., Thalheim, B., Šaloun, P. (eds.) ADBIS 2016. LNCS, vol. 9809, pp. 60–73. Springer, Cham (2016). doi:10.1007/978-3-319-44039-2_5
2. Barrios, J.M., Bustos, B., Skopal, T.: Analyzing and dynamically indexing the query set. Inf. Syst. **45**, 37–47 (2014)
3. Batko, M., Falchi, F., Lucchese, C., Novak, D., Perego, R., Rabitti, F., Sedmidubsky, J., Zezula, P.: Building a web-scale image similarity search system. Multimedia Tools Appl. **47**(3), 599–629 (2010)
4. Braunmuller, B., Ester, M., Kriegel, H.P., Sander, J.: Multiple similarity queries: a basic DBMS operation for mining in metric databases. IEEE TKDE **13**(1), 79–95 (2001)
5. Brisaboa, N.R., Cerdeira-Pena, A., Gil-Costa, V., Marin, M., Pedreira, O.: Efficient similarity search by combining indexing and caching strategies. In: Italiano, G.F., Margaria-Steffen, T., Pokorný, J., Quisquater, J.-J., Wattenhofer, R. (eds.) SOFSEM 2015. LNCS, vol. 8939, pp. 486–497. Springer, Heidelberg (2015). doi:10.1007/978-3-662-46078-8_40
6. Budikova, P., Batko, M., Zezula, P.: Evaluation platform for content-based image retrieval systems. In: Gradmann, S., Borri, F., Meghini, C., Schuldt, H. (eds.) TPDL 2011. LNCS, vol. 6966, pp. 130–142. Springer, Heidelberg (2011). doi:10.1007/978-3-642-24469-8_15
7. Chung, Y.C., Su, I.F., Lee, C., Liu, P.C.: Multiple k nearest neighbor search. In: World Wide Web, pp. 1–28 (2016)
8. Falchi, F., Lucchese, C., Orlando, S., Perego, R., Rabitti, F.: Similarity caching in large-scale image retrieval. Inf. Process. Manag. **48**(5), 803–818 (2012)
9. Jia, Y., Shelhamer, E., Donahue, J., Karayev, S., Long, J., Girshick, R., Guadarrama, S., Darrell, T.: Caffe: convolutional architecture for fast feature embedding. In: Proceedings of the ACM International Conference on Multimedia, pp. 675–678. ACM (2014)
10. Nalepa, F., Batko, M., Zezula, P.: Enhancing similarity search throughput by dynamic query reordering. In: Hartmann, S., Ma, H. (eds.) DEXA 2016. LNCS, vol. 9828, pp. 185–200. Springer, Cham (2016). doi:10.1007/978-3-319-44406-2_14
11. Novak, D., Batko, M., Zezula, P.: Metric index: an efficient and scalable solution for precise and approximate similarity search. Inf. Syst. **36**(4), 721–733 (2011)
12. Pandey, S., Broder, A., Chierichetti, F., Josifovski, V., Kumar, R., Vassilvitskii, S.: Nearest-neighbor caching for content-match applications. In: Proceedings of the 18th International Conference on World Wide Web, pp. 441–450. ACM (2009)
13. Skopal, T., Lokoc, J., Bustos, B.: D-cache: universal distance cache for metric access methods. IEEE Trans. Knowl. Data Eng. **24**(5), 868–881 (2012)
14. Solar, R., Gil-Costa, V., Marín, M.: Evaluation of static/dynamic cache for similarity search engines. In: Freivalds, R.M., Engels, G., Catania, B. (eds.) SOFSEM 2016. LNCS, vol. 9587, pp. 615–627. Springer, Heidelberg (2016). doi:10.1007/978-3-662-49192-8_50

15. Sundaram, N., Turmukhametova, A., Satish, N., Mostak, T., Indyk, P., Madden, S., Dubey, P.: Streaming similarity search over one billion tweets using parallel locality-sensitive hashing. Proc. VLDB Endowment **6**(14), 1930–1941 (2013)

16. Zezula, P., Amato, G., Dohnal, V., Batko, M.: Similarity Search: The Metric Space Approach, vol. 32. Springer, Boston (2006)

ANN-Benchmarks: A Benchmarking Tool for Approximate Nearest Neighbor Algorithms

Martin Aumüller[1]([✉]), Erik Bernhardsson[2], and Alexander Faithfull[1]

[1] IT University of Copenhagen, Copenhagen, Denmark
maau@itu.dk, alef@itu.dk
[2] Better, New York, USA
mail@erikbern.com

Abstract. This paper describes ANN-Benchmarks, a tool for evaluating the performance of in-memory approximate nearest neighbor algorithms. It provides a standard interface for measuring the performance and quality achieved by nearest neighbor algorithms on different standard data sets. It supports several different ways of integrating k-NN algorithms, and its configuration system automatically tests a range of parameter settings for each algorithm. Algorithms are compared with respect to many different (approximate) quality measures, and adding more is easy and fast; the included plotting front-ends can visualise these as images, LATEX plots, and websites with interactive plots. ANN-Benchmarks aims to provide a constantly updated overview of the current state of the art of k-NN algorithms. In the short term, this overview allows users to choose the correct k-NN algorithm and parameters for their similarity search task; in the longer term, algorithm designers will be able to use this overview to test and refine automatic parameter tuning. The paper gives an overview of the system, evaluates the results of the benchmark, and points out directions for future work. Interestingly, very different approaches to k-NN search yield comparable quality-performance trade-offs. The system is available at http://sss.projects.itu.dk/ann-benchmarks/.

1 Introduction

Nearest neighbor search is one of the most fundamental tools in many areas of computer science, such as image recognition, machine learning, and computational linguistics. For example, one can use nearest neighbor search on image descriptors such as MNIST [17] to recognize handwritten digits, or one can find semantically similar phrases to a given phrase by applying the word2vec embedding [22] and finding nearest neighbors. The latter can, for example, be used to tag articles on a news website and recommend new articles to readers that have shown an interest in a certain topic. In some cases, a generic nearest neighbor

---

The research of the first and third authors has received funding from the European Research Council under the European Union's 7th Framework Programme (FP7/2007-2013)/ERC grant agreement no. 614331.

C. Beecks et al. (Eds.): SISAP 2017, LNCS 10609, pp. 34–49, 2017.
DOI: 10.1007/978-3-319-68474-1_3

search under a suitable distance or measure of similarity offers surprising quality improvements [7].

In many applications, the data points are described by high-dimensional vectors, usually ranging from 100 to 1000 dimensions. A phenomenon called the *curse of dimensionality*, the existence of which is also supported by popular algorithmic hardness conjectures (see [2,28]), tells us that to obtain the true nearest neighbors, we have to use either linear time (in the size of the dataset) or time/space that is exponential in the dimensionality of the dataset. In the case of *massive* high-dimensional datasets, this rules out *efficient* and *exact* nearest neighbor search algorithms.

To obtain efficient algorithms, research has focused on allowing the returned neighbors to be an *approximation* of the true nearest neighbors. Usually, this means that the answer to finding the nearest neighbors to a query point is judged by how *close* (in some technical sense) the result set is to the set of true nearest neighbors.

There exist many different algorithmic techniques for finding approximate nearest neighbors. Classical algorithms such as kd-trees [4] or M-trees [8] can simulate this by terminating the search early, for example shown by Zezula et al. [29] for M-trees. Other techniques [20,21] build a graph from the dataset, where each vertex is associated with a data point, and a vertex is adjacent to its true nearest neighbors in the data set. Others involve projecting data points into a lower-dimensional space using hashing. A lot of research has been conducted with respect to locality-sensitive hashing (LSH) [14], but there exist many other techniques that rely on hashing for finding nearest neighbors; see [27] for a survey on the topic. We note that, in the realm of LSH-based techniques, algorithms guarantee sublinear query time, but solve a problem that is only distantly related to finding the k nearest neighbors of a query point. In practice, this could mean that the algorithm runs *slower* than a linear scan through the data, and countermeasures have to be taken to avoid this behavior [1,25].

Given the difficulty of the problem of finding nearest neighbors in high-dimensional spaces and the wide range of different solutions at hand, it is natural to ask how these algorithms perform in empirical settings. Fortunately, many of these techniques already have good implementations: see, e.g., [5,19,23] for tree-based, [6,10] for graph-based, and [3] for LSH-based solutions. This means that a new (variant of an existing) algorithm can show its worth by comparing itself to the many previous algorithms on a collection of standard benchmark datasets with respect to a collection of quality measures. What often happens, however, is that the evaluation of a new algorithm is based on a small set of competing algorithms and a small number of selected datasets. This approach poses problems for everyone involved:

(i) *The algorithm's authors*, because competing implementations might be unavailable, they might use other conventions for input data and output of results, or the original paper might omit certain required parameter settings (and, even if these are available, exhaustive experimentation can take lots of CPU time).

(ii) *Their reviewers and readers*, because experimental results are difficult to reproduce and the selection of datasets and quality measures might appear selective.

This paper proposes a way of standardizing benchmarking for nearest neighbor search algorithms, taking into account their properties and quality measures. Our benchmarking framework provides a unified approach to experimentation and comparison with existing work. The framework has already been used for experimental comparison in other papers [20] (to refer to parameter choice of algorithms) and algorithms have been contributed by the community, e.g., by the authors of NMSLib [6] and FALCONN [3]. An earlier version of our framework is already widely used as a benchmark referred to from other websites, see, e.g., [3,5,6,10,19].

Related work. Generating reproducible experimental results is one of the greatest challenges in many areas of computer science, in particular in the machine learning community. As an example, openml.org [26] and codalab.org provide researchers with excellent platforms to share reproducible research results.

The automatic benchmarking system developed in connection with the mlpack machine learning library [9,12] shares many characteristics with our framework: it automates the process of running algorithms with preset parameters on certain datasets, and can visualize these results. However, the underlying approach is very different: it calls the algorithms natively and parses the standard output of the algorithm for result metrics. Consequently, the system relies solely on the correctness of the algorithms' own implementations of quality measures, and adding new quality measures needs changes in *every single* algorithm implementation. Very recently, Li et al. [18] presented a comparison of many approximate nearest neighbor algorithms, including many algorithms that are considered in our framework as well. Their approach is to take existing algorithm implementations and to heavily modify them to fit a common style of query processing, in the process changing compiler flags (and sometimes even core parts of the implementation). There is no general framework, and including new features again requires manual changes in each single algorithm.

Our benchmarking framework does not aim to replace these tools; instead, it complements them by taking a different approach. We require that algorithms expose a simple programmatic interface, which is only required to return the set of nearest neighbors of a given query, after preprocessing the set of data points. All the timing and quality measure computation is conducted within our framework, which lets us add new metrics without rerunning the algorithms, if the metric can be computed from the set of returned elements. Moreover, we benchmark each implementation as intended by the author. That means that we benchmark *implementations*, rather than *algorithmic ideas* [16].

Contributions. We describe our system for benchmarking approximate nearest neighbor algorithms with the general approach described in Sect. 3. The system allows for easy experimentation with k-NN algorithms, and visualizes algorithm runs in an approachable way. Moreover, in Sect. 4 we use our benchmark suite to overview the performance and quality of current state-of-the-art k-NN algorithms. This allows us to identify areas that already have competitive algorithms, to compare different methodological approaches to nearest neighbor search, but also to point out challenging datasets and metrics, where good implementations are missing or do not take full advantage of properties of the underlying metric. Having this overview has immediate practical benefits, as users can select the right combination of algorithm and parameters for their application. In the longer term, we expect that more algorithms will become able to tune their own parameters according to the user's needs, and our benchmark suite will also serve as a testbed for this automatic tuning.

2 Problem Definition and Quality Measures

We assume that we want to find nearest neighbors in a space X with a distance measure dist: $X \times X \to \mathbb{R}$, for example the d-dimensional Euclidean space $\mathbb{R}^d$ under Euclidean distance (l_2 norm), or Hamming space $\{0,1\}^d$ under Hamming distance.

An algorithm $\mathcal{A}$ for nearest neighbor search builds a data structure $\mathrm{DS}_\mathcal{A}$ for a data set $S \subset X$ of n points. In a preprocessing phase, it creates $\mathrm{DS}_\mathcal{A}$ to support the following type of queries: For a query point $q \in X$ and an integer k, return a *result tuple* $\pi = (p_1, \ldots, p_{k'})$ of $k' \leq k$ distinct points from S that are "close" to the query q. Nearest neighbor search algorithms generate π by refining a set $C \subseteq S$ of candidate points w.r.t. q by choosing the k closest points among those using distance computations. The size of C (and thus the number of distance computations) is denoted by N. We let $\pi^* = (p_1^*, \ldots, p_k^*)$ denote the tuple containing the true k nearest neighbors for q in S (where ties are broken arbitrarily). We assume in the following that all tuples are sorted according to their distance to q.

2.1 Quality Measures

We use different notions of "recall" as a measure of the quality of the result returned by the algorithm. Intuitively, recall is the ratio of the number of points in the result tuple that are true nearest neighbors to the number k of true nearest neighbors. However, this intuitive definition is fragile when distances are not distinct or when we try to add a notion of approximation to it. To avoid these issues, we use the following distance-based definitions of recall and $(1+\varepsilon)$-approximative recall, that take the distance of the k-th true nearest neighbor as threshold distance.

$$\mathrm{recall}(\pi, \pi^*) = \frac{|\{p \text{ contained in } \pi \mid \mathrm{dist}(p,q) \leq \mathrm{dist}(p_k^*,q)\}|}{k}$$

$$\mathrm{recall}_\varepsilon(\pi, \pi^*) = \frac{|\{p \text{ contained in } \pi \mid \mathrm{dist}(p,q) \leq (1+\varepsilon)\mathrm{dist}(p_k^*,q)\}|}{k}, \quad \text{for } \varepsilon > 0.$$

(If all distances are distinct, $\mathrm{recall}(\pi, \pi^*)$ matches the intuitive notion of recall.)

Table 1. Performance measures used in the framework.

Name of measure	Computation of measure
Index size of DS	Size of DS after preprocessing finished (in kB)
Index build time DS	Time it took to build DS (in seconds)
Number of distance computations	N
Time of a query	Time it took to run the query and generate result tuple π

We note that (approximate) recall in high dimensions is sometimes criticised; see, for example, [6, Sect. 2.1]. We investigate the impact of approximation as part of the evaluation in Sect. 4, and plan to include other quality measures such as position-related measures [29] in future work.

2.2 Performance Measures

With regard to the performance, we use the performance measures defined in Table 1, which are divided into measures of the performance of the preprocessing step, i.e., generation of the data structure, and measures of the performance of the query algorithm. With respect to the query performance, different communities are interested in different cost values. Some rely on actual timings of query times, where others rely on the number of distance computations. The framework can take both of these measures into account. However, none of the currently included algorithms report the number of distance computations.

3 System Design

ANN-Benchmarks is implemented as a Python library with several different front-ends: one script for running experiments and a handful of others for working with and plotting results. It is designed to be run in a virtual machine or Docker container, and so comes with shell scripts for automatically installing algorithm implementations, dependencies, and datasets.

The experiment front-end has some parameters of its own that influence what algorithm implementations will be tested: the dataset to be searched (and an optional dataset of query points), the number of neighbours to search for, and the distance metric to be used to compare points. The plotting front-ends are also aware of these parameters, which are used to select and label plots.

This section gives only a high-level overview of the system; see http://sss.projects.itu.dk/ann-benchmarks/ for more detailed technical information.

3.1 Installing Algorithms and Datasets

Each dataset and library has a shell script that downloads, builds and installs it. These scripts are built on top of a shell function library that defines a few common operations, like cloning and patching a Git repository or downloading a dataset and checking its integrity. Datasets may also need to be converted; we include Python scripts for converting a few commonly-used formats into the plain-text format used by our system, and the shell scripts make use of these.

Although we hope that algorithm libraries will normally bundle their own Python bindings, our shell function library can also apply a patch series to an implementation once it has been downloaded, allowing us to (temporarily) carry patches for bindings that will later be moved upstream.

Adding support for a new algorithm implementation to ANN-Benchmarks is as easy as writing a script to install it and its dependencies, making it available to Python by writing a wrapper (or by reusing an existing one), and adding the parameters to be tested to the configuration files. Most of the installation scripts fetch the latest version of their library from its Git repository, but there is no requirement to do this; indeed, installing several different versions of a library would make it possible to use the framework for regression testing.

Algorithm wrappers. To be usable by our system, each of the implementations to be tested must have some kind of Python interface. Many libraries already provide their own Python wrappers, either written by hand or automatically generated using a tool like SWIG; others are implemented partly or entirely in Python.

To bring implementations that do not provide a Python interface into the framework, we specify a simple text-based protocol that supports the few operations we care about: parameter configuration, sending training data, and running queries. The framework comes with a wrapper that communicates with external programs using this protocol. In this way, experiments can be run in external front-end processes implemented in any programming language.

The protocol has been designed to be easy to implement. Every message is a line of text that will be split into tokens according to the rules of the POSIX shell, good implementations of which are available for most programming languages. The protocol is flexible and extensible: front-ends are free to include extra information in replies, and they can also implement special configuration options that cause them to diverge from the protocol's basic behaviour. As an example, we provide a simple C implementation that supports an alternative query mode in which parsing and preparing a query data point and running a query are two different commands. (As the overhead of parsing a string representation of a data point is introduced by the use of the protocol, removing it makes the timings more representative.)

Comparing implementations. We note at this point that we are explicitly comparing algorithm *implementations*. Implementations make many different decisions that will affect their performance and two implementations of the same algorithm can have somewhat different performance characteristics [16]. Our framework supports other quality measures, such as the number of distance computations, which is more suited for comparing algorithms on a more abstract level; however, the implementations we consider do not yet support this measure.

3.2   Loading Datasets and Computing Ground Truth

Once we have datasets available, we must load them and compute the ground truth for the query set: the true nearest neighbours for each query point, along with their distances. This ground truth is passed, along with the values obtained by each experiment, to the functions used by the plotting scripts to calculate the quality metrics.

The query set for a dataset is, by default, a pseudorandomly-selected set of ten thousand entries separated from the rest of the training data. If this behavior is not wanted, datasets can declare a different number of queries in their metadata, or the user can provide an explicit query set instead.

Depending on the values of the system's own configuration parameters, many different sets may have to be computed. Each of these is stored in a separate cache file.

3.3   Creating Algorithm Instances

After loading the dataset, the framework moves on to creating the algorithm instances. It does so based on a YAML configuration file that specifies a hierarchy of dictionaries: the first level specifies the point type, the second the distance metric, and the third each algorithm implementation to be tested. Each implementation gives the name of its wrapper's Python constructor; a number of other entries are then expanded to give the arguments to that constructor. Figure 1 shows an example of this configuration file.

```
float:
  any:
    annoy:
      constructor: Annoy
      base-args: ["@metric"]
      run-groups:
        one-or-two-hundred-trees:
          args: [[100, 200], [100, 200, 400, 1000]]
        four-hundred-trees:
          args: [400, [1000, 2000, 4000, 10000]]
```

Fig. 1. An example of a fragment of an algorithm configuration file.

The base-args list consists of those arguments that should be prepended to every invocation of the constructor. Figure 1 also shows one of the special keywords, "@metric", that is used to pass one of the framework's configuration parameters to the constructor.

Algorithms must specify one or more "run groups", each of which will be expanded into one or more lists of constructor arguments. The args entry completes the argument list, but not directly: instead, the Cartesian product of all of its entries is used to generate *many* lists of arguments. The annoy entry in Fig. 1, for example, expands into twelve different algorithm instances, from Annoy("euclidean", 100, 100) to Annoy("euclidean", 400, 10000).

3.4   The Experiment Loop

Once the framework knows what instances should be run, it moves on to the experiment loop. (Figure 2 gives an overview of the loop.) Each algorithm instance is run in a separate subprocess. This makes it easy to clean up properly after each run: simply destroying the subprocess takes care of everything. This approach also gives us a simple and implementation-agnostic way of computing the memory usage of an implementation: the subprocess takes a snapshot of its memory consumption before and after initialising the algorithm instance's data structures and compares the two.

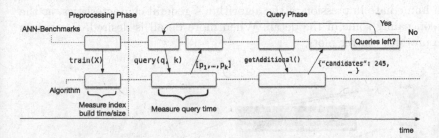

Fig. 2. Overview of the interaction between ANN-Benchmarks and an algorithm instance under test. The instance builds an index data structure for the dataset X in the preprocessing phase. During the query phase, it returns k data points for each query point; after answering a query, it can also report any extra information it might have, such as the size of the candidate set.

The complete results of each run are sent back to the main process using a pipe. The main process performs a blocking, timed wait on the other end of the pipe, and will destroy the subprocess if the user-configurable timeout is exceeded before any results are available.

3.5   Results and Metrics

For each run, we store the full name – including the parameters – of the algorithm instance, the time it took to evaluate the training data, and the near neighbours returned by the algorithm, along with their distance from the query point. (To avoid affecting the timing of algorithms that do not indicate the distance of a result, the experiment loop independently re-computes distance values after the run is otherwise finished.) Each run is stored in a separate file in a directory hierarchy that encodes the framework's configuration. Keeping runs in separate files makes them easy to compress, easy to enumerate, and easy to re-run, and individual results – or sets of results – can easily be shared to make results more transparent.

Metric functions are passed the ground truth and the results for a particular run; they can then compute their result however they see fit. Adding a new quality metric is a matter of writing a short Python function and adding it to an internal data structure; the plotting scripts query this data structure and will automatically support the new metric.

3.6   Frontend

ANN-Benchmarks provides two options to evaluate the results of the experiments: a script to generate individual plots using Python's matplotlib and a script to generate a website that summarizes the results and provides interactive plots with the option to export the plot as LATEX code using pgfplots. See Fig. 3 for an example. Plots depict the Pareto frontier over all runs of an algorithm; this gives an immediate impression of the algorithm's general characteristics, at the cost of concealing some of the detail. When more detail is desired, the scripts can also produce scatter plots.

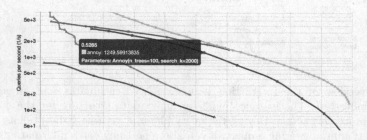

Fig. 3. Interactive plot screen from framework's website (cropped). Plot shows "Queries per second" (*y*-axis, log-scaled) against "Recall" (*x*-axis, not shown). Highlighted data point corresponds to a run of Annoy with parameters as depicted, giving about 1249 queries per second for a recall of about 0.52.

4   Evaluation

In this section we present a short evaluation of our findings from running benchmarks in the benchmarking framework.

Experimental setup. All experiments were run in Docker containers on *Amazon EC2 c4.2xlarge* instances that are equipped with Intel Xeon E5-2666v3 processors (4 cores available, 2.90 GHz, 25.6 MB Cache) and 15 GB of RAM running Amazon Linux. We ran a single experiment multiple times to verify that performance was reliable, and compared the experiments results with a 4-core Intel Core i7-4790 clocked at 3.6 GHz with 32 GB RAM. While the latter was a little faster, the relative order of algorithms remained stable. For each parameter setting and dataset, the algorithm was given thirty minutes to build the index and answer the queries.

Tested Algorithms. Table 2 summarizes the algorithms that are used in the evaluation; see the references provided for details. More implementations are

Table 2. Overview of tested algorithms (abbr. in parentheses). Implementations in *italics* have "recall" as quality measure provided as an input parameter.

Principle	Algorithms
k-NN graph	KGraph (KG) [10], SWGraph (SWG) [6,21], HNSW [6,20]
Tree-based	*FLANN* [23], *BallTree (BT)* [6]
LSH	FALCONN (FAL) [3], *MPLSH* [6,11]
Random-proj. forest	Annoy (A) [5], RPForest (RPF) [19]
Other	Multi-Index Hashing (MIH) [24] (exact Hamming search)

included in the framework, but turned out to be non-competitive (details can be found on the framework's website). The scripts that set up the framework automatically fetch the most current version found in each algorithm's repository.

Datasets. The datasets used in this evaluation are summarized in Table 3. Results for other datasets are found on the framework's website. The NYTimes dataset was generated by building tf-idf descriptors from the bag-of-words version, and embedding them into a lower dimensional space using the Johnson-Lindenstrauss Transform [15]. Hamming space versions have been generated by applying Spherical Hashing [13] using the implementation provided by the authors of [13]. The random dataset Rand-Angular (where data points lie on the surface of the d-dimensional unit sphere) is generated by choosing 500 query points at random and putting clusters of 500 points at distance around $\alpha\sqrt{2}/3$, where α grows linearly from 0 to 1 with step size 1/500. Each cluster has 500 points at distance around $2\alpha\sqrt{2}/3$ added. The rest of the dataset consists of random data points, 500 of which are chosen as the other set of query points (with closest neighbors expected to be at distance $\sqrt{2}$).

Table 3. Datasets under consideration

Dataset	Data/Query points	Dimensionality	Metric
SIFT	1 000 000/10 000	128	Euclidean
GLOVE	1 183 514/10 000	100	Angular/Cosine
NYTimes	234 791/10 000	256	Euclidean
Rand-Angular	1 000 000/1 000	128	Angular/Cosine
SIFT-Hamming	1 000 000/10 000	256	Hamming
NYTimes-Hamming	234 791/10 000	128	Hamming

Parameters of Algorithms. Most algorithms do not allow the user to explicitly specify a quality target—in fact, only three implementations from Table 2 provide "recall" as an input parameter. We used our framework to test many parameter settings at once. The detailed settings tested for each algorithm can be found on the framework's website.

4.1   Objectives of the Experiments

We used the benchmarking framework to find answers to the following questions:

(Q1) Performance. Given a dataset, a quality measure and a number k of nearest neighbors to return, how do algorithms compare to each other with respect to different performance measures, such as query time or index size?

(Q2) Robustness. Given an algorithm $\mathcal{A}$, how is its performance and result quality influenced by the dataset and the number of returned neighbors?

(Q3) Approximation. Given a dataset, a number k of nearest neighbors to return, and an algorithm $\mathcal{A}$, how does its performance improve when the returned neighbors can be an approximation? Is the effect comparable for different algorithms?

(Q4) Embeddings. Equipped with a framework with many different datasets and distance metrics, we can try interesting combinations. How do algorithms targeting Euclidean space or Cosine similarity perform in, say, Hamming space? How does replacing the internals of an algorithm with Hamming space related techniques improve its performance?

The following discussion is based on a combination of the plots found on the framework's website; see the website for more complete and up-to-date results.

4.2   Discussion

(Q1) Performance. Figure 4 shows the relationship between an algorithm's achieved recall and the number of queries it can answer per second (its QPS) on the two datasets GLOVE (Cosine similarity) and SIFT (Euclidean distance) for 10- and 100-nearest neighbor queries.

For GLOVE, we observe that the graph-based algorithms HNSW and SWGraph, the LSH-based FALCONN, and the "random-projection forest"-based Annoy algorithm are fastest. For high recall values, HNSW is fastest, while for lower recall values, FALCONN achieves highest QPS. We can also observe the importance of implementation decisions: although Annoy and RPForest are both built upon the same algorithmic idea, they have very different performance characteristics.

On SIFT, all tested algorithms can achieve close to perfect recall. In particular, the graph-based algorithms (along with KGraph) are fastest, followed by Annoy. FALCONN, BallTree, and FLANN have very similar performance.

Very few of these algorithms can tune themselves to produce a particular recall value. In particular, the fastest algorithms on the GLOVE dataset expose many parameters, leaving the user to find the combination that works best. The KGraph algorithm, on the other hand, uses only a single parameter, which—even in its "smallest" choice—still guarantees recall at least 0.9 on SIFT. FLANN manages to tune itself for a particular recall value on the SIFT dataset; for GLOVE with high recall values, however, the tuning does not complete within the time limit, especially with 100-NN.

Figure 5 relates an algorithm's performance to its index size. High recall can be achieved with small indexes by probing many points; however, this probing

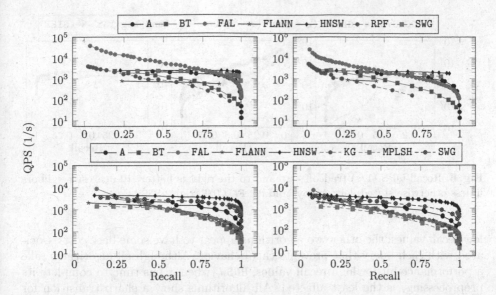

Fig. 4. Recall-QPS (1/s) tradeoff - up and to the right is better. Top: GLOVE, bottom: SIFT; left: 10-NN, right: 100-NN.

is expensive, and so the QPS drops dramatically. To reflect this performance cost, we scale the size of the index by the QPS it achieves. This reveals that, on SIFT, SWGraph and FLANN achieve good recall values with small indexes. Both BallTree and HNSW show a similar behavior to each other. Annoy and FALCONN need rather large indexes to achieve high QPS. The picture is very different on GLOVE, where FALCONN provides the best ratio up to recall 0.8, only losing to the graph-based approaches at higher recall values.

(Q2) Robustness. Figure 6 plots recall against QPS for Annoy, FALCONN, and HNSW with fixed parameters over a range of datasets. Each algorithm has a distinct performance curve. In particular, FALCONN has very fast query times for

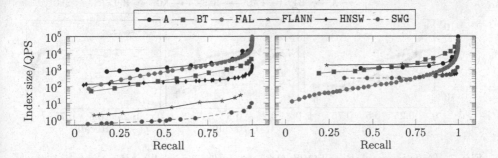

Fig. 5. Recall-Index size (kB)/QPS (s) tradeoff - down and to the right is better. Left: SIFT ($k = 100$), right: GLOVE ($k = 10$).

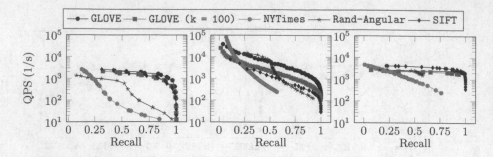

Fig. 6. Recall-QPS (1/s) tradeoff - up and to the right is better, 10-nearest neighbors unless otherwise stated, left: `Annoy`, middle: `FALCONN`, right: `HNSW`.

low recall values; the other two algorithms appear to have some base cost associated with each query that prevents this behavior. Although all algorithms take a performance hit for high recall values, `HNSW` (when it has time to complete its preprocessing) is the least affected. All algorithms show a sharp transition for the random dataset; this is to be expected based on the dataset's composition (cf. **Datasets** above).

(Q3) Approximation. Figure 7 relates achieved QPS to the (approximate) recall of an algorithm. The plots show results on the NYTimes dataset for recall with no approximation and approximation factors of 1.01 and 1.1. The dataset is notoriously difficult; with no approximation, only a handful of algorithms can achieve a recall above 0.98. However, we know the candidate sets of most algorithms are very close to the true nearest neighbors, as even a very small approximation factor of 1.01 improves the situation drastically: all algorithms get more than 0.9 recall. Allowing for an approximation of 1.1 yields very high performance for most algorithms, although some benefit more than others: `FALCONN`, for example, now always outperforms `HNSW`, while `Annoy` suddenly leaps ahead of its competitors.

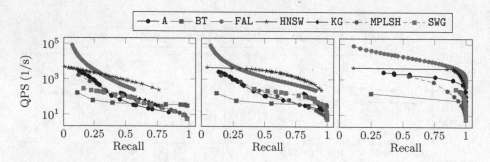

Fig. 7. (Approximate) Recall-QPS (1/s) tradeoff - up and to the right is better, NYTimes dataset; left: $\varepsilon = 0$, middle: $\varepsilon = 0.01$, right: $\varepsilon = 0.1$.

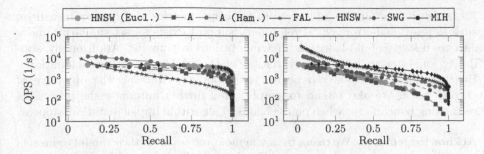

Fig. 8. Recall-QPS (1/s) tradeoff - up and to the right is better, 10-nearest neighbors, left: SIFT-Hamming, right: NYTimes-Hamming.

(Q4) Embeddings. Figure 8 shows a comparison between selected algorithms on the binary versions of SIFT and NYTimes. The performance plot for HNSW in the original Euclidean-space version is also shown. On SIFT, algorithms perform very similarly to the original Euclidean-space version (see Fig. 4), which indicates that the queries are as difficult to answer in the embedded space as they are in the original space. The behavior is very different on NYTimes, where all algorithms improve their speed *and* quality. The only dedicated Hamming space algorithm shown here, exact multi-index hashing, shows good performance at around 180 QPS on SIFT and 400 QPS on NYTimes.

As an experiment, we created a Hamming space-aware version of Annoy, using popcount for distance computations, and sampling single bits (as in Bitsampling LSH [14]) instead of choosing hyperplanes. This version is an order of magnitude faster on NYTimes; on SIFT, the running times converge for high recall values.

The embedding into Hamming space does have some consistent benefits that we do not show here. Hamming space-aware algorithms should always have smaller index sizes, for example, due to the compactness of bit vectors stored as binary strings.

5   Conclusion and Further Work

We introduced ANN-Benchmarks, an automated benchmarking system for approximate nearest neighbor algorithms. We described the system and used it to evaluate existing algorithms. Our evaluation showed that well-engineered solutions for Euclidean and Cosine distance exist, and many techniques allow for fast nearest neighbor search algorithms. At the moment, graph-based approaches such as HNSW or KGraph outperform the other approaches for very high recalls, whereas LSH-based solutions such as FALCONN yield very high performance at lower recall values. Index building for graph-based approaches takes a long time for datasets with difficult queries. We did not find solutions targeting Hamming space under Hamming distance, but showed that substituting Hamming space-specific techniques into more general algorithms can greatly improve their running time.

In future, we aim to add support for other metrics and quality measures, such as positional errors [29]. Preliminary support exists for set similarity under Jaccard distance, but algorithm implementations are missing. Additionally, similarity joins are an interesting variation of the problem worth benchmarking. Benchmarking GPU-powered nearest neighbor algorithms is the objective of current work. We also intend to simplify and further automate the process of re-running benchmarks when new versions of algorithm implementations appear.

Acknowledgements. We thank the anonymous reviewers for their careful comments that allowed us to improve the paper. The first and third authors thank all members of the algorithm group at ITU Copenhagen for fruitful discussions.

References

1. Ahle, T.D., Aumüller, M., Pagh, R.: Parameter-free locality sensitive hashing for spherical range reporting. In: SODA 2017, pp. 239–256
2. Alman, J., Williams, R.: Probabilistic polynomials and hamming nearest neighbors. In: FOCS 2015, pp. 136–150
3. Andoni, A., Indyk, P., Laarhoven, T., Razenshteyn, I.P., Schmidt, L.: Practical and optimal LSH for angular distance. In: NIPS 2015, pp. 1225–1233. https://falconn-lib.org/
4. Bentley, J.L.: Multidimensional binary search trees used for associative searching. Commun. ACM **18**(9), 509–517 (1975)
5. Bernhardsson, E.: Annoy. https://github.com/spotify/annoy
6. Boytsov, L., Naidan, B.: Engineering efficient and effective non-metric space library. In: Brisaboa, N., Pedreira, O., Zezula, P. (eds.) SISAP 2013. LNCS, vol. 8199, pp. 280–293. Springer, Heidelberg (2013). doi:10.1007/978-3-642-41062-8_28
7. Boytsov, L., Novak, D., Malkov, Y., Nyberg, E.: Off the beaten path: let's replace term-based retrieval with k-NN search. In: CIKM 2016, pp. 1099–1108
8. Ciaccia, P., Patella, M., Zezula, P.: M-tree: an efficient access method for similarity search in metric spaces. In: VLDB 1997, pp. 426–435 (1997)
9. Curtin, R.R., Cline, J.R., Slagle, N.P., March, W.B., Ram, P., Mehta, N.A., Gray, A.G.: MLPACK: a scalable C++ machine learning library. J. Mach. Learn. Res. **14**, 801–805 (2013)
10. Dong, W.: KGraph. https://github.com/aaalgo/kgraph
11. Dong, W., Wang, Z., Josephson, W., Charikar, M., Li, K.: Modeling LSH for performance tuning. In: CIKM 2008, pp. 669–678. ACM. http://lshkit.sourceforge.net/
12. Edel, M., Soni, A., Curtin, R.R.: An automatic benchmarking system. In: NIPS 2014 Workshop on Software Engineering for Machine Learning (2014)
13. Heo, J.P., Lee, Y., He, J., Chang, S.F., Yoon, S.E.: Spherical hashing: binary code embedding with hyperspheres. IEEE TPAMI **37**(11), 2304–2316 (2015)
14. Indyk, P., Motwani, R.: Approximate nearest neighbors: towards removing the curse of dimensionality. In: STOC 1998, pp. 604–613
15. Johnson, W.B., Lindenstrauss, J., Schechtman, G.: Extensions of Lipschitz maps into Banach spaces. Isr. J. Math. **54**(2), 129–138 (1986)
16. Kriegel, H., Schubert, E., Zimek, A.: The (black) art of runtime evaluation: are we comparing algorithms or implementations? Knowl. Inf. Syst. **52**(2), 341–378 (2017)

17. LeCun, Y., Bottou, L., Bengio, Y., Haffner, P.: Gradient-based learning applied to document recognition. Proc. IEEE **86**(11), 2278–2324 (1998)
18. Li, W., Zhang, Y., Sun, Y., Wang, W., Zhang, W., Lin, X.: Approximate nearest neighbor search on high dimensional data - experiments, analyses, and improvement (v1.0). CoRR abs/1610.02455 (2016). http://arxiv.org/abs/1610.02455
19. Lyst Engineering: Rpforest. https://github.com/lyst/rpforest
20. Malkov, Y.A., Yashunin, D.A.: Efficient and robust approximate nearest neighbor search using Hierarchical Navigable Small World graphs. ArXiv e-prints, March 2016
21. Malkov, Y., Ponomarenko, A., Logvinov, A., Krylov, V.: Approximate nearest neighbor algorithm based on navigable small world graphs. Inf. Syst. **45**, 61–68 (2014)
22. Mikolov, T., Sutskever, I., Chen, K., Corrado, G.S., Dean, J.: Distributed representations of words and phrases and their compositionality. In: NIPS 2013, pp. 3111–3119
23. Muja, M., Lowe, D.G.: Fast approximate nearest neighbors with automatic algorithm configuration. In: VISSAPP 2009, pp. 331–340. INSTICC Press
24. Norouzi, M., Punjani, A., Fleet, D.J.: Fast search in hamming space with multi-index hashing. In: CVPR 2012, pp. 3108–3115. IEEE
25. Pham, N.: Hybrid LSH: faster near neighbors reporting in high-dimensional space. In: EDBT 2017, pp. 454–457
26. van Rijn, J.N., Bischl, B., Torgo, L., Gao, B., Umaashankar, V., Fischer, S., Winter, P., Wiswedel, B., Berthold, M.R., Vanschoren, J.: OpenML: a collaborative science platform. In: Blockeel, H., Kersting, K., Nijssen, S., Železný, F. (eds.) ECML PKDD 2013. LNCS, vol. 8190, pp. 645–649. Springer, Heidelberg (2013). doi:10.1007/978-3-642-40994-3_46
27. Wang, J., Shen, H.T., Song, J., Ji, J.: Hashing for similarity search: a survey. CoRR abs/1408.2927 (2014). http://arxiv.org/abs/1408.2927
28. Williams, R.: A new algorithm for optimal 2-constraint satisfaction and its implications. Theor. Comput. Sci. **348**(2–3), 357–365 (2005)
29. Zezula, P., Savino, P., Amato, G., Rabitti, F.: Approximate similarity retrieval with M-Trees. VLDB J. **7**(4), 275–293 (1998)

Improving Similarity Search Methods and Applications

Sketches with Unbalanced Bits
for Similarity Search

Vladimir Mic$^{(\boxtimes)}$, David Novak, and Pavel Zezula

Masaryk University, Brno, Czech Republic
xmic@fi.muni.cz

Abstract. In order to accelerate efficiency of similarity search, compact bit-strings compared by the Hamming distance, so called sketches, have been proposed as a form of dimensionality reduction. To maximize the data compression and, at the same time, minimize the loss of information, sketches typically have the following two properties: (1) each bit divides datasets approximately in halves, i.e. bits are *balanced*, and (2) individual bits have low pairwise *correlations*, preferably zero. It has been shown that sketches with such properties are minimal with respect to the retained information. However, they are very difficult to index due to the dimensionality curse – the range of distances is rather narrow and the distance to the nearest neighbour is high. We suggest to use sketches with *unbalanced* bits and we analyse their properties both analytically and experimentally. We show that such sketches can achieve practically the same quality of similarity search and they are much easier to index thanks to the decrease of distances to the nearest neighbours.

1 Introduction

Treating data objects according to their pairwise similarity closely corresponds to the human perception of reality, thus it represents an important field of data processing. Features of complex objects are typically characterized by *descriptors*, which are often high dimensional vectors. These descriptors can be bulky and evaluation of their pairwise similarity may be computationally demanding [16,21]. Thus techniques to process them efficiently are needed. In this paper we consider one to one mapping between objects and descriptors, thus we do not distinguish these terms and we use just term *object*. One of the state-of-the-art approaches allowing to search big datasets efficiently is based on object transformation to short binary strings – *sketches*. The objective of a *sketching technique* is to construct the binary strings so that they, together with *Hamming distance h*, preserve pairwise similarity relations between objects as much as possible. Thanks to their compact size and computational efficiency of the Hamming distance, sketches have been used by several authors who report promising results for different data types, dimensions, and similarity functions [5,7,15,19].

Many sketching techniques were proposed and majority of them produce sketches $sk(o)$ with *balanced* bits with *low correlations* [12,13,15,19], because these properties are reported to support the quality of similarity approximation:

© Springer International Publishing AG 2017
C. Beecks et al. (Eds.): SISAP 2017, LNCS 10609, pp. 53–63, 2017.
DOI: 10.1007/978-3-319-68474-1_4

- Bit i is balanced (with respect to dataset X) iff it is set to 1 in one half of all sketches $sk(o), o \in X$.
- Bit correlations are investigated in pairwise manner over all pairs of bits of sketches $sk(o), o \in X$.

To the best of our knowledge, there is no prior work discussing disadvantages arising from these properties. In this paper, we analyse their pros and cons and we further focus on sketches with bits balanced to a given ratio b:

- Bit i is balanced to ratio b (with respect to dataset X) iff it is set to 1 in $b \cdot |X|$ sketches $sk(o), o \in X$. Without loss of generality, we assume $0.5 \leq b \leq 1$, since the opposite case is symmetric.

We denote S_b the set of all sketches $sk(o), o \in X$ with bits balanced to b.

We show that the Hamming distance distribution on sketches $S_{0.5}$ (i.e. with balanced bits) with low pairwise bit correlations makes an efficient indexing practically impossible. The main contribution of this paper is analytical and experimental investigation of sketches with unbalanced bits, which shows that they can achieve practically the same quality of the similarity search but they are significantly easier to index.

2   Background

To formalize the concept of similarity, we adopt the model of *metric space* $M = (D, d)$, where D is a domain of objects and $d : D \times D \mapsto \mathbb{R}$ is a *distance function* which determines the dissimilarity of objects [21]. Further we consider a finite dataset $X \subseteq D$. The goal of this section is to provide basic observations about the sketches, which influence their indexability and ability to preserve similarity relationships of objects. First, let us focus on Hamming distance density of sketches S_b with length λ.

Lemma 1 (Mean value of Hamming distance). *Let us have set S_b of sketches $sk(o), o \in X$ with length λ. The mean value of Hamming distance on S_b is $2\lambda \cdot b \cdot (1 - b)$ regardless of pairwise bit correlations.*

Proof. Let us consider one bit i of the sketches. The Hamming distance h of sketches $sk(o_1)$, $sk(o_2)$ on bit i is 1 iff $sk(o_1)$, $sk(o_2)$ have different values in bit i. Considering all $|X|$ sketches, it happens in $2b|X| \cdot (1 - b)|X|$ cases. Thus sum of all Hamming distances over λ bits is $2\lambda b|X| \cdot (1 - b)|X|$, and the mean Hamming distance is $2\lambda b(1 - b)$ as we summed $|X|^2$ distances in the previous step.

The mean of Hamming distance is maximized for $b = 0.5$, i.e. for the balanced bits.

Next, we focus on the bit correlation, which we express with the Pearson correlation coefficient. According to Theorem 2 in [12], the variance of Hamming distance on sketches $S_{0.5}$ decreases with decreasing absolute value of the average pairwise bit correlation; it is minimized for uncorrelated bits.

In case of uncorrelated bits, the Hamming distance density of sketches S_b has binomial distribution, thus for variance σ^2 of Hamming distances holds: $\sigma^2 = \lambda b(1 - b)$. In other words, (1) sketches with balanced bits have maximum mean distance and, (2) for these sketches, minimization of the pairwise bit correlations means minimization of the variance of the Hamming distance, which is maximization of all distances lower than the mean distance. Clearly, maximizing values of the smallest inter-object distances violates the key objective of the data transformation for the similarity indexing: distances $h(sk(o_1), sk(o_2)), o_1, o_2 \in X$ for very similar objects o_1, o_2 are desired to be small [5]. Moreover these consequences imply a problem known as the dimensionality curse [18].

A formalised view is provided by the *intrinsic dimensionality* of sketches. Intrinsic dimensionality (*iDim*) expresses "*the minimum number of parameters needed to account for the observed properties of the data*" [6]. We use the formula proposed by Chavez and Navarro [2] for the estimation of *iDim*:

$$iDim \approx \frac{\mu^2}{2 \cdot \sigma^2}, \tag{1}$$

where μ is the mean of distance density, and σ^2 is its variance. In compliance with the previous paragraph, it has been proven that:

- for uncorrelated bits, *iDim* is maximized iff they are balanced [18],
- for balanced bits, *iDim* is maximized iff they are uncorrelated [12].

In the field of similarity search, *iDim* expresses "the difficulty" of data indexing [17]. Thus techniques which produce sketches $S_{0.5}$ with bit correlations close to zero produce hard-to-index sketches. Moreover indexing techniques typically assume at least a few objects in small distances from the query object [14,16].

2.1 Observations on GHP Sketches

We illustrate our findings on a sketching technique based on the *generalized hyperplane partitioning* (GHP) [21]. Bit i of all sketches $sk(o), o \in X$ is determined using a pair of pivoting objects p_{i0}, p_{i1}, which splits objects $o \in X$ by comparing distances $d(o, p_{i0}), d(o, p_{i1})$; value of bit $sk_i(o)$ expresses which pivot is closer to o. This technique is described in detail e.g. in [12].

Let us consider query object $q \in D$ and its most similar object $o_{q1} \in X$, $o_{q1} \neq q$. As we have explained, the Hamming distance $h(sk(q), sk(o_{q1}))$ is high on average on sketches $S_{0.5}$ with low pairwise bit correlations. It means that many hyperplanes separate q and o_{q1}. On the contrary, in case of sketches with unbalanced bits, e.g. $S_{0.8}$, the distance $h(sk(q), sk(o_{q1}))$ should be lower.

For motivation, consider the situation in 2D Euclidean space shown in Fig. 1. In case of hyperplanes dividing dataset into halves (Fig. 1a), the Hamming distances between originally close objects are high which suggests that many hyperplanes split dense subspaces of dataset X. On the other hand, in case of sketches with unbalanced bits (Fig. 1b), the Hamming distances are smaller not only for

(a) $b = 0.5$ (b) $b = 0.8$

Fig. 1. Hyperplanes producing sketches with bits balanced to different balance b

originally close objects, but also for more distant ones; as shown later, this draw-back can be compensated by using longer sketches. Please, note that this is only an artificial example in 2D, but these properties are implied by values of mean and variance of the Hamming distance and thus hold even for real-life, high dimensional data.

Figure 1b suggests, that unbalanced bits may lead to many objects with all bits set to 1 (the "center" of the figure). However, our practical experience with sketches in high dimensional space show that there is just a few such objects. In particular, we conducted experiments with $b \in \{0.85, 0.9\}$ and $\lambda = 205$ (see Sect. 4 for details on the dataset). We realized that there was no sketch with all bits set to 1 in case of $b = 0.85$, and only eight out of one million in case of $b = 0.9$.

2.2 Related Work

Charikar has introduced in his pioneering work [1] the idea of using random hyperplanes to summarize objects in a multi-dimensional vector spaces in such a way that the resulting bit strings respect the cosine distance. Lv et al. [11] have proposed a sketching technique for spaces of vectors compared by (weighted) L_1 distance function. Their method is based on *thresholding*; a threshold is determined for each dimension of the original space. Individual bits of the sketches are set according to values in corresponding dimensions and compressed. Pagh et al. propose *odd sketches* [14] – short binary strings created as a transformation of original vector space, based on *min-hashing* [9,10]. Odd sketches are compared by the Hamming distance, and they suffer a lot from the curse of dimensionality. Daugman [3] uses bit strings to describe human irises to identify people. His method is based on encoding shades of grey colour shown in images of irises in UV light, and it constitutes the most widely used approach of human irises recognition.

3   Analysis of Bits Balanced to b

The objective of this section is to quantify all trends mentioned above. More specifically, we analytically derive the influence of the ratio b on bit correlations, Hamming distance distribution, and *iDim* of sketches.

Consider sketches S_b and their two arbitrarily selected bits i, j, $0 \leq i < \lambda$, $0 \leq j < \lambda$. Set S_b can be split into four parts according to combination of values in bits i, j. Let us denote $\#_{11}, \#_{10}, \#_{01}, \#_{00}$ the relative numbers of the sketches in these four parts. It holds that $\#_{10} = \#_{01}$ regardless of correlation of bits i and j. Denoting $sk_i(o)$ the ith bit of sketch $sk(o)$, the Pearson correlation coefficient of bits i, j can be simplified:

$$Corr(i, j) = \frac{\sum\limits_{o \in X} (sk_i(o) - b)(sk_j(o) - b)}{\sqrt{\sum\limits_{o \in X} (sk_i(o) - b)^2 \sum\limits_{o \in X} (sk_j(o) - b)^2}} = \frac{\#_{11} - b^2}{b(1 - b)}. \tag{2}$$

Let us point out one difference between balanced and unbalanced bits: if we switch all values in arbitrary bit i in case of balanced bits, only the sign of correlations $Corr(i, j)$ (with all other bits j) changes. Thus only the absolute values of pairwise bit correlations matter [12]. On the other hand, in case of unbalanced bits ($b \neq 0.5$), the sign of correlation matters, as opposite correlations express different space partitioning. For example for $b = 0.8$, correlation -0.25 means object distribution $\#_{11} = 60\%$, $\#_{10} = \#_{01} = 20\%$, $\#_{00} = 0\%$ while correlation $+0.25$ means distribution $\#_{11} = 68\%$, $\#_{10} = \#_{01} = 12\%$, $\#_{00} = 8\%$. Therefore, we keep the same bit orientation for all bits (specifically, 1 in $b \cdot |X|$ sketches).

It has been shown [12], that high intrinsic dimensionality $iDim$ increases potential of sketches with balanced bits to well approximate similarity relationships of objects. So, let us analyse the $iDim$ of sketches with bits balanced to b. We denote H_i the list of all $|X|^2$ Hamming distances measured just on bit i of sketches $sk(o), o \in X$. Then $Corr(H_i, H_j)$ is the Pearson correlation of lists H_i, H_j, and $Corr_{Avg}$ is the average pairwise correlation over all lists $H_i, H_j, 0 \leq i < j < \lambda$. We have derived in [12] the variance σ^2 of Hamming distance on sketches $S_{0.5}$. Using Lemma 7 from that paper and analogous approach, it is possible to derive σ^2 for sketches S_b:

$$\sigma^2 = 2b(1 - b) \cdot [1 - 2b(1 - b)] \cdot [\lambda + (\lambda^2 - \lambda) \cdot Corr_{Avg}]. \tag{3}$$

Using Lemma 1 and Eq. 3, the $iDim$ of sketches with bits balanced to b is:

$$iDim \approx \frac{\mu^2}{2\sigma^2} = \frac{b \cdot (1 - b) \cdot \lambda}{(2b^2 - 2b + 1) \cdot [1 + (\lambda - 1) Corr_{Avg}]}. \tag{4}$$

Therefore $iDim$ of sketches increases with decreasing $Corr_{Avg}$. In the following, we lower bound average correlation $Corr_{Avg}$, which implies the upper bound for $iDim$ of sketches with bits balanced to b.

Minimum average correlation $Corr_{Avg}$ occurs iff all pairwise correlations of lists H_i, H_j are minimal. Thus, we focus on $Corr(H_i, H_j)$:

$$Corr(H_i, H_j) = \frac{2(\#_{11} \cdot \#_{00} + \#_{10}^2) - [2b(1 - b)]^2}{2b(1 - b) - [2b(1 - b)]^2}. \tag{5}$$

Using this equation, it is possible to express values $\#_{11}, \#_{10}$ and $\#_{00}$ implying minimum value of $Corr(H_i, H_j)$. Please, notice that all fractions $\#_{11}, \#_{10}$ and $\#_{00}$ must be non-negative.

Theorem 1. *Maximum iDim of sketches S_b with bits balanced to b occurs iff for all pairs of bits $0 \leq i < j < \lambda$ holds: $\#_{00} = \max(0, 3/4-b)$, $\#_{10} = \min(1/4, 1-b)$ and $\#_{11} = \max(2b-1, b-1/4)$.*

Proof. Theorem holds as a consequence of Eqs. 2, 4 and 5.

Values $\#_{00}, \#_{10}$ and $\#_{11}$ implying maximum $iDim$ of sketches, imply negative pairwise correlations $Corr(i, j)$ for $b > 0.5$, which bring a problem: it is not possible to create meaningful sketches for similarity search with significantly negative pairwise bit correlations. Considering a given ratio $b > 0.5$ and negative bit correlations, each zero in an arbitrary bit i of any sketch $sk_i(o_1)$ pushes other values $sk_i(o_2), o_2 \in X \wedge o_2 \neq o_1$ to be 1. However the number of ones is given by ratio b.

In case of $b \geq 0.75$, maximum $iDim$ occurs iff $\forall 0 \leq i < j \leq \lambda : \#_{00} = 0$. In this case, each sketch contains exactly one or none bit set to 0, and therefore at most $\lambda + 1$ different sketches of length λ exist (including one with all bits set to 1). In the other words, an effort to minimize ratio $\#_{00}$ leads to extremely long sketches.

In practice when a realistic sketch length λ is preserved, higher $iDim$ of sketches S_b may be achieved with an effort to produce uncorrelated bits rather than negatively correlated. The reason is, that few significant negative correlations usually cause higher increase of other correlations which leads to an increase of average pairwise correlation above zero. We illustrate these statements in an experiment in Sect. 4.

As a result of provided analysis and experiments, we propose to search for uncorrelated unbalanced bits, i.e. for sketches with binomial distance distribution, but with lower mean value than in case of balanced bits, which is favourable for indexing of sketches.

4   Evaluation

At first, we run an experiment to confirm the suitability of producing sketches with uncorrelated rather than negatively correlated bits. Then we focus on the quality of similarity search with unbalanced sketches and their indexability. Let us briefly describe the testing data and sketching technique:

Testing Data

The experiments are conducted on a real-life data collection consisting of visual descriptors extracted from images. More specifically, we use *DeCAF* descriptors [4] – 4096-dimensional vectors taken as an output from the last hidden layer of a deep convolutional neural network [8]. These descriptors were extracted from a 1M subset of the *Profiset collection*[1]. The DeCAF descriptors are compared by the Euclidean distance to form a metric space.

Sketching Technique

In order to create sketches, we randomly select a set of 512 pivots and we investigate all $\binom{512}{2}$ pivot pairs. We use a random subset of 100,000 data objects and analyse the balance b of generalized hyperplane partitioning (GHP) defined by each pair of pivots (see Sect. 2.1 for examples of GHP). From pivot pairs implying a proper balance b (which is about 8,000–15,000 pairs) we further select those, producing sketches with low correlated bits using our heuristic. Description of this heuristic is available online[2].

4.1   Searching for Negatively Correlated Bits

Table 1 contains evaluated properties of sketches created by the sketching technique, which tried to (1) find sketches with uncorrelated bits, and (2) find as negatively correlated bits as possible. Ratio b was 0.8 in these experiments, and results for four sketch lengths λ are presented. The average pairwise bit correlation is lower in case of searching for uncorrelated bits, rather then negatively correlated, in three cases. The numbers of negative and positive pairwise bit correlations confirms these results as well. There is an exception in Table 1, the sketch length $\lambda = 205$ for which average bit correlation is lower when searched for negatively correlated bits. However, observed difference is tiny in this case.

Table 1. Sketching technique: searching for uncorrelated and negatively correlated unbalanced bits, $b = 0.8$

	Searching for uncorrelated			Searching for negative correlations		
λ	Average corr	# positive	# negative	Average corr	# positive	# negative
64	+0.0019	1,000	1,016	+0.0024	1,032	984
128	+0.0046	4,066	4,062	+0.0053	4,106	4,022
205	+0.0064	10,494	10,416	+0.0063	10,469	10,441
256	+0.0072	16,406	16,234	+0.0077	16,436	16,204

The reasons of observed tendencies are discussed in theoretical Sect. 3.

---

[1] http://disa.fi.muni.cz/profiset/.
[2] http://www.fi.muni.cz/~xmic/sketches/AlgSelectLowCorBits.pdf.

4.2   Quality of Sketches

The most important requirement for sketches with unbalanced bits is that they have to provide acceptable quality of the similarity search in comparison to sketches with balanced bits. In the following experiments, we use k-recall@k' of approximate kNN search using sketches. More specifically, for each query object q, we compare the set of k most similar objects from X found by the sequential scan of X (denoted as $Prec(q)$) with k objects found by the *filter and refine* approach based on sketches: First, in the filtering phase, we select k' objects $o \in X$, $k' \geq k$ with smallest Hamming distances $h(sk(q), sk(o))$. Then these k' objects o are refined by evaluating distances $d(q, o)$ in order to identify approximate kNN answer denoted as $Ans(q, k')$. The ability of sketches to approximate similarity relationships of objects $o \in X$ is expressed by measure k-recall@k':

$$k\text{-recall@}k' = \frac{Prec(q) \cap Ans(q, k')}{k}. \tag{6}$$

In the following, we present results only for $k = 10$, because trends observed in these experiments are the same even for other values of k. Size of dataset X in the following experiments is $|X| = 1{,}000{,}000$. All results are averages over 1,000 randomly selected queries q.

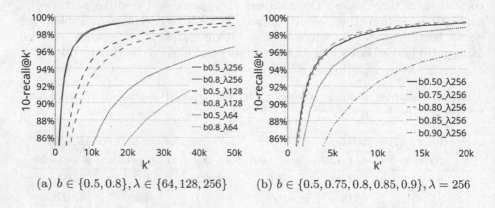

(a) $b \in \{0.5, 0.8\}, \lambda \in \{64, 128, 256\}$ (b) $b \in \{0.5, 0.75, 0.8, 0.85, 0.9\}, \lambda = 256$

Fig. 2. Quality of approx. similarity search with sketches balanced to different b

We demonstrate by Fig. 2a that the difference in 10-recall@k' using sketches $S_{0.5}$ and $S_{0.8}$ is relatively high in case of short sketches, however with increasing length λ it is becoming negligible (in our case, this happens approximately for $\lambda \geq 200$). Figure 2b depicts 10-recall@k' for sketch length $\lambda = 256$ and different ratio b. Using $b \in \{0.5, 0.75, 0.8\}$, the results are practically the same; decrease is noticeable in case of $b = 0.85$: for example about 2.3 percentage points for k' = 5,000 (i.e. 0.5 % of $|X|$) and it is significant for $b = 0.9$: e.q. 9.2 percentage points for k' = 5,000.

Table 2. Sketches with $\lambda = 256$: 10-recall@k', *iDim* and avg. Hamming distances to k'th closest sketch

b	10-recall@k'		*iDim*	$h(sk(q), sk(o_{qk'}))$		
	k'=2,500	k'=10,000		k'=1	k'=100	k'=10,000
0.50	93.20 %	98.41 %	29.4	43.1	58.9	83,6
0.75	93.43 %	98.66 %	24.1	32.0	44.3	63.2
0.80	92.79 %	98.52 %	19.6	27.4	38.0	54.0
0.85	89.19 %	97.30 %	13.5	21.1	29.7	42.2
0.90	80.31 %	92.46 %	9.0	13.8	19.9	28.3

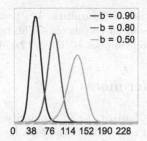

Fig. 3. Ham. distance densities for $\lambda = 256$

4.3 Indexability of Sketches

The indexability of sketches is illustrated by their *iDim* and by the average Hamming distances $h(sk(q), sk(o_{qk'}))$ between $sk(q)$ and its k'th nearest sketch for $k' \in \{1; 100; 10,000\}$. We show results for $b \in \{0.5, 0.75, 0.8, 0.85, 0.9\}$ in Table 2. These results make possible to utilize techniques for bit-strings indexing and other processing [16, 20]

As expected, the *iDim* of sketches decreases as ratio b grows (for $b \geq 0.5$). For instance the *iDim* of sketches $S_{0.5}$ and $S_{0.8}$ differs about one third for $\lambda = 256$. In order to remind results from Sect. 4.2, we show 10-recall@k' for two selected k': the difference of 10-recall@k' for $b \in [0.5, 0.8]$ is negligible. It confirms, that properly unbalanced sketches can be used as a full-fledged but easily indexable alternative to sketches with balanced bits. Better indexability is confirmed by the decrease of distances to the k' nearest sketches (shown in last three columns of Table 2), and by distribution of Hamming distance densities presented in Fig. 3. All these measurements confirm the analytic results from Sects. 2 and 3.

5 Conclusions

We have investigated sketches – bit strings created by such transformation of data objects, which should preserve the similarity relationships between the objects. Sketching techniques proposed so far usually aim at producing bit strings with balanced and low correlated bits. Sketches with these properties have been reported to provide the best trade-off between their length and ability to approximate similarity relationships between objects. In this paper, we studied one drawback of such sketches: these properties lead to maximization of the *intrinsic dimensionality* of the set of sketches making them hard-to-index (because of the *dimensionality curse*). We thus focus on sketches with bits balanced to some given ratio b and we derive various theoretical properties of such sketches. Further, we show on a real life dataset that the proposed approach can achieve practically the same quality of the similarity search, but with sketches having *iDim* about one third lower than sketches with balanced bits.

Acknowledgements. We thank to Matthew Skala for his advices about negative pairwise bit correlations. This research was supported by the Czech Science Foundation project number GBP103/12/G084.

References

1. Charikar, M.S.: Similarity estimation techniques from rounding algorithms. In: Proceedings of the 34th Annual ACM Symposium on Theory of Computing. ACM, New York (2002)
2. Chávez, E., Navarro, G., Baeza-Yates, R., Marroquín, J.L.: Searching in metric spaces. ACM Comput. Surv. **33**(3) (2001)
3. Daugman, J.: The importance of being random: statistical principles of iris recognition. Pattern Recognit. **36**(2) (2003)
4. Donahue, J., Jia, Y., Vinyals, O., Hoffman, J., Zhang, N., Tzeng, E., Darrell, T.: Decaf: a deep convolutional activation feature for generic visual recognition. In: ICML 2014, vol. 32, pp. 647–655 (2014)
5. Dong, W., Charikar, M., Li, K.: Asymmetric distance estimation with sketches for similarity search in high-dimensional spaces. In: Proceedings of the 31st Annual International ACM SIGIR Conference on Research and Development in Information Retrieval. ACM (2008)
6. Fukunaga, K.: Introduction to Statistical Pattern Recognition. Academic Press, San Diego (2013)
7. Jegou, H., Douze, M., Schmid, C.: Hamming embedding and weak geometric consistency for large scale image search. In: Forsyth, D., Torr, P., Zisserman, A. (eds.) ECCV 2008. LNCS, vol. 5302, pp. 304–317. Springer, Heidelberg (2008). doi:10. 1007/978-3-540-88682-2_24
8. Krizhevsky, A., Sutskever, I., Hinton, G.E.: Imagenet classification with deep convolutional neural networks. In: Advances in Neural Information Processing Systems (2012)
9. Leskovec, J., Rajaraman, A., Ullman, J.D.: Mining of Massive Datasets. Cambridge University Press, Cambridge (2014)
10. Li, P., König, A.C.: Theory and applications of b-bit minwise hashing. Commun. ACM **54**(8), 101–109 (2011)
11. Lv, Q., Charikar, M., Li, K.: Image similarity search with compact data structures. In: Proceedings of the 13th ACM International Conference on Information and Knowledge Management, pp. 208–217. ACM (2004)
12. Mic, V., Novak, D., Zezula, P.: Designing sketches for similarity filtering. In: 2016 IEEE 16th International Conference on Data Mining Workshops (ICDMW), pp. 655–662, December 2016
13. Mic, V., Novak, D., Zezula, P.: Speeding up similarity search by sketches. In: Amsaleg, L., Houle, M.E., Schubert, E. (eds.) SISAP 2016. LNCS, vol. 9939, pp. 250–258. Springer, Cham (2016). doi:10.1007/978-3-319-46759-7_19
14. Mitzenmacher, M., Pagh, R., Pham, N.: Efficient estimation for high similarities using odd sketches. In: Proceedings of the 23rd International Conference on World Wide Web, pp. 109–118. ACM (2014)
15. Muller-Molina, A.J., Shinohara, T.: Efficient similarity search by reducing i/o with compressed sketches. In: Proceedings of the 2nd International Workshop on Similarity Search and Applications, pp. 30–38 (2009)

16. Pagh, R.: Locality-sensitive hashing without false negatives. In: Proceedings of the Twenty-Seventh Annual ACM-SIAM Symposium on Discrete Algorithms, pp. 1–9. Society for Industrial and Applied Mathematics (2016)
17. Skala, M.: Measuring the difficulty of distance-based indexing. In: Consens, M., Navarro, G. (eds.) SPIRE 2005. LNCS, vol. 3772, pp. 103–114. Springer, Heidelberg (2005). doi:10.1007/11575832_12
18. Skala, M.A.: Aspects of Metric Spaces in Computation. Ph.D. thesis, University of Waterloo (2008)
19. Wang, Z., Dong, W., Josephson, W., Lv, Q., Charikar, M., Li, K.: Sizing sketches: a rank-based analysis for similarity search. SIGMETRICS Perform. Eval. Rev. **35**(1), 157–168 (2007)
20. Zezula, P., Rabitti, F., Tiberio, P.: Dynamic partitioning of signature files. ACM Trans. Inf. Syst. **9**(4), 336–367 (1991)
21. Zezula, P., Amato, G., Dohnal, V., Batko, M.: Similarity Search: The Metric Space Approach, vol. 32. Springer, Boston (2006)

Local Intrinsic Dimensionality I: An Extreme-Value-Theoretic Foundation for Similarity Applications

Michael E. Houle[✉]

National Institute of Informatics,
2-1-2 Hitotsubashi, Chiyoda-ku, Tokyo 101-8430, Japan
meh@nii.ac.jp

Abstract. Researchers have long considered the analysis of similarity applications in terms of the intrinsic dimensionality (ID) of the data. This theory paper is concerned with a generalization of a discrete measure of ID, the expansion dimension, to the case of smooth functions in general, and distance distributions in particular. A local model of the ID of smooth functions is first proposed and then explained within the well-established statistical framework of extreme value theory (EVT). Moreover, it is shown that under appropriate smoothness conditions, the cumulative distribution function of a distance distribution can be completely characterized by an equivalent notion of data discriminability. As the local ID model makes no assumptions on the nature of the function (or distribution) other than continuous differentiability, its extreme generality makes it ideally suited for the non-parametric or unsupervised learning tasks that often arise in similarity applications. An extension of the local ID model is also provided that allows the local assessment of the rate of change of function growth, which is then shown to have potential implications for the detection of inliers and outliers.

1 Introduction

In an attempt to alleviate the effects of high dimensionality, and thereby improve the discriminability of data, simpler representations of data are often sought by means of a number of supervised or unsupervised learning techniques. One of the earliest and most well-established simplification strategies is dimensional reduction, which seeks a projection to a lower-dimensional subspace that minimizes the distortion of the data according to a given criterion. In general, dimensional reduction requires that an appropriate dimension for the reduced space (or approximating manifold) be either supplied or learned, ideally so as to minimize the error or loss of information incurred. The dimension of the surface that best approximates the data can be regarded as an indication of the intrinsic dimensionality (ID) of the data set, or of the minimum number of latent variables needed to represent the data. Intrinsic dimensionality thus serves as an important natural measure of the complexity of data.

© Springer International Publishing AG 2017
C. Beecks et al. (Eds.): SISAP 2017, LNCS 10609, pp. 64–79, 2017.
DOI: 10.1007/978-3-319-68474-1_5

1.1   Characterizations of Intrinsic Dimensionality

Over the past decades, many characterizations of ID have been proposed. The earliest theoretical measures of ID such as the classical Hausdorff dimension, Minkowski-Bouligand or 'box counting' dimension, and packing dimension, all associate a non-negative real number to metric spaces in terms of their covering or packing properties (for a general reference, see [1]). Although they are of significant theoretical importance, they are impractical for direct use in similarity applications, as the value of such measures is zero for any finite set. However, these theoretical measures have served as the foundation of practical methods for finite data samples, including the correlation dimension [2], and 'fractal' methods which estimate ID from the space-filling capacity or self-similarity properties of the data [3,4]. Other practical techniques for the estimation of ID include the topological approaches, which estimate the basis dimension of the tangent space of a data manifold from local samples (see for example [5]). In their attempt to determine lower-dimensional projective spaces or surfaces that approximate the data with minimum error, projection-based learning methods such as PCA can produce as a byproduct an estimate of the ID of the data. Parametric modeling and estimation of distribution often allow for estimators of ID to be derived [6].

An important family of dimensional models, including the minimum neighbor distance (MiND) models [5], the expansion dimension (ED) [7], generalized expansion dimension (GED) [8], and the local intrinsic dimension (LID) [9], quantify the ID in the vicinity of a point of interest in the data domain. More precisely, expansion models of dimensionality assess the rate of growth in the number of data objects encountered as the distance from the point increases. For example, in Euclidean spaces the volume of an m-dimensional set grows proportionally to r^m when its size is scaled by a factor of r — from this rate of volume growth with distance, the dimension m can be deduced. Expansion models of dimensionality provide a local view of the dimensional structure of the data, as their estimation is restricted to a neighborhood of the point of interest. They hold an advantage over parametric models in that they require no explicit knowledge of the underlying global data distribution. Expansion models also have the advantage of computational efficiency: as they require only an ordered list of the neighborhood distance values, no expensive vector or matrix operations are required for the computation of estimates. Expansion models have seen applications in the design and analysis of index structures for similarity search [7,10–14], and heuristics for anomaly detection [15], as well as in manifold learning.

1.2   Local Intrinsic Dimensionality and Extreme Value Theory

With one exception, the aforementioned expansion models assign a measure of intrinsic dimensionality to specific sets of data points. The exception is the local intrinsic dimension ('local ID', or 'LID'), which extends the GED model to a statistical setting that assumes an underlying (but unknown) distribution of distances from a given reference point [9]. Here, each object of the data set induces a distance to the reference point; together, these distances can be regarded as samples from the distribution. The only assumptions made on the nature of the distribution are those of smoothness.

In [9], the local intrinsic dimension is shown to be equivalent to a notion of discriminability of the distance measure, as reflected by the growth rate of the cumulative distribution function. For a random distance variable $\mathbf{X}$, with a continuous cumulative distribution function $F_\mathbf{X}$, the k-nearest neighbor distance within a sample of n points is an estimate of the distance value r for which $F_\mathbf{X}(r) = k/n$. If k is fixed, and n is allowed to tend to infinity, the indiscriminability of $F_\mathbf{X}$ at the k-nearest neighbor distance tends to the local intrinsic dimension. The local intrinsic dimension can thus serve to characterize the degree of difficulty in performing similarity-based operations within query neighborhoods using the underlying distance measure, asymptotically as the sample size (that is, the data set size) scales to infinity.

From the perspective of a given query point, the smallest distances encountered in a query result could be regarded as 'extreme events' associated with the lower tail of the underlying distance distribution [16]. The modeling of neighborhood distance values can thus be investigated from the viewpoint of extreme value theory (EVT), a statistical discipline concerned with the extreme behavior of stochastic processes. One of the pillars of EVT, a theorem independently proven by Balkema and de Haans [17] and by Pickands [18], states that under very reasonable assumptions, the tails of continuous probability distributions converge to a form of power-law distribution, the Generalized Pareto Distribution (GPD) [19]. In an equivalent (and much earlier) formulation of EVT due to Karamata [20], the cumulative distribution function of a tail distribution can be represented in terms of a 'regularly varying' function whose dominant factor is a polynomial in the distance [19]; the degree (or 'index') of this polynomial factor determines the shape parameter of the associated GPD. The index has been interpreted as a form of dimension within statistical contexts [19]. Many practical methods have been developed for the estimation of the index, including the well-known Hill estimator and its variants (for a survey, see [21]).

In a recent paper, Amsaleg et al. [22] developed estimators of local ID through a heuristic approximation of the true underlying distance distribution by a transformed GPD. The scale parameter of the GPD was shown to determine the local ID value. Estimators of the scale parameter of the GPD were then considered as candidates for the heuristic estimation of the local ID of the true distance distribution. Of these, the Hill estimator [23] has recently been used for ID estimation in the context of reverse k-NN search [14] and the analysis of non-functional dependencies among data features [24].

1.3   Contributions

In this paper, we revisit the intrinsic dimensionality model proposed in [9] so as to establish a firm theoretical connection between LID and EVT. The specific contributions of the paper include the following:

1. In Sect. 2.2, an overview of the LID model, extended so as to cover not only the cumulative distribution functions of distance distributions, but also a more general class of functions satisfying certain smoothness conditions.
2. In Sect. 3, a theoretical result demonstrating that any smooth functions can be fully represented in terms of an associated LID discriminability function. When applied to distance distributions, the result implies that the cumulative distribution function can be characterized entirely in terms of its discriminability, with no explicit knowledge of probability densities.
3. In Sect. 4, the development of a second-order theory of local intrinsic dimensionality that captures the growth rates within the discriminability measure itself. In the context of distance distributions, the second-order LID is shown to be a natural measure of the inlierness or outlierness of the underlying data distribution.
4. In Sect. 5, the theory developed in Sect. 3 is revealed to be a reworking of extreme value theory from first principles, for the growth rate of smooth functions from the origin. Rather than relying on the heuristic asymptotic connection to the generalized Pareto distribution that was identified in [22], we show that the LID characterization theorem is a more precise statement of the Karamata representation for the case of short-tailed distributions, with all elements of the Karamata representation being given an interpretation in terms of LID. A well-studied second-order EVT parameter governing the convergence rate of extreme values is also given an interpretation in terms of higher-order LID.

2   Background and Preliminaries

In this section, we give an overview of the LID model of [9], extended to account for a more general class of smooth functions (and not just cumulative distribution functions over the non-negative real domain). We begin the discussion with an overview of the expansion dimension and its applications.

2.1   Expansion Dimension

For the Euclidean distance metric in $\mathbb{R}^m$, increasing the radius of a ball by a factor of Δ would increase its volume by a factor of Δ^m. Were we inclined to measure the volumes V_1 and V_2 of two balls of radii r_1 and r_2, with $r_2 > r_1 > 0$, taking the logarithm of their ratios would reveal the dimension m:

$$\frac{V_2}{V_1} = \left(\frac{r_2}{r_1}\right)^m \implies m = \frac{\ln(V_2/V_1)}{\ln(r_2/r_1)}. \tag{1}$$

The *generalized expansion dimension* (GED) can be regarded as the smallest upper bound on the values of m that would be produced over a set of allowable ball placements and ball radii [8]; Karger and Ruhl's original expansion dimension (ED) further constrained r_2 to be double the value of r_1 [7].

The ED and GED have also appeared in the complexity analyses of several other similarity search structures [10, 12, 25]. The GED has also been successfully applied to guide algorithmic decisions at runtime for a form of adaptive search, the so-called *multi-step* similarity search problem [11, 13, 14, 26]. In [15], a heuristic for outlier detection was presented in which approximations of the well-known local outlier factor (LOF) score [27] were calculated after projection to a lower-dimensional space. The quality of the approximation was shown to depend on a measure of expansion dimension, in which the ratio of the ball radii relates to a targeted error bound.

2.2   Intrinsic Dimensionality of Distance Distributions

If one accepts the observed data set as indicative of an underlying generation process, the generalized expansion dimension can be regarded as an attempt to model the worst-case growth characteristics of the distribution of distances to generated objects, as measured from a reference object drawn from $\mathcal{U}$. When the reference object $q \in \mathcal{U}$ is fixed, a supplied data set S thus gives rise to a sample of values drawn from the distance distribution associated with q.

For finite data sets, GED formulations are obtained by estimating the volume of balls by the numbers of points they enclose [8]. In contrast, for continuous real-valued random distance variables, the notion of volume is naturally analogous to that of probability measure. As shown in [9], the generalized expansion dimension can thus be adapted for distance distributions by replacing the notion of ball set size by that of the probability measure of lower tails of the distribution. As in Eq. 1, intrinsic dimensionality can then be modeled as a function of distance $\mathbf{X} = x$, by letting the radii of the two balls be $r_1 = x$ and $r_2 = (1 + \epsilon)x$, and letting $\epsilon \to 0$. The following definition (adapted from [9]) generalizes this notion even further, to any real-valued function that is non-zero in the vicinity of $x \neq 0$.

Definition 1. *Let F be a real-valued function that is non-zero over some open interval containing $x \in \mathbb{R}$, $x \neq 0$. The* intrinsic dimensionality of F at x is *defined as*

$$\mathrm{IntrDim}_F(x) \triangleq \lim_{\epsilon \to 0} \frac{\ln\left(F((1+\epsilon)x)/F(x)\right)}{\ln\left((1+\epsilon)x/x\right)} = \lim_{\epsilon \to 0} \frac{\ln\left(F((1+\epsilon)x)/F(x)\right)}{\ln(1+\epsilon)},$$

whenever the limit exists.

Using the same assumptions on the distance distribution, [9] also proposed a natural measure of the discriminability of a random distance variable $\mathbf{X}$, in terms of the relative rate at which its cumulative distance function $F_{\mathbf{X}}$ increases as the distance increases. If $\mathbf{X}$ is discriminative at a given distance r, then expanding the distance by some small factor should incur a small increase in probability measure as a proportion of the value of $F_{\mathbf{X}}(r)$ (or, expressed in terms of a data sample, the proportional expansion in the expected number of data points in the neighborhood of the reference point q). Conversely, if the distance variable $\mathbf{X}$ is indiscriminative at distance r, then the proportional increase in probability measure would be large. Accordingly, [9] defined the indiscriminability of the distance variable as the limit of the ratio of two quantities: the proportional rate of increase of probability measure, and the proportional rate of increase in distance. As with the intrinsic dimensionality formulation of Definition 1, we generalize the notion of a cumulative distribution function to any real-valued function $F(x)$ that is non-zero in the vicinity of x.

Definition 2. *Let F be a real-valued function that is non-zero over some open interval containing $x \in \mathbb{R}$, $x \neq 0$. The* indiscriminability *of F at x is defined as*

$$\mathrm{InDiscr}_F(x) \triangleq \lim_{\epsilon \to 0} \left[\frac{(F((1+\epsilon)x) - F(x))}{F(x)} \bigg/ \frac{(1+\epsilon)x - x}{x} \right]$$

$$= \lim_{\epsilon \to 0} \frac{F((1+\epsilon)x) - F(x)}{\epsilon \cdot F(x)},$$

whenever the limit exists.

When F satisfies certain smoothness conditions in the vicinity of $x > 0$, the intrinsic dimensionality and the indiscriminability of F both exist at x, and are equivalent. Once again, we generalize the original statement appearing in [9] so as to apply not only to distance distributions, but also to any general function $F : \mathbb{R} \to \mathbb{R}$ at values for which F is both non-zero and continuously differentiable. The proof follows from applying l'Hôpital's rule to the numerator and denominator in the limits of $\mathrm{IntrDim}_F$ and $\mathrm{InDiscr}_F$; since it is essentially the same as the version in [9], we omit it here.

Theorem 1. *Let F be a real-valued function that is non-zero over some open interval containing $x \in \mathbb{R}$, $x \neq 0$. If F is continuously differentiable at x, then*

$$\mathrm{IntrDim}_F(x) = \mathrm{InDiscr}_F(x) = \frac{x \cdot F'(x)}{F(x)}.$$

This equivalence can be extended to those cases where $x = 0$ or $F(x) = 0$ by taking the limit of $\mathrm{IntrDim}_F(t) = \mathrm{InDiscr}_F(t)$ as $t \to x$, wherever the limit exists.

Corollary 1. *Let F be a real-valued function that is non-zero and continuously differentiable over some open interval containing $x \in \mathbb{R}$, except perhaps at x itself. Then*

$$\mathrm{ID}_F(x) \triangleq \lim_{t \to x} \frac{t \cdot F'(t)}{F(t)} = \lim_{t \to x} \mathrm{IntrDim}_F(t) = \lim_{t \to x} \mathrm{InDiscr}_F(t),$$

whenever the limits exist.

For values of x at which $\mathrm{ID}_F(x)$ exists, we observe that $\mathrm{ID}_F(x) = \mathrm{ID}_{-F}(x)$; the LID model therefore expresses the local growth rate relative to the magnitude of F, regardless of its sign. Although in general ID_F is negative whenever $|F|$ is decreasing, if F is a cumulative distribution function, ID_F must be non-negative whenever it exists.

ID_F can be viewed interchangeably as both the intrinsic dimensionality and the indiscriminability of F at x. However, we will henceforth refer to $\mathrm{ID}_F(x)$ as the *indiscriminability of* F at x whenever $x \neq 0$, and to $\mathrm{ID}_F^* \triangleq \mathrm{ID}_F(0)$ as the *local intrinsic dimension of* F.

3   ID-Based Representation of Smooth Functions

The LID formula $\mathrm{ID}_F(x) = x \cdot F'(x)/F(x)$ established in Corollary 1 simultaneously expresses the notions of local intrinsic dimensionality and indiscriminability. In general, the formula measures the instantaneous rate of change $F'(x)$ normalized by the cumulative rate of change $F(x)/x$. When F is the cumulative distribution function of a distance distribution, the formula can be interpreted as a normalization of the probability density $F'(x)$ with respect to the cumulative density $F(x)/x$. The following theorem states conditions for which the indiscriminability ID_F fully characterizes F.

Theorem 2 (Local ID Representation). *Let* $F : \mathbb{R} \to \mathbb{R}$ *be a real-valued function, and let* $v \in \mathbb{R}$ *be a value for which* $\mathrm{ID}_F(v)$ *exists. Let* x *and* w *be values for which* x/w *and* $F(x)/F(w)$ *are both positive. If* F *is non-zero and continuously differentiable everywhere in the interval* $[\min\{x, w\}, \max\{x, w\}]$, *then*

$$\frac{F(x)}{F(w)} = \left(\frac{x}{w}\right)^{\mathrm{ID}_F(v)} \cdot G_{F,v,w}(x), \text{ where}$$

$$G_{F,v,w}(x) \triangleq \exp\left(\int_x^w \frac{\mathrm{ID}_F(v) - \mathrm{ID}_F(t)}{t}\, \mathrm{d}t\right),$$

whenever the integral exists.

Proof. For any x and w for which x/w and $F(x)/F(w)$ are both positive,

$$F(x) = F(w) \cdot \exp\left(\ln(F(x)/F(w))\right)$$
$$= F(w) \cdot \exp\left(\mathrm{ID}_F(v)\ln(x/w) + \mathrm{ID}_F(v)\ln(w/x) + \ln(F(x)/F(w))\right)$$
$$= F(w) \cdot \left(\frac{x}{w}\right)^{\mathrm{ID}_F(v)} \cdot \exp\left(\mathrm{ID}_F(v)\ln(w/x) - \ln(F(w)/F(x))\right)$$
$$= F(w) \cdot \left(\frac{x}{w}\right)^{\mathrm{ID}_F(v)} \cdot \exp\left(\mathrm{ID}_F(v)\int_x^w \frac{1}{t}\, \mathrm{d}t - \int_x^w \frac{F'(t)}{F(t)}\, \mathrm{d}t\right),$$

since F is differentiable within the range of integration. Furthermore, since F is also non-zero over the range, and since F' is continuous, Corollary 1 implies that $F'(t)/F(t)$ can be substituted by $\mathrm{ID}_F(t)/t$. Combining the two integrals, the result follows.   □

The representation formula in Theorem 2 can be used to characterize the behavior of the function F in the vicinity of a given reference value v.

To see why, let us consider the value of the function at a point w that is tending towards v. The following theorem shows that when x is restricted to lie not too far from w, the exponential factor $G_{F,v,w}(x)$ eventually vanishes: in other words, the relationship stated in Theorem 2 tends asymptotically towards $F(x)/F(w) = (x/w)^{\mathrm{ID}_F(v)}$. This asymptotic relationship fits the intuition presented in Eq. 1 of Sect. 2.1, where the dimension is revealed by the ratios of the volumes and the radii of two balls. Here, as per the definitions of local intrinsic dimensionality and indiscriminability in Sect. 2, the role of volume is played by probability measure, and the dimension is the local ID. The asymptotic relationship is formalized in the following theorem.

Theorem 3. *Let $F : \mathbb{R} \to \mathbb{R}$ be a real-valued function, and let $v \in \mathbb{R}$ be a value for which $\mathrm{ID}_F(v)$ exists. Assume that there exists an open interval containing v for which F is non-zero and continuously differentiable, except perhaps at v itself. For any fixed $c > 1$, if $v \neq 0$, then*

$$\lim_{\substack{w \to v \\ |x-v| \le c|w-v|}} G_{F,v,w}(x) = 1;$$

otherwise, if $v = 0$, then

$$\lim_{\substack{w \to 0^+ \\ 0 < 1/c \le x/w \le c}} G_{F,0,w}(x) = \lim_{\substack{w \to 0^- \\ 0 < 1/c \le x/w \le c}} G_{F,0,w}(x) = 1.$$

Proof. For each case, it suffices to show that $\int_x^w (\mathrm{ID}_F(v) - \mathrm{ID}_F(t))/t\, dt \to 0$. First we consider the case where $v = 0$. Since $\mathrm{ID}_F(v)$ is assumed to exist, for any real value $\epsilon > 0$ there must exist a value $0 < \delta < 1$ such that $|t - v| < \delta$ implies that $|\mathrm{ID}_F(t) - \mathrm{ID}_F(v)| < \epsilon$. Therefore, when $|w - v| < \delta$,

$$\left| \int_x^w \frac{\mathrm{ID}_F(v) - \mathrm{ID}_F(t)}{t}\, dt \right| \le \epsilon \cdot \left| \int_x^w \frac{1}{t}\, dt \right| = \epsilon \ln \frac{w}{x}. \tag{2}$$

Since we have that $0 < 1/c \le w/x \le c$, $\ln(w/x)$ is bounded from above and below by constants. Therefore, since ϵ can be made arbitrarily small, the limit is indeed 0, and the result follows for the case $v = 0$.

Next, we consider the case where $v \neq 0$. The argument is the same as when $v = 0$, except that δ is chosen such that $0 < \delta < |v|/c$. Again, when $|w - v| < \delta$, Inequality 2 holds. Moreover, since by assumption $|x - v| \le c|w - v| < c\delta < |v|$, we have $||v| - |x|| < c\delta$ and $||w| - |v|| < \delta$. Together, these inequalities imply that

$$0 < \frac{\delta(c-1)}{2|v|} < \frac{|v| - \delta}{|v| + c\delta} < \frac{w}{x} = \frac{|w|}{|x|} < \frac{|v| + \delta}{|v| - c\delta}.$$

Since $\ln(w/x)$ is once again bounded from above and below by positive constants, the limits in this case exist and are 0, and the result follows. □

For the case when $v = 0$, x can be allowed to range over an arbitrarily large range relative to the magnitude of w, by choosing c sufficiently large. However,

x and w must be of the same sign (either both strictly positive or both strictly negative). When $v \neq 0$, the separation between x and v can be much greater than that between w and v, provided that the ratio of the two separations remains bounded by a constant — the constant can be chosen to be arbitrarily large, but once fixed, it cannot be changed.

Given a random distance variable **X**, its cumulative distribution function satisfies the conditions of Theorem 2 with $v = 0$, provided that it is strictly positive and continuously differentiable within some open interval of distances with lower endpoint 0. The ID representation expresses the behavior of the entire distribution in terms of the local intrinsic dimensionality and the indiscriminability function, without the explicit involvement of a probability density function. In this sense, the indiscriminability function holds all the information necessary to reconstruct the distribution.

Taken together, Theorems 2 and 3 show that within the extreme lower tail of a smooth distance distribution, ratios of probability measure tend to a polynomial function of the corresponding ratios in distance, with degree equal to the local ID of the cumulative distribution function. If the distances were generated from a reference point in the relative interior of a local manifold to points selected uniformly at random within the manifold, the polynomial growth rate would simply be the dimension of the manifold. However, it should be noted that in general, data distributions may not be perfectly modelled by a manifold, in which case the growth rate (and intrinsic dimensionality) may not necessarily be an integer.

4   Second-Order Local ID

In the previous section, we saw that a smooth function F can be represented in terms of its indiscriminability function ID_F. Here, we show that a representation formula for ID_F can be obtained for the second-order LID function $\mathrm{ID}_{\mathrm{ID}_F}(x)$ from the first-order representation formulae for F and F'.

4.1   Second-Order ID Representation

For the proof of the representation formula for ID_F, we require two technical lemmas. The first of the two lemmas shows that the second-order LID function $\mathrm{ID}_{\mathrm{ID}_F}(x)$ can be expressed in terms of the difference between the indiscriminabilities of F and F'. The proof is omitted due to space limitations.

Lemma 1. *Let F be a real-valued function over the interval $I = (0, z)$, for some choice of $z > 0$ (possibly infinite). If F is twice differentiable at some distance $x \in I$ for which $F(x) \neq 0$ and $F'(x) \neq 0$, then $\mathrm{ID}_F(x)$, $\mathrm{ID}_{F'}(x)$ and $\mathrm{ID}'_F(x)$ all exist, and*

$$\mathrm{ID}_{\mathrm{ID}_F}(x) = \frac{x \cdot \mathrm{ID}'_F(x)}{\mathrm{ID}_F(x)} = \mathrm{ID}_{F'}(x) + 1 - \mathrm{ID}_F(x).$$

The next technical lemma shows that the second-order LID converges to 0 as $x \to 0$. Again, the proof is omitted due to space limitations.

Lemma 2. *Let F be a real-valued function over the interval $I = (0, z)$, for some choice of $z > 0$ (possibly infinite). If F and F' are twice-differentiable and either positive everywhere or negative everywhere on I, if $F(x) \to 0$ as $x \to 0$, and if ID_F^* exists, then $\mathrm{ID}_{F'}^*$ also exists, and*

$$\mathrm{ID}_{\mathrm{ID}_F}^* = \mathrm{ID}_{F'}^* + 1 - \mathrm{ID}_F^* = 0.$$

We are now in a position to state and prove a characterization of the first-order LID function in terms of the second-order LID function.

Theorem 4 (Second-Order ID Representation). *Let F be a real-valued function over the interval $I = (0, z)$, for some choice of $z > 0$ (possibly infinite). Also, assume that F and F' are twice-differentiable and either positive everywhere or negative everywhere on I. Given any distance values $x, w \in (0, z)$, $\mathrm{ID}_F(x)$ admits the following representation:*

$$\mathrm{ID}_F(x) = \mathrm{ID}_F(w) \cdot \exp\left(-\int_x^w \frac{\mathrm{ID}_{\mathrm{ID}_F}(t)}{t}\, dt\right).$$

Furthermore, if $F(x) \to 0$ as $x \to 0$, and if ID_F^* exists and is non-zero, then the representation is also valid for $x = 0$.

Proof. The assumptions on F and F', together with Lemma 1, imply that ID_F, $\mathrm{ID}_{F'}$, ID_F' and $\mathrm{ID}_{\mathrm{ID}_F}$ exist everywhere, and that $\mathrm{ID}_F(x)$ and $\mathrm{ID}_F(w)$ are non-zero and share the same sign. We can therefore establish the result for the case where $x > 0$, as follows:

$$\mathrm{ID}_F(x)/\mathrm{ID}_F(w) = \exp\ln\left(\mathrm{ID}_F(x)/\mathrm{ID}_F(w)\right)$$

$$= \exp\left(-\int_x^w \frac{\mathrm{ID}_F'(t)}{\mathrm{ID}_F(t)}\, dt\right) = \exp\left(-\int_x^w \frac{\mathrm{ID}_{\mathrm{ID}_F}(t)}{t}\, dt\right),$$

where the last step follows from Theorem 1. If $F(x) \to 0$ as $x \to 0$, and if ID_F^* exists and is non-zero, by Lemma 2 we have that $\mathrm{ID}_{F'}^*$ exists, and that $\mathrm{ID}_{\mathrm{ID}_F}^* = 0$. Since $\mathrm{ID}_F(w)$ is also non-zero, the integral in the representation formula must converge, and therefore the representation is valid for $x = 0$ as well.   □

4.2   Inlierness, Outlierness and LID

Local manifold learning techniques such as Locally-Linear Embedding [28] typically model data dimensionality as the dimension of a manifold that well approximates the data within a region of interest. Under these assumptions, with respect to given reference point $\mathbf{q}$ on the manifold, the model assumes that the data distribution within a neighborhood of $\mathbf{q}$ tends to uniformity as the radius of the neighborhood tends to zero. The local ID of the manifold at $\mathbf{q}$ is simply the value of ID_F^*, where F is the cumulative distribution function for the induced distance distribution from $\mathbf{q}$. In addition, the indiscriminability function ID_F can indicate whether $\mathbf{q}$ should be regarded as an inlier or as an outlier relative to its locality within the manifold, as the following argument shows.

If $\text{ID}_F(x) < \text{ID}_F^*$ throughout a neighborhood of $\mathbf{q}$ of radius $0 < x < \epsilon$ (where $\epsilon > 0$ is chosen to be sufficiently small), then from the local ID representation formula of Theorem 2, we observe that $G_{F,0,\epsilon}(x)$ is greater than 1, and that $F(\epsilon)/F(x) < (\epsilon/x)^{\text{ID}_F^*}$. Consequently, the growth rate in probability measure within distance x from $\mathbf{q}$ is less than would be expected for a locally-uniform distribution of points within a manifold of dimension ID_F^*. The drop in indiscriminability (or rise in discriminability) indicates a decrease in local density as the distance from $\mathbf{q}$ increases. Under this interpretation, the relationship between $\mathbf{q}$ and its neighborhood can be deemed to be that of an *inlier*.

By similar arguments, if instead $\text{ID}_F(x) > \text{ID}_F^*$, then a rise in indiscriminability (or drop in discriminability) would indicate an increase in local density as the distance from $\mathbf{x}$ increases, in which case $\mathbf{q}$ would be an *outlier* with respect to its neighborhood.

Within a small local neighborhood $0 < x < \epsilon$, the condition $\text{ID}_F(x) < \text{ID}_F^*$ is equivalent to that of $\text{ID}_F'(x) < 0$, and the condition $\text{ID}_F(x) > \text{ID}_F^*$ is equivalent to that of $\text{ID}_F'(x) > 0$. The strength of the inlierness or outlierness of $\mathbf{q}$ can be judged according to the magnitude $|\text{ID}_F'(x)|$. However, for ease of comparison across manifolds of different intrinsic dimensions, and across different distances x, $|\text{ID}_F'(x)|$ should be normalized with respect to these two quantities. The second-order LID function $\text{ID}_{\text{ID}_F}(x) = x \cdot \text{ID}_F'(x)/\text{ID}_F(x)$ can thus be viewed as a natural measure of the inlierness (when negative) or outlierness (when positive) of $\mathbf{q}$, one that normalizes the relative rate of change of the LID function with respect to the average (radial) rate of change of LID within distance x of $\mathbf{q}$, namely $\text{ID}_F(x)/x$.

As an illustration of the ability of second-order LID to naturally determine the inlierness or outlierness of a point with respect to a data distribution, let us consider a Gaussian distribution in $\mathbb{R}^m$ generated as a vector of normally distributed random variables with means μ_i and variances σ_i^2, for $1 \le i \le m$. Then the normalized distance from the origin to a point $\mathbf{X} = (X_1, X_2, \ldots, X_m)$, defined as $Z = \sqrt{\sum_{i=1}^m (X_i^2/\sigma_i^2)}$, follows a noncentral chi distribution. Although the details are omitted due to the complexity of the derivations, Theorem 1 can be applied to the probability density function for Z to show that

$$\text{ID}_{F_Z}^* = m\,, \quad \text{and} \quad \text{ID}_{\text{ID}_{F_Z}}^* = 2$$

where $\lambda = \sqrt{\sum_{i=1}^m (\mu_i^2/\sigma_i^2)}$ is a distributional parameter representing the normalized distance between the Gaussian mean and the origin. Moreover, as z tends to 0, the sign of $\text{ID}_{\text{ID}_{F_Z}}(z)$ is positive when $\lambda > \sqrt{m}$, and negative when $0 \le \lambda < \sqrt{m}$, indicating 'outlierness' of the tail region of the Gaussian beyond the inflection boundary $\lambda = \sqrt{m}$, and 'inlierness' of the central region. It is worth noting that the strength of outlierness or inlierness is a constant value,

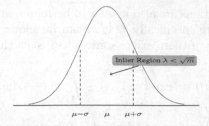

Fig. 1. The inlier region of a 1-dimensional Gaussian distribution. The boundary between the inlier (central) region and the outlier (tail) regions is at normalized distance $\lambda = \sqrt{m} = 1$, or equivalently at $|x - \mu| = \sigma$.

regardless of the actual dimension m, or of the (normalized) distance λ to the Gaussian center. The 1-dimensional case is illustrated in Fig. 1.

5 Local ID and Extreme Value Theory

The characterization of continuous distance distributions established from first principles in Sect. 3 can be regarded as an elucidation of extreme value theory (EVT) in the setting of short-tailed distributions. Several mutually-equivalent formulations of EVT exist; here, the formulation with which we will concern ourselves is that of regularly varying functions, pioneered by Karamata in the 1930s. There is a vast literature on EVT and its applications, the majority of which involve the upper tails of distributions. For a detailed account of regular variation and EVT, see (for example) [29].

5.1 First-Order EVT

Let F be a function that is continuously differentiable and strictly positive over the open interval $I = (0, z)$ for some $z > 0$. Although Karamata's original representation theorem [20] deals with the behavior of smooth functions as they diverge to (positive) infinity, the theorem can be reformulated by applying a reciprocal transformation of the function domain $(1/z, \infty)$ into the interval I; this yields the result that the function F restricted to I can be expressed in the form $F(x) = x^{\gamma} \ell(1/x)$ for some constant γ, where ℓ is differentiable and *slowly varying* (at infinity); that is, for all $c > 0$, ℓ satisfies

$$\lim_{u \to \infty} \frac{\ell(cu)}{\ell(u)} = 1.$$

The function F restricted to I is itself said to be *regularly varying* with index γ.

Note that the slowly-varying component $\ell(u)$ is not necessarily constant as $u \to \infty$. However, the slowly-varying condition ensures that the derivative $\ell'(u)$ is bounded, and that the following auxiliary function tends to 0:

$$\varepsilon(u) \triangleq \frac{u \ell'(u)}{\ell(u)}, \qquad \lim_{u \to \infty} \varepsilon(u) \to 0.$$

Slowly varying functions are also known to be representable in terms of their auxiliary function. More specifically, $\ell(1/x)$ can be shown to be slowly varying as $1/x \to \infty$ if and only if there exists some $w > 0$ such that

$$\ell(1/x) = \exp\left(\eta(1/x) + \int_{1/w}^{1/x} \frac{\varepsilon(u)}{u}\, du\right),$$

where η and ε are measurable and bounded functions such that $\eta(1/x)$ tends to a constant, and $\varepsilon(1/t)$ tends to 0, as x and t tend to 0. Note that under the substitution $t = 1/u$, the slowly-varying component can be expressed as

$$\ell(1/x) = \exp\left(\eta(1/x) + \int_x^w \frac{\varepsilon(1/t)}{t}\, dt\right).$$

Thus the formula $F(x) = x^\gamma \ell(1/x)$ can easily be verified to fit the form of the representation given in Theorem 2, with the following choices:

$$\gamma = \mathrm{ID}_F^*\,; \quad \eta(1/x) = \ln F(w) - \mathrm{ID}_F^* \ln w\,; \quad \varepsilon(1/t) = \mathrm{ID}_F^* - \mathrm{ID}_F(t)\,.$$

5.2   Second-Order EVT

An issue of great importance and interest in the design and performance of semi-parametric EVT estimators is the speed of convergence of extreme values to their limit [30]. As is the case with first-order EVT, many approaches to the estimation of second-order parameters have been developed [21].

Here, we will follow the formulation appearing in [31] using second-order regular variation. In their paper, de Haan and Resnick proved the equivalence of two conditions regarding the derivatives of regularly varying functions, which can be stated as follows. Let $\phi : (0, \infty) \to \mathbb{R}$ be twice differentiable, with $\phi'(t)$ eventually positive as $t \to \infty$, and let $\gamma \in \mathbb{R}$. Consider a function $A(t)$ whose absolute value is regularly varying with index $\rho \leq 0$, such that $A(t) \to 0$ as $t \to \infty$ with $A(t)$ either eventually positive or eventually negative. Then the condition

$$A(t) \triangleq \frac{t \cdot \phi''(t)}{\phi'(t)} - \gamma + 1$$

is equivalent to ϕ' having the following representation for some non-zero constant k:

$$\phi'(t) = k \cdot t^{\gamma-1} \cdot \exp\left(\int_1^t \frac{A(u)}{u}\, du\right).$$

As in the discussion of first-order EVT in Sect. 5.1, we apply a reciprocal transform of the domain to an interval of the form $I = (0, w)$, by setting $t = 1/x$ and $\phi'(t) = F'(x)$. Noting that $F''(x) = -t^2 \phi''(t)$, and defining $B(x) \triangleq A(t)$, the first condition can be shown to be

$$B(x) \triangleq 1 - \gamma - \frac{x \cdot F''(x)}{F'(x)} = 1 - \gamma - \mathrm{ID}_{F'}(x),$$

and, under the substitution $u = 1/y$, the second condition can be shown to be

$$F'(x) = k \cdot x^{1-\gamma} \cdot \exp\left(\int_x^1 \frac{B(y)}{y} \, dy\right).$$

Thus these equivalent conditions can be verified to fit the form of the representation given in Theorem 2, with $w = 1$, $v = 0$, and

$$k = F'(1);$$
$$\gamma = 1 - \mathrm{ID}_{F'}^* = 2 - \mathrm{ID}_F^*;$$
$$B(x) = 1 - \gamma - \mathrm{ID}_{F'}(x) = \mathrm{ID}_F^* - 1 - \mathrm{ID}_{F'}(x).$$

Second-order EVT is largely concerned with the estimation of the parameter ρ. The following theorem establishes that the two functions $B(x)$ and $\mathrm{ID}_{\mathrm{ID}_{F'}}(x)$ both have as their index of regular variation the non-negative value $-\rho$.

Theorem 5. *Let F be a function that is twice differentiable over the interval $I = (0, z)$, for some choice of $z > 0$ (possibly infinite). Furthermore, assume that F' and F'' are positive everywhere or negative everywhere over I, that $F'(x) \to 0$ as $x \to 0$, and that ID_F^* exists. Let $B(x) = \mathrm{ID}_{F'}^* - 1 - \mathrm{ID}_{F'}(x)$. Then $B(x)$ and $B_*(x) \triangleq \mathrm{ID}_{\mathrm{ID}_{F'}}(x)$ are both regularly varying with index $-\rho$. Furthermore, if B_* is continuously differentiable, then $-\rho = \mathrm{ID}_{B_*}^*$.*

The proof relies heavily on Lemma 2 and Theorem 4. However, due to space limitations, the details are omitted in this version of the paper.

6   Conclusion

Among the implications of the extreme-value-theoretic foundation introduced in this paper, perhaps the one with the greatest potential impact for similarity applications is that intrinsic dimensionality reveals the interchangeability between probability and distance. For distance distributions, the ID representation formula of Theorem 2 essentially states that the ratio of the expected numbers of points in neighborhoods of different radii asymptotically tends to the ratio of the neighborhood radii themselves, raised to the power of the intrinsic dimension. Knowledge of any 4 of these 5 quantities would help to determine the value of the unknown quantity. Indeed, this relationship among probability, distance and ID has already been successfully exploited to improve the accuracy/time tradeoff of certain similarity search tasks, via dimensional testing [11–14].

To realize the full potential of the theory of local intrinsic dimensionality for similarity applications, it is essential that accurate and efficient estimators be available. Estimators for the first-order EVT scale parameter have been developed within the EVT community; generally, they require on the order of 100 neighborhood distance samples in order to converge [22]. However, second-order EVT estimators generally require many thousands of neighbors for convergence [32]. Reducing the sample size for both first- and second-order LID/EVT estimation would be a worthwhile target.

78   M.E. Houle

Another important future research direction is that of feature selection and metric learning. The LID model provides a natural measure of data discriminability that could in principle be used to guide the selection of features, or the learning of similarity measures. Towards this goal, in a companion paper [33], a theoretical investigation is made into how the local IDs of distance distributions can change as their cumulative distribution functions are combined.

Acknowledgments. The author gratefully acknowledges the financial support of JSPS Kakenhi Kiban (A) Research Grant 25240036 and JSPS Kakenhi Kiban (B) Research Grant 15H02753.

References

1. Falconer, K.: Fractal Geometry: Mathematical Foundations and Applications. Wiley, Hoboken (2003)
2. Grassberger, P., Procaccia, I.: Measuring the strangeness of strange attractors. Physica D **9**(1–2), 189–208 (1983)
3. Camastra, F., Vinciarelli, A.: Estimating the intrinsic dimension of data with a fractal-based method. IEEE TPAMI **24**(10), 1404–1407 (2002)
4. Gupta, A., Krauthgamer, R., Lee, J.R.: Bounded geometries, fractals, and low-distortion embeddings. In: FOCS, pp. 534–543 (2003)
5. Rozza, A., Lombardi, G., Ceruti, C., Casiraghi, E., Campadelli, P.: Novel high intrinsic dimensionality estimators. Mach. Learn. J. **89**(1–2), 37–65 (2012)
6. Larrañaga, P., Lozano, J.A.: Estimation of Distribution Algorithms: A New Tool for Evolutionary Computation, vol. 2. Springer, Heidelberg (2002)
7. Karger, D.R., Ruhl, M.: Finding nearest neighbors in growth-restricted metrics. In: STOC, pp. 741–750 (2002)
8. Houle, M.E., Kashima, H., Nett, M.: Generalized expansion dimension. In: ICDMW, pp. 587–594 (2012)
9. Houle, M.E.: Dimensionality, discriminability, density & distance distributions. In: ICDMW, pp. 468–473 (2013)
10. Beygelzimer, A., Kakade, S., Langford, J.: Cover trees for nearest neighbors. In: ICML, pp. 97–104 (2006)
11. Houle, M.E., Ma, X., Nett, M., Oria, V.: Dimensional testing for multi-step similarity search. In: ICDM, pp. 299–308 (2012)
12. Houle, M.E., Nett, M.: Rank-based similarity search: reducing the dimensional dependence. IEEE TPAMI **37**(1), 136–150 (2015)
13. Houle, M.E., Ma, X., Oria, V., Sun, J.: Efficient similarity search within user-specified projective subspaces. Inf. Syst. **59**, 2–14 (2016)
14. Casanova, G., Englmeier, E., Houle, M.E., Kröger, P., Nett, M., Zimek, A.: Dimensional testing for reverse k-nearest neighbor search. PVLDB **10**(7), 769–780 (2017)
15. de Vries, T., Chawla, S., Houle, M.E.: Density-preserving projections for large-scale local anomaly detection. Knowl. Inf. Syst. **32**(1), 25–52 (2012)
16. Furon, T., Jégou, H.: Using Extreme Value Theory for Image Detection. Research report RR-8244, INRIA, February 2013
17. Balkema, A.A., de Haan, L.: Residual life time at great age. Ann. Probab. **2**, 792–804 (1974)
18. Pickands, J.: Statistical inference using extreme order statistics. Ann. Stat. **3**, 119–131 (1975)

19. Coles, S.: An Introduction to Statistical Modeling of Extreme Values. Springer, London (2001)
20. Karamata, J.: Sur un mode de croissance réguliere des fonctions. Mathematica (Cluj) **4**, 38–53 (1930)
21. Gomes, M.I., Canto e Castro, L., Fraga Alves, M.I., Pestana, D.: Statistics of extremes for IID data and breakthroughs in the estimation of the extreme value index: Laurens de Haan leading contributions. Extremes **11**, 3–34 (2008)
22. Amsaleg, L., Chelly, O., Furon, T., Girard, S., Houle, M.E., Kawarabayashi, K., Nett, M.: Estimating local intrinsic dimensionality. In: KDD, pp. 29–38 (2015)
23. Hill, B.M.: A simple general approach to inference about the tail of a distribution. Ann. Stat. **3**(5), 1163–1174 (1975)
24. Romano, S., Chelly, O., Nguyen, V., Bailey, J., Houle, M.E.: Measuring dependency via intrinsic dimensionality. In: ICPR, pp. 1207–1212 (2016)
25. Krauthgamer, R., Lee, J.R.: Navigating nets: simple algorithms for proximity search. In: SODA, pp. 798–807 (2004)
26. Houle, M.E., Ma, X., Oria, V.: Effective and efficient algorithms for flexible aggregate similarity search in high dimensional spaces. IEEE TKDE **27**(12), 3258–3273 (2015)
27. Breunig, M.M., Kriegel, H.P., Ng, R.T., Sander, J.: LOF: identifying density-based local outliers. SIGMOD Rec. **29**(2), 93–104 (2000)
28. Roweis, S.T., Saul, L.K.: Nonlinear dimensionality reduction by locally linear embedding. Science **290**(5500), 2323–2326 (2000)
29. Beirlant, J., Goegebeur, Y., Segers, J., Teugels, J.: Statistics of Extremes: Theory and Applications. Wiley, Hoboken (2004)
30. de Haan, L., Stadtmüller, U.: Generalized regular variation of second order. J. Aust. Math. Soc. (Series A) **61**(3), 381–395 (1996)
31. de Haan, L., Resnick, S.: Second-order regular variation and rates of convergence in extreme-value theory. Ann. Probab. **24**(1), 97–124 (1996)
32. Fraga Alves, M.I., de Haan, L., Lin, T.: Estimation of the parameter controlling the speed of convergence in extreme value theory. Math. Methods Stat. **12**(2), 155–176 (2003)
33. Houle, M.E.: Local intrinsic dimensionality II: multivariate analysis and distributional support. In: SISAP, pp. 1–16 (2017)

Local Intrinsic Dimensionality II: Multivariate Analysis and Distributional Support

Michael E. Houle[✉]

National Institute of Informatics, 2-1-2 Hitotsubashi,
Chiyoda-ku, Tokyo 101-8430, Japan
meh@nii.ac.jp

Abstract. Distance-based expansion models of intrinsic dimensionality have had recent application in the analysis of complexity of similarity applications, and in the design of efficient heuristics. This theory paper extends one such model, the local intrinsic dimension (LID), to a multivariate form that can account for the contributions of different distributional components towards the intrinsic dimensionality of the entire feature set, or equivalently towards the discriminability of distance measures defined in terms of these feature combinations. Formulas are established for the effect on LID under summation, product, composition, and convolution operations on smooth functions in general, and cumulative distribution functions in particular. For some of these operations, the dimensional or discriminability characteristics of the result are also shown to depend on a form of distributional support. As an example, an analysis is provided that quantifies the impact of introduced random Gaussian noise on the intrinsic dimension of data. Finally, a theoretical relationship is established between the LID model and the classical correlation dimension.

1 Introduction

In such areas as search and retrieval, machine learning, data mining, multimedia, recommendation systems, and bioinformatics, the efficiency and efficacy of many fundamental operations commonly depend on the interplay between measures of data similarity and the choice of features by which objects are represented. Similarity search, perhaps the most fundamental operation involving similarity measures, is ubiquitous in data analysis tasks such as clustering, k-nearest-neighbor classification, and anomaly detection, as well as content-based multimedia applications.

One of the most common strategies employed in similarity search is that of neighborhood expansion, in which the radius of the search (or, equivalently, the number of points visited) is increased until a neighborhood of the desired size has been identified. Even when this radius is known in advance, the actual number of points visited can be considerably larger than the target neighborhood size, particularly if the similarity measure is not discriminative. A highly indiscriminative similarity measure is more susceptible to measurement error,

© Springer International Publishing AG 2017
C. Beecks et al. (Eds.): SISAP 2017, LNCS 10609, pp. 80–95, 2017.
DOI: 10.1007/978-3-319-68474-1_6

and (in the case of distance metrics) is less suited to classical methods for search path pruning based on the triangle inequality.

1.1 Discriminability and Dimensionality

In the design and analysis of similarity applications, measures or criteria often directly or indirectly express some notion of the discriminability of similarity measures within neighborhoods. Examples include spectral feature selection criteria [1], such as the Laplacian score [2], that measure the discriminative power of candidate feature in terms of the variance of feature values. The distance ratio (aspect ratio), defined as the ratio between the largest and smallest pairwise distances within a data set, has been applied to the analysis of nearest-neighbor search [3]. As a way of characterizing the dimensionality of data, data concentration has been measured in terms of the relationship between the mean and variance of pairwise distance values [4]; in the context of similarity indexing, data concentration has been linked to the theory of VC dimensionality [5]. Disorder inequalities, relaxations of the usual metric triangle inequality, have been proposed for the analysis of combinatorial search algorithms making use of rankings of data points with respect to a query [6]. The degree of relaxation of the disorder inequality can be regarded as a measure of the discriminability of the data.

In an attempt to alleviate the effects of high dimensionality, and thereby improve the discriminability of data, simpler representations are often sought by means of a number of supervised or unsupervised learning techniques. One of the earliest and most well-established simplification strategies is dimensional reduction, which seeks a projection to a lower-dimensional subspace that minimizes data distortion. In general, dimensional reduction requires that an appropriate dimension for the reduced space (or approximating manifold) must be either supplied or learned, ideally so as to minimize the error or loss of information incurred. The dimension of the surface that best approximates the data can be regarded as an indication of the intrinsic dimensionality (ID) of the data set, or of the minimum number of latent variables needed to represent the data. Intrinsic dimensionality thus serves as an important natural measure of the complexity of data.

An important family of dimensional models, including the minimum neighbor distance (MiND) models [7], the expansion dimension (ED) [8,9], and the local intrinsic dimension (LID) [10], quantify the ID in the vicinity of a point of interest in the data domain, by assessing the rate of growth in the number of data objects encountered as the distance from the point increases. For example, in Euclidean spaces the volume of an m-dimensional set grows proportionally to r^m when its size is scaled by a factor of r — from this rate of volume growth with distance, the dimension m can be deduced. These so-called expansion models provide a local view of the dimensional structure of the data, as their estimation is restricted to a neighborhood of the point of interest. They hold an advantage over parametric models in that they require no explicit knowledge of the underlying global data distribution. Expansion models have been successfully applied to the design and analysis of index structures for similarity search [8,11–15], and heuristics for anomaly detection [16], as well as in manifold learning.

With one exception, the aforementioned expansion models assign a measure of intrinsic dimensionality to specific sets of data points. The exception is the local intrinsic dimension ('local ID', or 'LID'), which extends the ED model to a statistical setting that assumes an underlying (but unknown) distribution of distances from a given reference point [10]. Here, each object of the data set induces a distance to the reference point; together, these distances can be regarded as samples from the distribution. The only assumptions made on the nature of the distribution are those of smoothness.

In [10], the local intrinsic dimension is shown to be equivalent to a notion of discriminability of the distance measure, as reflected by the growth rate of the cumulative distribution function. For a random distance variable $\mathbf{X}$ with a continuous cumulative distribution function $F_{\mathbf{X}}$, the k-nearest neighbor distance within a sample of n points is an estimate of the distance value r for which $F_{\mathbf{X}}(r) = k/n$. If k is fixed, and n is allowed to tend to infinity, the indiscriminability of $F_{\mathbf{X}}$ at the k-nearest neighbor distance tends to the local intrinsic dimension. The local intrinsic dimension can thus serve to characterize the degree of difficulty in performing similarity-based operations within query neighborhoods using the underlying distance measure, asymptotically as the sample size (that is, the data set size) scales to infinity.

A strong connection has been shown between local intrinsic dimensionality and the scale parameter of the generalized Pareto distribution [17], and to the index of regularly varying functions in the Karamata representation from extreme value theory [18]. Estimators developed within the extreme value theory research community have been shown to be effective for local ID [17]. Of these, the Hill estimator [19] has recently been used for ID estimation in the context of reverse k-NN search [15] and the analysis of non-functional dependencies among data features [20].

1.2   Contributions

This work is a companion paper to [18], which establishes a formal connection between the LID model and the statistical discipline of extreme value theory (EVT) [21,22]. In this theoretical paper, we extend the LID model to a multivariate form that can account for contributions from different distributions towards an overall intrinsic dimensionality; or equivalently towards the discriminability of distance measures defined in terms of feature combinations.

The paper is organized as follows: in the next section, the local ID model of [10,18] is reintroduced in a multivariate framework, and an equivalence is shown between the notions of local intrinsic dimensionality and indiscriminability. In Sect. 3 we introduce the notion of local intrinsic dimensionality weighted by a measure of the support for the function or distribution to which it is associated. In Sect. 4, we derive and state rules by which support-weighted and unweighted local ID change as functions or distributions are combined. The effect on ID of adding noise to feature values is investigated in Sect. 5, where we develop a rule for the convolution of probability density functions. Using this rule, we consider as an example the effect of introducing random Gaussian noise. A relationship between the local intrinsic dimensionality and the correlation dimension is established in Sect. 6. The paper concludes in Sect. 7.

2 Intrinsic Dimensionality and Indiscriminability

In this section, we rework the local ID model from [10] in a multivariate framework. The model as originally proposed considered only the cumulative distribution functions of distance distributions. Here, we extend the notions of intrinsic dimensionality and indiscriminability to a more general class of smooth functions, and show that these notions remain equivalent within the new framework.

2.1 Multivariate Formulation of Local ID

The definitions of intrinsic dimensionality and indiscriminability introduced in [10] and further developed in [18] can be extended to the multivariate case in a natural way, by replacing the magnitude of the independent variable from the univariate case with the multivariate vector norm.

Definition 1. *Let F be a real-valued multivariate function over a normed vector space $(\mathbb{R}^m, \|\cdot\|)$ that is non-zero at $\mathbf{x} \neq \mathbf{0} \in \mathbb{R}^m$. The intrinsic dimensionality of F at $\mathbf{x}$ is defined as*

$$\mathrm{IntrDim}_F(\mathbf{x}) \triangleq \lim_{\epsilon \to 0} \frac{\ln\left(F((1+\epsilon)\mathbf{x})/F(\mathbf{x})\right)}{\ln\left(\|(1+\epsilon)\mathbf{x}\|/\|\mathbf{x}\|\right)} = \lim_{\epsilon \to 0} \frac{\ln\left(F((1+\epsilon)\mathbf{x})/F(\mathbf{x})\right)}{\ln(1+\epsilon)},$$

wherever the limit exists.

Definition 1 expresses the local intrinsic dimensionality of F as the growth rate (or 'scale') in function value over the growth rate in the norm of its variable vector. F can be regarded as playing the role of the volume of a ball in some space, with $\|\mathbf{x}\|$ playing the role of its radius — in Euclidean space, the quotient appearing in the limit would reveal the dimension of the space [8–10,18].

Definition 2 expresses the indiscriminability of F as a ratio between the proportional increase in value of the function versus the proportional increase in the norm of its variable vector, as the vector expands [10,18].

Definition 2. *Let F be a real-valued multivariate function over a normed vector space $(\mathbb{R}^m, \|\cdot\|)$ that is non-zero at $\mathbf{x} \neq \mathbf{0} \in \mathbb{R}^m$. The indiscriminability of F at $\mathbf{x}$ is defined as*

$$\mathrm{InDiscr}_F(\mathbf{x}) \triangleq \lim_{\epsilon \to 0} \left(\frac{F((1+\epsilon)\mathbf{x}) - F(\mathbf{x})}{F(\mathbf{x})} \middle/ \frac{(1+\epsilon)\|\mathbf{x}\| - \|\mathbf{x}\|}{\|\mathbf{x}\|} \right)$$

$$= \lim_{\epsilon \to 0} \frac{F((1+\epsilon)\mathbf{x}) - F(\mathbf{x})}{\epsilon \cdot F(\mathbf{x})},$$

wherever the limit exists.

If the function F depends only on the norm of $\mathbf{x}$ (that is, if $F(\mathbf{x}) = F(\mathbf{y})$ when $\|\mathbf{x}\| = \|\mathbf{y}\|$), then F can be expressed as a univariate function of a single variable, $\|\mathbf{x}\|$. In this case, the definitions of the multivariate intrinsic dimensionality and indiscriminability stated above are equivalent to the original univariate versions from [10]. In particular, the indiscriminability of F at $\mathbf{x}$ (or equivalently, the distance $\|\mathbf{x}\|$ from the origin) would account for the growth rate of F over expanding balls centered at the origin, when measured at radius $\|\mathbf{x}\|$.

In general, however, F may not depend only on the norm of $\mathbf{x}$, and the dependencies of the growth rate of F on its variables may differ greatly from variable to variable.

2.2 Equivalence of Indiscriminability and Intrinsic Dimensionality

The following result generalizes the main theorem of [10] for univariate cumulative distribution functions to the case of multivariate real functions $F : \mathbb{R}^m \to \mathbb{R}$, for those locations (with the exception of the origin) at which the gradient $\nabla F = (\frac{\partial F}{\partial x_i})_{i=1}^m$ is defined and is continuous.

Theorem 1. *Let F be a real-valued multivariate function over a normed vector space $(\mathbb{R}^m, \|\cdot\|)$, and let $\mathbf{x} \neq \mathbf{0} \in \mathbb{R}^m$ be a vector of positive norm. If there exists an open interval $I \subseteq \mathbb{R}$ containing 0 such that F is non-zero and continuously differentiable at $(1 + \epsilon)\mathbf{x}$ for all $\epsilon \in I$, then*

$$\mathrm{ID}_F(\mathbf{x}) \triangleq \frac{\mathbf{x} \cdot \nabla F(\mathbf{x})}{F(\mathbf{x})} = \mathrm{IntrDim}_F(\mathbf{x}) = \mathrm{InDiscr}_F(\mathbf{x}) \,.$$

Proof. By assumption, F is non-zero and differentiable at $(1+\epsilon)\mathbf{x}$ for all real values of ϵ within an open interval containing 0. We may therefore apply l'Hôpital's rule to the limits of Definitions 1 and 2 along this interval. Using the chain rule for multivariate differentiation, and letting $\mathbf{y} = (1 + \epsilon)\mathbf{x}$, the indiscriminability of F thus becomes

$$\mathrm{InDiscr}_F(\mathbf{x}) = \lim_{\epsilon \to 0} \left[\sum_{i=1}^m \left(\frac{\partial F(\mathbf{y})}{\partial x_i} \cdot \frac{\partial x_i}{\partial y_i} \cdot \frac{\partial y_i}{\partial \epsilon} \right) \middle/ \frac{\partial(\epsilon \cdot F(\mathbf{x}))}{\partial \epsilon} \right]$$

$$= \frac{1}{F(\mathbf{x})} \cdot \lim_{\epsilon \to 0} \sum_{i=1}^m \left(\frac{\partial F(\mathbf{y})}{\partial x_i} \cdot \frac{\partial(y_i/(1+\epsilon))}{\partial y_i} \cdot \frac{\partial((1+\epsilon)x_i)}{\partial \epsilon} \right)$$

$$= \frac{1}{F(\mathbf{x})} \cdot \lim_{\epsilon \to 0} \left[\frac{1}{1+\epsilon} \cdot \sum_{i=1}^m \left(\frac{\partial F((1+\epsilon)\mathbf{x})}{\partial x_i} \cdot x_i \right) \right]$$

$$= \frac{1}{F(\mathbf{x})} \cdot \sum_{i=1}^m \left(x_i \cdot \frac{\partial F(\mathbf{x})}{\partial x_i} \right) = \frac{\mathbf{x} \cdot \nabla F(\mathbf{x})}{F(\mathbf{x})} \,;$$

the limit holds due to the assumption of continuity for each of the partial derivatives of F, together with the assumption that $F(\mathbf{x}) \neq 0$.

To complete the proof, we observe that the multivariate intrinsic dimensionality can be derived in a similar way. (The details are omitted due to space limitations.) □

The equivalence between the multivariate indiscriminability and intrinsic dimensionality can be extended to the origin $\mathbf{0}$, or to points $\mathbf{x} \neq \mathbf{0}$ where F is zero, by taking limits of $\mathrm{IntrDim}_F$ and $\mathrm{InDiscr}_F$ radially, along a line passing through $\mathbf{x}$ and the origin. When taken at $\mathbf{x}$, the limit will be referred to as the *indiscriminability of F at $\mathbf{x}$*, which we will denote by $\mathrm{ID}_F(\mathbf{x})$. Otherwise, if the limit is taken at the origin, $\mathbf{x}$ is interpreted as the direction vector of a line of points along which $\mathrm{IntrDim}_F$ and $\mathrm{InDiscr}_F$ are evaluated as the distance to the origin tends to zero. We will refer to this limit as the *local intrinsic dimensionality of F in the direction of $\mathbf{x}$*, and denote it by $\mathrm{ID}_F^*(\mathbf{x})$.

Corollary 1. *Let F be a real-valued multivariate function over a normed vector space $(\mathbb{R}^m, \|\cdot\|)$, and let $\mathbf{x} \neq \mathbf{0} \in \mathbb{R}^m$ be a vector of positive norm. If there exists an open interval $I \subseteq \mathbb{R}$ containing 0 such that F is non-zero and continuously differentiable at $(1 + \epsilon)\mathbf{x}$ for all $\epsilon \in I$, then*

$$\mathrm{ID}_F(\mathbf{x}) \triangleq \lim_{\epsilon \to 0} \frac{(1 + \epsilon)\mathbf{x} \cdot \nabla F((1 + \epsilon)\mathbf{x})}{F((1 + \epsilon)\mathbf{x})}$$
$$= \lim_{\epsilon \to 0} \mathrm{IntrDim}_F((1 + \epsilon)\mathbf{x}) = \lim_{\epsilon \to 0} \mathrm{InDiscr}_F((1 + \epsilon)\mathbf{x}),$$

whenever the limits exist.

Alternatively, if F is non-zero and continuously differentiable at $\epsilon\mathbf{x}$ for all $\epsilon \geq 0 \in I$, then

$$\mathrm{ID}_F^*(\mathbf{x}) \triangleq \lim_{\epsilon \to 0^+} \frac{\epsilon\mathbf{x} \cdot \nabla F(\epsilon\mathbf{x})}{F(\epsilon\mathbf{x})} = \lim_{\epsilon \to 0^+} \mathrm{IntrDim}_F(\epsilon\mathbf{x}) = \lim_{\epsilon \to 0^+} \mathrm{InDiscr}_F(\epsilon\mathbf{x}),$$

whenever the limits exist.

If $\mathbf{x} = (x_i)_{i=1}^m$ is such that $x_i = 0$ for some choices of $1 \leq i \leq m$, Corollary 1 holds within the subspace spanned by those variables that are non-zero at $\mathbf{x}$. No contributions are made by the ith coordinate towards the values of $\mathrm{ID}_F(\mathbf{x})$ or ID_F^*.

In the univariate case, if the function F is such that $F((1 + \epsilon)x) = 0$ for all $\epsilon \in I$, the values of the derivative $F'((1 + \epsilon)x)$ are also 0, and $\mathrm{ID}_F(0)$ is in indeterminate form. However, for the remainder of the paper, $F((1 + \epsilon)x) \cdot \mathrm{ID}_F((1 + \epsilon)x)$ can safely be deemed to be 0 whenever $(1 + \epsilon)x \cdot F'((1 + \epsilon)x) = 0$.

2.3 Transformations of Variable

In some similarity applications, transformations of the underlying distance measures are sometimes sought so as to improve the overall performance of tasks that depend upon them. In the context of distance distributions with smooth cumulative distribution functions, for certain smooth transformations of the underlying distance measure, it was shown in [10] that the indiscriminability of a cumulative distribution function after transformation can be decomposed into two factors: the indiscriminability of the cumulative distribution function before transformation, and the indiscriminability of the transform itself. Here, we restate the theorem for the case when F is not necessarily a cumulative distribution function.

Theorem 2 ([10]). *Let g be a real-valued function that is non-zero and contin-uously differentiable over some open interval containing $x \in \mathbb{R}$, except perhaps at x itself. Let f be a real-valued function that is non-zero and continuously dif-ferentiable over some open interval containing $g(x) \in \mathbb{R}$, except perhaps at $g(x)$ itself. Then*

$$\mathrm{ID}_{f \circ g}(x) = \mathrm{ID}_g(x) \cdot \mathrm{ID}_f(g(x))$$

whenever $\mathrm{ID}_g(x)$ and $\mathrm{ID}_f(g(x))$ are defined. If $x = f(x) = g(x) = 0$, then

$$\mathrm{ID}^*_{f \circ g} = \mathrm{ID}^*_f \cdot \mathrm{ID}^*_g$$

whenever ID^*_f and ID^*_g are defined.

Note that as with Corollary 1, the theorem also holds for one-sided limits, and for boundary points whenever the domain(s) of the functions are restricted to intervals of the real line.

3   Support-Weighted Local Intrinsic Dimensionality

When considering the intrinsic dimensionality of a function of several variables, or a distance distribution on a space of many features, it is natural to ask which variables or features are contributing most to the overall discriminability of the function or cumulative distribution function (as the case may be). Two variables or features with the same local ID value taken individually may not necessarily have the same impact on the overall ID value when taken together. To see this, let Φ and Ψ be the respective cumulative distribution functions of two univariate distance distributions on distance variable x. The indiscriminability $\mathrm{ID}_\Phi(x)$ can be thought of as having a 'support' equal to the probability measure associated with distance x — namely, $\Phi(x)$; similarly, the support for $\mathrm{ID}_\Psi(x)$ would be $\Psi(x)$. Even when the indiscriminabilities $\mathrm{ID}_\Phi(x)$ and $\mathrm{ID}_\Psi(x)$ are equal, if (say) the support $\Phi(x)$ greatly exceeded $\Psi(x)$, one would be forced to conclude that the features associated with ID_Φ are more significant than those of ID_Ψ, at least within the neighborhood of radius x, if Φ and Ψ were to be combined.

Accordingly, we define a 'support-weighted' local intrinsic dimension as follows.

Definition 3 (Support-Weighted ID). *Let F be a real-valued multivariate function over a normed vector space $(\mathbb{R}^m, \| \cdot \|)$, and let $\mathbf{x} \neq \mathbf{0} \in \mathbb{R}^m$ be a vector of positive norm. The support-weighted indiscriminability of F at $\mathbf{x}$ is defined as*

$$\mathrm{wID}_F(\mathbf{x}) \triangleq F(\mathbf{x}) \, \mathrm{ID}_F(\mathbf{x}) = \mathbf{x} \cdot \nabla F(\mathbf{x}).$$

Theorem 1 implies that the support-weighted ID is equivalent to $\mathbf{x} \cdot \nabla F(\mathbf{x})$, the directional derivative of $F(\mathbf{x})$ in the direction of $\mathbf{x}$ itself. It can be regarded as a measure of the growth rate F over an expanding ball, that rewards increases in the magnitudes of both the rate of change of F (from the gradient $\nabla F(\mathbf{x})$) and the radius of the ball itself (from $\mathbf{x}$).

For examples in which the function represents the cumulative distribution of a distance distribution, estimating support-weighted ID for the purpose of assessing indiscriminability can be complicated by the need to standardize the distance within which the indiscriminabilities are measured — otherwise, if the local ID of each feature were assessed at widely-varying distances, there would be no basis for the comparison of performance. In practice, however, estimation of ID requires samples that are the result of a k-nearest neighbor query on the underlying data set. Nevertheless, in the univariate case, standardization can be achieved using the local ID representation theorem from [18]:

Theorem 3 (Local ID Representation [18]**).** *Let $\Phi : \mathbb{R} \to \mathbb{R}$ be a real-valued function, and let $v \in \mathbb{R}$ be a value for which $\mathrm{ID}_{\Phi}(v)$ exists. Let x and w be values for which x/w and $\Phi(x)/\Phi(w)$ are both positive. If Φ is non-zero and continuously differentiable everywhere in the interval $[\min\{x, w\}, \max\{x, w\}]$, then*

$$\frac{\Phi(x)}{\Phi(w)} = \left(\frac{x}{w}\right)^{\mathrm{ID}_{\Phi}(v)} \cdot G_{\Phi,v,w}(x), \text{ where}$$

$$G_{\Phi,v,w}(x) \triangleq \exp\left(\int_x^w \frac{\mathrm{ID}_{\Phi}(v) - \mathrm{ID}_{\Phi}(t)}{t}\,\mathrm{d}t\right),$$

whenever the integral exists.

For a univariate cumulative distribution function Φ at distance x, we can use the theorem with $v = 0$ to relate the support $\Phi(x)$ with the support at another desired distance w. If n is the size of the data set that we are given, we choose the distance at which over n selection trials one would expect k samples to fall within the neighborhood — that is, w would satisfy $\Phi(w) = k/n$. The support-weighted ID would thus be:

$$\mathrm{wID}_{\Phi}(x) = \Phi(x)\,\mathrm{ID}_{\Phi}(x) = \frac{k\,\mathrm{ID}_{\Phi}(x)}{n} \cdot \left(\frac{x}{w}\right)^{\mathrm{ID}_{\Phi}^*} \cdot G_{\Phi,0,w}(x).$$

In [18] it is shown that (under certain mild assumptions) the function $G_{\Phi,0,w}(x)$ tends to 1 as $x, w \to 0$ (or equivalently, as $n \to \infty$); also, $\mathrm{ID}_{\Phi}(x)$ would tend to ID_{Φ}^*, for which reliable estimators are known [17,19]. Thus, for reasonably large data set sizes, one could use the following approximation:

$$\mathrm{wID}_{\Phi}(x) \approx \frac{k\,\mathrm{ID}_{\Phi}^*}{n} \cdot \left(\frac{x}{w}\right)^{\mathrm{ID}_{\Phi}^*}.$$

4   Composition Rules for Local ID

The support-weighted ID can be interpreted as the radially-directed directional derivative $\mathbf{x} \cdot \nabla F(\mathbf{x})$. Each contribution to the inner product has the form of a univariate directional derivative, which Theorem 1 tells us is equivalent to a univariate support-weighted ID. In this section, the properties of the multivariate directional derivative will be exploited to prove a number of composition rules.

Theorem 4 (General Decomposition Rule). *Let F be a real-valued multivariate function over a normed vector space $(\mathbb{R}^m, \|\cdot\|)$, and let $\mathbf{x} \neq \mathbf{0} \in \mathbb{R}^m$ be a vector of positive norm. If there exists an open interval $I \subseteq \mathbb{R}$ containing 0 such that F is non-zero and continuously differentiable at $(1+\epsilon)\mathbf{x}$ for all $\epsilon \in I$, then*

$$\mathrm{wID}_F(\mathbf{x}) = \sum_{i=1}^m \mathrm{wID}_{F_i}(x_i),$$

where for all $1 \leq i \leq m$, F_i is any univariate antiderivative of the partial derivative $\partial F/\partial x_i$ obtained by treating the other variables x_j $(j \neq i)$ as constants.

Alternatively, if F is non-zero and continuously differentiable at $\epsilon\mathbf{x}$ for all $\epsilon \geq 0 \in I$, then

$$\mathrm{ID}_F^*(\mathbf{x}) = \sum_{i=1}^m \omega_i^* \, \mathrm{ID}_{F_i}^*, \qquad \omega_{i.}^* \triangleq \lim_{\epsilon \to 0^+} \frac{F_i(\epsilon x_i)}{F(\epsilon\mathbf{x})},$$

whenever the limits ω_i^* exist.

Proof. With the assumption of the continuous differentiablity of F, together with the definition of F_i, Theorem 1 can be applied twice to give

$$F(\mathbf{x})\,\mathrm{ID}_F(\mathbf{x}) = \sum_{i=1}^m \left(x_i \cdot \frac{\partial F(\mathbf{x})}{\partial x_i} \right) = \sum_{i=1}^m x_i \, F_i'(x_i) = \sum_{i=1}^m F_i(x_i)\,\mathrm{ID}_{F_i}(x_i).$$

Replacing $\mathbf{x}$ by $\epsilon\mathbf{x}$ and then taking the limit as $\epsilon \to 0^+$, we obtain

$$\mathrm{ID}_F^*(\mathbf{x}) = \lim_{\epsilon \to 0^+} \mathrm{ID}_F(\epsilon\mathbf{x}) = \lim_{\epsilon \to 0^+} \sum_{i=1}^m \left(\frac{F_i(\epsilon x_i)}{F(\epsilon\mathbf{x})} \cdot \mathrm{ID}_{F_i}(\epsilon\mathbf{x}) \right) = \sum_{i=1}^m \omega_i^* \, \mathrm{ID}_{F_i}^*.$$

$\square$

The antiderivative F_i is not unique — each choice satisfying the conditions of Theorem 4 would be associated with its own support-weighted ID function (although the sum would be the same). If, however, each F_i is constructed from F by fixing all variable values to those of a specific point $\mathbf{x}$, we find that the local ID values taken at $\mathbf{x}$ would themselves sum to the total local ID. Such a collection F_i will be referred to as the *canonical decomposition* of the support-weighted ID.

Corollary 2 (Canonical Decomposition Rule). *Let F satisfy the conditions of Theorem 4 at $\mathbf{x}$, with the additional constraint that $x_i \neq 0$ for all $1 \leq i \leq m$. Let F_i be the univariate function on x_i obtained directly from F by treating the variable x_j as a constant, for all choices $j \neq i$. Then*

$$\mathrm{ID}_F(\mathbf{x}) = \sum_{i=1}^m \mathrm{ID}_{F_i}(x_i) \quad and \quad \mathrm{ID}_F^*(\mathbf{x}) = \sum_{i=1}^m \mathrm{ID}_{F_i}^*.$$

Proof. Since the definition of F_i implies that $F_i(x_i) = F(\mathbf{x}) \neq 0$, we have that $F_i'(x_i) = \partial F(\mathbf{x})/\partial x_i$. Theorem 4 can therefore be applied to give

$$\mathrm{ID}_F(\mathbf{x}) = \sum_{i=1}^{m} \left(\frac{F_i(x_i)}{F(\mathbf{x})} \cdot \mathrm{ID}_{F_i}(x_i) \right) = \sum_{i=1}^{m} \mathrm{ID}_{F_i}(x_i).$$

To prove the second statement of the theorem, it suffices to replace $\mathbf{x}$ by $\epsilon\mathbf{x}$, and let $\epsilon \to 0$. $\square$

Corollary 2 can be used to show that whenever a function can be expressed as the product of univariate functions (each on their own variable), the local ID of the product is the sum of the individual IDs of each univariate function. The proof is omitted due to space limitations.

Corollary 3 (Product Rule). *Let $F(\mathbf{x}) = \prod_{i=1}^{m} \Phi_i(x_i)$ be a function satisfying the conditions of Theorem 4 at $\mathbf{x}$. If we have that $\mathrm{ID}_{\Phi_i}(x_i)$ exists for all $1 \leq i \leq m$, or alternatively that $\mathrm{ID}_{\Phi_i}^*$ exists, then*

$$\mathrm{ID}_F(\mathbf{x}) = \sum_{i=1}^{m} \mathrm{ID}_{\Phi_i}(x_i), \quad \text{or alternatively,} \quad \mathrm{ID}_F^*(\mathbf{x}) = \sum_{i=1}^{m} \mathrm{ID}_{\Phi_i}^*.$$

Theorem 4 can be used directly to show that whenever a function can be expressed as the sum of univariate functions (each on their own variable), the support-weighted ID of the sum is the sum of the individual support-weighted IDs of each univariate function. Again, we omit the proof.

Corollary 4 (Summation Rule). *Let $F(\mathbf{x}) = \sum_{i=1}^{m} \Phi_i(x_i)$ be a function satisfying the conditions of Theorem 4 at $\mathbf{x}$. Then*

$$\mathrm{wID}_F(\mathbf{x}) = \sum_{i=1}^{m} \mathrm{wID}_{\Phi_i}(x_i), \quad \text{or alternatively,}$$

$$\mathrm{ID}_F^*(\mathbf{x}) = \sum_{i=1}^{m} \omega_i^* \, \mathrm{ID}_{\Phi_i}^*, \quad \omega_i^* \triangleq \lim_{\epsilon \to 0^+} \frac{\Phi_i(\epsilon x_i)}{F(\epsilon\mathbf{x})},$$

whenever the limits ω_i^* exist.

Note that for the summation rule, the sum of the weights ω_i^* must equal 1. Intuitively, whenever a function can be expressed as the sum of univariate functions (each on their own variable), the local ID of the sum is a weighted sum of the individual IDs of each univariate function — the weighting being determined by the proportional support of each univariate function.

5   Convolution and Support-Weighted ID

As a motivating example, let us consider a situation in which multivariate data is perturbed through the addition of random noise. Given a distance of interest

$x > 0$, if before the perturbation the distribution of a distance feature is associated with a cumulative distribution function Φ, the probability of the feature value lying in the range $[0, x]$ would be $\Phi(x)$; if Φ were smooth at x, the distance distribution would have a probability density of $\Phi'(x)$. However, if the value were perturbed by an amount t by the addition of noise, the probability density associated with this event would be $\Phi'(x - t)$. In general, the overall probability density at x would be determined by the convolution

$$F'(x) = \int_{-\infty}^{\infty} \phi(x - t)\,\psi(t)\,dt\,,$$

where ϕ and ψ are the probability density functions associated with the original distance distribution and the additive noise, respectively.

In this situation, one may wonder how the noise affects the intrinsic dimensionality of the distribution. Some insight into this issue can be gained from an analysis of support-weighted ID. Before attempting an answer, we will state and prove the following general result (not necessarily restricted to probability density functions of distance distributions).

Theorem 5 (Convolution Rule). *Let F be the real-valued univariate convolution function over a normed vector space $(\mathbb{R}, \|\cdot\|)$, defined as follows:*

$$F(x) = \int_0^x \int_{-\infty}^{\infty} \phi(s - t)\,\psi(t)\,dt\,ds\,,$$

where ϕ and ψ have the antiderivatives Φ and Ψ, respectively, defined almost everywhere. Consider the value $x \neq 0$. If there exists an open interval $I \subseteq \mathbb{R}$ containing 0 such that F is non-zero and continuously differentiable at $(1 + \epsilon)x$ for all $\epsilon \in I$, then

$$\mathrm{wID}_F(x) = \int_{-\infty}^{\infty} \mathrm{wID}_\Phi(x - t)\,\psi(t)\,dt + \int_{-\infty}^{\infty} \mathrm{wID}_\Psi(x - t)\,\phi(t)\,dt\,.$$

Proof. With the assumption of the continuous differentiablity of F, Theorem 1 can be applied twice to give

$$\mathrm{wID}_F(x) = F(x)\,\mathrm{ID}_F(x) = x \cdot F'(x) = x \cdot \int_{-\infty}^{\infty} \phi(x - t)\,\psi(t)\,dt$$

$$= \int_{-\infty}^{\infty} (x - t)\,\phi(x - t)\,\psi(t)\,dt + \int_{-\infty}^{\infty} t\,\phi(x - t)\,\psi(t)\,dt\,.$$

After the substitution $u = x - t$ in the second integral, and applying Theorem 1, the result follows:

$$\mathrm{wID}_F(x) = \int_{-\infty}^{\infty} (x - t)\,\Phi'(x - t)\,\psi(t)\,dt + \int_{-\infty}^{\infty} (x - u)\,\Psi'(x - u)\,\phi(u)\,du$$

$$= \int_{-\infty}^{\infty} \Phi(x - t)\,\mathrm{ID}_\Phi(x - t)\,\psi(t)\,dt + \int_{-\infty}^{\infty} \Psi(x - t)\,\mathrm{ID}_\Psi(x - t)\,\phi(t)\,dt\,.$$

$\square$

The convolution rule has a natural interpretation when ϕ and ψ are both probability density functions: the support-weighted ID of the convolution (wID$_F$) is the sum of the expected values of the support-weighted ID of each function (wID$_\Phi$ and wID$_\Psi$), each taken with respect to each other's distribution (using the probability densities ψ and ϕ, respectively), shifted by the value of x and reversed in orientation.

Returning to our example involving the perturbation of the data by noise, let us assume that a certain amount of random Gaussian noise (whose probability density is ψ) is added to the value of each variable of the data (whose probability density is ϕ).

If the original distribution is associated with a real data set, its contribution to the overall support-weighted ID is likely to be low. Real data sets typically have intrinsic dimensionalities much lower than the representational dimension — typical values range up to approximately 20, even when the number of features are in the thousands or more [17]. On the other hand, the contribution of the Gaussian noise is likely to be much higher, particularly when the variance is high: the central portion of a high-variance Gaussian would have high support as well as an ID value approaching 1. (As the probability density of the Gaussian distribution is well known, this assertion can be verified analytically; however, the details are omitted due to space limitations.) In short, the convolution rule indicates that great care should be used when introducing additive (Gaussian) noise into data sets, as they can produce sets with unrealistically high complexity as measured by support-weighted or unweighted local ID.

6   Local ID and the Correlation Dimension

In [23], Pesin explored the relationship between several formulations of the correlation dimension with other measures of intrinsic dimensionality, such as the Hausdorff dimension, the 'box counting' dimension, and the Rényi dimension. Rather than assuming any smoothness conditions (such as continuous differentiability), Pesin used upper limits (lim sup) and lower limits (lim inf) to define the various measures of dimension. A partial ordering of these measures was then exhibited. For some of the theoretical arguments, the Hausdorff and certain other dimensional variants were characterized in terms of the extreme values of a local ID formulation closely related to ID*.

A full accounting of the interrelationships between the various measures of intrinsic dimensionality is beyond the scope of this paper — instead, the interested reader is referred to [20,23] for more information. Here, we will show that for smooth distance distributions, the correlation dimension of a domain can be expressed as a form of weighted averaging of the local intrinsic dimensionalities associated with the distributions of distances from each of the locations within the domain.

We begin with a formal definition of the correlation dimension. Consider an infinite sequence of random point samples from a domain $\mathcal{D}$ for which a probability measure μ and distance measure $d : \mathcal{D} \times \mathcal{D} \to \mathbb{R}^{\geq 0}$ are defined.

Given a positive distance threshold $r > 0$, the *correlation integral* $C(r)$ is given by the limit

$$C(r) \triangleq \lim_{n \to \infty} \rho_n(r), \quad \rho_n(r) = 2g_n(r)/(n^2 - n),$$

where $g_n(r)$ is the number of unordered pairs of points of the nth sample having mutual distance at most r. As r tends to 0, if the correlation integral asymptotically tends to a polynomial form $C(r) \to r^\nu$, the value of ν for which this relationship holds is the *correlation dimension* [24]. More formally, the correlation dimension is given by

$$\nu = \lim_{r \to 0^+} \log C(r)/\log r.$$

In practice, the correlation dimension ν is estimated from values of $\rho_n(r)$ over a large sample. The simplest method for estimating ν is to plot $\rho_n(r)$ against r in log-log scale, to then fit a line to the lower tail, and determine its slope [25]. More sophisticated estimators have been proposed by Takens and Theiler [26].

When C is smooth in the vicinity of $r = 0$, l'Hôpital's rule can be used to show that the correlation dimension can be expressed in terms of the local ID of C.

Definition 4. *Whenever $C(r)$ is non-zero and continuously differentiable in some open interval bounded below by $r = 0$, the* correlation dimension *can be expressed as*

$$\mathrm{CD}^* \triangleq \lim_{r \to 0^+} \frac{\log C(r)}{\log r} = \lim_{r \to 0^+} \frac{r \cdot C'(r)}{C(r)} = \lim_{r \to 0^+} \mathrm{ID}_C(r).$$

In addition, $\mathrm{CD}(r) \triangleq \mathrm{ID}_C(r)$ *will be referred to as the* correlation indiscriminability *at radius r wherever $\mathrm{ID}_C(r)$ is well-defined.*

The quantity $\rho_n(r)$ can be regarded as an estimate of the probability that two points independently drawn from $\mathcal{D}$ have mutual distance at most r. The correlation integral $C(r)$ can thus be represented as the following Lebesgue integral:

$$C(r) = \int_{\mathcal{D}} F_r \, d\mu = \iint_{\mathcal{D}} \theta(r - d(\mathbf{x}, \mathbf{y})) \, d\mu(\mathbf{y}) \, d\mu(\mathbf{x}), \tag{1}$$

where $F_r(\mathbf{x})$ is the probability of drawing a sample from $\mathcal{D}$ within distance r of $\mathbf{x}$, and where $\theta(x)$ equals 1 if $x \geq 1$ and equals 0 otherwise.

With respect to any given fixed location $\mathbf{x}$, the random selection of points from $\mathcal{D}$ induces a distribution of non-negative values, determined by the distances of selected points from $\mathbf{x}$. The value of the cumulative distribution function of this distribution at distance r is precisely the integrand $F_r(\mathbf{x})$ appearing in Eq. 1. To avoid notational confusion, we will denote the cumulative distribution function for distances relative to $\mathbf{x}$ by $F_{\mathbf{x}}(r) \triangleq F_r(\mathbf{x})$.

The following theorem uses this notion of distance distributions relative to the locations in $\mathcal{D}$ to establish a relationship between the correlation indiscriminability and the indiscriminabilities of these distributions.

Theorem 6. *Let $\mathcal{D}$ be a domain for which a probability measure μ and distance measure $d : \mathcal{D} \times \mathcal{D} \to \mathbb{R}^{\geq 0}$ are defined. For a given point $\mathbf{x} \in \mathcal{D}$ and a positive distance value $r > 0$, let $F_r(\mathbf{x}) = F_{\mathbf{x}}(r)$ be the probability associated with the ball $\{\mathbf{y} \in \mathcal{D} \,|\, d(\mathbf{y}, \mathbf{x}) \leq r\}$ centered at $\mathbf{x}$ with radius r, as defined above. If the correlation indiscriminability $\mathrm{CD}(r)$ exists at r, and if the indiscriminability $\mathrm{ID}_{F_r}(\mathbf{x}) \triangleq \mathrm{ID}_{F_{\mathbf{x}}}(r)$ exists almost everywhere (that is, for all $\mathbf{x} \in \mathcal{D}$ except possibly over a subset of $\mathcal{D}$ of measure zero), then*

$$\mathrm{CD}(r) \cdot \int_{\mathcal{D}} F_r \, \mathrm{d}\mu = \int_{\mathcal{D}} F_r \cdot \mathrm{ID}_{F_r} \, \mathrm{d}\mu \,.$$

Furthermore, if the intrinsic dimensionality $\mathrm{ID}^*_F(\mathbf{x}) \triangleq \mathrm{ID}_{F_{\mathbf{x}}}(0)$ exists almost everywhere, then

$$\mathrm{CD}^* = \lim_{r \to 0} \left(\int_{\mathcal{D}} F_r \cdot \mathrm{ID}^*_F \, \mathrm{d}\mu \, \Big/ \int_{\mathcal{D}} F_r \, \mathrm{d}\mu \right) ,$$

wherever the limit exists.

Proof. Rewriting the left-hand side of the first equation in terms of the correlation integral C, and applying Corollary 1, we obtain

$$\mathrm{CD}(r) \cdot \int_{\mathcal{D}} F_r \, \mathrm{d}\mu = \mathrm{ID}_C(r) \cdot C(r) = r \cdot C'(r) = \int_{\mathcal{D}} r \cdot f_r \, \mathrm{d}\mu \,,$$

where $f_r(\mathbf{x}) = F'_{\mathbf{x}}(r)$ is the derivative of the cumulative distribution function $F_{\mathbf{x}}$ at r (which must exist for almost all $\mathbf{x} \in \mathcal{D}$, since $\mathrm{ID}_{F_{\mathbf{x}}}(r)$ is assumed to exist almost everywhere). Applying Corollary 1 again to the integrand $r \cdot f_r$, the first equation follows.

The second equation of the theorem statement follows from the first, by taking the limit of the ratio of the two integrals as r tends to zero, and noting that the local intrinsic dimensionality ID^*_F is assumed to exist almost everywhere. $\square$

Intuitively, Theorem 6 extends the finite summation rule of Corollary 4 to the possibly infinite collection of distance distributions induced by the locations of the domain $\mathcal{D}$. Here, the correlation indiscriminability and correlation dimension can be viewed as a form of expectation of the local ID, taken across all locations of $\mathcal{D}$ according to a weighting function that depends on the local cumulative distribution function. If one accepts ID^*_F and $\mathrm{ID}_F(r)$ as the natural measures of intrinsic dimensionality and indiscriminability for a smooth local distance distribution, then the correlation dimension is a natural measure of global intrinsic dimensionality within a domain where almost every location is associated with a smooth distance distribution.

7   Conclusion

In 2005, Yang and Wu [27] identified ten challenging problems in data mining, by consulting 14 of the most active researchers in data mining as to what they considered to be the most important topics for future research. The first challenging problem listed was that of developing a unifying theory of data mining, for which the authors stated:

> *Several respondents feel that the current state of the art of data mining research is too 'ad-hoc'. Many techniques are designed for individual problems, such as classification or clustering, but there is no unifying theory. However, a theoretical framework that unifies different data mining tasks including clustering, classification, association rules, etc., as well as different data mining approaches (such as statistics, machine learning, database systems, etc.), would help the field and provide a basis for future research.*

Although a comprehensive theory of data mining is not yet in sight, the similarity-based intrinsic dimensional framework presented in this paper and others can constitute an early step, as it provides a framework for the formal study of such fundamental notions as similarity measure, data density, data discriminability, intrinsic dimensionality, and (in [18]) local inlierness and outlierness. Preliminary work on the use of support-weighted local ID in the selection of features for similarity graph construction [28] shows that this model also has promise for guiding practical applications.

Acknowledgments. The author gratefully acknowledges the financial support of JSPS Kakenhi Kiban (A) Research Grant 25240036 and JSPS Kakenhi Kiban (B) Research Grant 15H02753.

References

1. Zhao, Z., Liu, H.: Spectral feature selection for supervised and unsupervised learning. In: ICML, pp. 1151–1157 (2007)
2. He, X., Cai, D., Niyogi, P.: Laplacian score for feature selection. In: NIPS, pp. 507–514 (2005)
3. Clarkson, K.L.: Nearest neighbor queries in metric spaces. Discrete Comput. Geom. **22**, 63–93 (1999)
4. Chávez, E., Navarro, G., Baeza-Yates, R., Marroquín, J.L.: Searching in metric spaces. ACM Comput. Surv. **33**, 273–321 (2001)
5. Pestov, V.: Indexability, concentration, and VC theory. J. Discrete Algorithms **13**, 2–18 (2012)
6. Goyal, N., Lifshits, Y., Schütze, H.: Disorder inequality: a combinatorial approach to nearest neighbor search. In: WSDM, pp. 25–32 (2008)
7. Rozza, A., Lombardi, G., Ceruti, C., Casiraghi, E., Campadelli, P.: Novel high intrinsic dimensionality estimators. Mach. Learn. J. **89**(1–2), 37–65 (2012)
8. Karger, D.R., Ruhl, M.: Finding nearest neighbors in growth-restricted metrics. In: STOC, pp. 741–750 (2002)
9. Houle, M.E., Kashima, H., Nett, M.: Generalized expansion dimension. In: ICDMW, pp. 587–594 (2012)
10. Houle, M.E.: Dimensionality, discriminability, density & distance distributions. In: ICDMW, pp. 468–473 (2013)

11. Beygelzimer, A., Kakade, S., Langford, J.: Cover trees for nearest neighbors. In: ICML, pp. 97–104 (2006)
12. Houle, M.E., Ma, X., Nett, M., Oria, V.: Dimensional testing for multi-step similarity search. In: ICDM, pp. 299–308 (2012)
13. Houle, M.E., Nett, M.: Rank-based similarity search: reducing the dimensional dependence. IEEE TPAMI **37**(1), 136–150 (2015)
14. Houle, M.E., Ma, X., Oria, V., Sun, J.: Efficient similarity search within user-specified projective subspaces. Inf. Syst. **59**, 2–14 (2016)
15. Casanova, G., Englmeier, E., Houle, M.E., Kröger, P., Nett, M., Zimek, A.: Dimensional testing for reverse k-nearest neighbor search. PVLDB **10**(7), 769–780 (2017)
16. de Vries, T., Chawla, S., Houle, M.E.: Density-preserving projections for large-scale local anomaly detection. Knowl. Inf. Syst. **32**(1), 25–52 (2012)
17. Amsaleg, L., Chelly, O., Furon, T., Girard, S., Houle, M.E., Kawarabayashi, K., Nett, M.: Estimating local intrinsic dimensionality. In: KDD, pp. 29–38 (2015)
18. Houle, M.E.: Local intrinsic dimensionality I: An extreme-value-theoretic foundation for similarity applications. In: SISAP, pp. 1–16 (2017)
19. Hill, B.M.: A simple general approach to inference about the tail of a distribution. Ann. Stat. **3**(5), 1163–1174 (1975)
20. Romano, S., Chelly, O., Nguyen, V., Bailey, J., Houle, M.E.: Measuring dependency via intrinsic dimensionality. In: ICPR, pp. 1207–1212 (2016)
21. Coles, S.: An Introduction to Statistical Modeling of Extreme Values. Springer, London (2001)
22. Gomes, M.I., Canto e Castro, L., Fraga Alves, M.I., Pestana, D.: Statistics of extremes for IID data and breakthroughs in the estimation of the extreme value index: Laurens de Haan leading contributions. Extremes **11**, 3–34 (2008)
23. Pesin, Y.B.: On rigorous mathematical definitions of correlation dimension and generalized spectrum for dimensions. J. Stat. Phys. **71**(3–4), 529–547 (1993)
24. Grassberger, P., Procaccia, I.: Measuring the strangeness of strange attractors. Physica D **9**(1–2), 189–208 (1983)
25. Procaccia, I., Grassberger, P., Hentschel, V.G.E.: On the characterization of chaotic motions. In: Garrido, L. (ed.) Dynamical System and Chaos. Lecture Notes in Physics, vol. 179, pp. 212–221. Springer, Heidelberg (1983)
26. Theiler, J.: Lacunarity in a best estimator of fractal dimension. Phys. Lett. A **133**(4–5), 195–200 (1988)
27. Yang, Q., Wu, X.: 10 challenging problems in data mining research. Int. J. Inf. Technol. Decis. Making **5**(4), 597–604 (2006)
28. Houle, M.E., Oria, V., Wali, A.M.: Improving k-NN graph accuracy using local intrinsic dimensionality. In: SISAP, pp. 1–15 (2017)

High-Dimensional Simplexes
for Supermetric Search

Richard Connor[1]([✉]), Lucia Vadicamo[2], and Fausto Rabitti[2]

[1] Department of Computer and Information Sciences, University of Strathclyde,
Glasgow G1 1XH, UK
`richard.connor@strath.ac.uk`
[2] ISTI - CNR, Via Moruzzi 1, 56124 Pisa, Italy
`{lucia.vadicamo,fausto.rabitti}@isti.cnr.it`

Abstract. In a metric space, triangle inequality implies that, for any three objects, a triangle with edge lengths corresponding to their pairwise distances can be formed. The *n-point property* is a generalisation of this where, for any $(n + 1)$ objects in the space, there exists an n-dimensional simplex whose edge lengths correspond to the distances among the objects. In general, metric spaces do not have this property; however in 1953, Blumenthal showed that any semi-metric space which is isometrically embeddable in a Hilbert space also has the n-point property.

We have previously called such spaces *supermetric* spaces, and have shown that many metric spaces are also supermetric, including Euclidean, Cosine, Jensen-Shannon and Triangular spaces of any dimension.

Here we show how such simplexes can be constructed from only their edge lengths, and we show how the geometry of the simplexes can be used to determine lower *and upper* bounds on unknown distances within the original space. By increasing the number of dimensions, these bounds converge to the true distance.

Finally we show that for any Hilbert-embeddable space, it is possible to construct Euclidean spaces of arbitrary dimensions, from which these lower and upper bounds of the original space can be determined. These spaces may be much cheaper to query than the original. For similarity search, the engineering tradeoffs are good: we show significant reductions in data size and metric cost with little loss of accuracy, leading to a significant overall improvement in exact search performance.

Keywords: Supermetric space · Metric search · Metric embedding · Dimensionality reduction

1 Introduction

To set the context, we are interested in searching a (large) finite set of objects S which is a subset of an infinite set U, where (U, d) is a metric space. The general requirement is to efficiently find members of S which are similar to an

© Springer International Publishing AG 2017
C. Beecks et al. (Eds.): SISAP 2017, LNCS 10609, pp. 96–109, 2017.
DOI: 10.1007/978-3-319-68474-1_7

arbitrary member of U, where the distance function d gives the only way by which any two objects may be compared. There are many important practical examples captured by this mathematical framework, see for example [3,19]. Such spaces are typically searched with reference to a query object $q \in U$. A threshold search for some threshold t, based on a query $q \in U$, has the solution set $\{s \in S$ such that $d(q, s) \leq t\}$.

This becomes an interesting problem when exhaustive search is intractable, in which case the research problem is to find ways of pre-processing the collection ahead of query time in order to minimise the cost of query. There are three main problems with achieving efficiency. Most obviously, if the search space is very large, scalability is required. Less obviously, when the search space is large, semantic accuracy is important to avoid large numbers of false positive results – in the terminology of information retrieval, *precision* becomes relatively more important that *recall*. To achieve higher semantic accuracy will usually require more expensive metrics, and larger data representations.

Here, we present a new technique which can be used to address all three of these issues in supermetric spaces. Using properties of finite isometric embedding, we show a mechanism which allows spaces with certain properties to be translated into a second, smaller, space. For a metric space (U, d), we describe a family of functions ϕ_n which can be created by measuring the distances among n objects sampled from the original space, and which can then be used to create a *surrogate* space:

$$\phi_n : (U, d) \to (\mathbb{R}^n, \ell_2)$$

with the property

$$\ell_2(\phi_n(u_1), \phi_n(u_2)) \leq d(u_1, u_2) \leq g(\phi_n(u_1), \phi_n(u_2))$$

for an associated function g.

The advantages of the proposed technique are that (a) the ℓ_2 metric is very much cheaper than some Hilbert-embeddable metrics; (b) the size of elements of $\mathbb{R}^n$ may be much smaller than elements of U, and (c) in many cases we can achieve both of these along with an *increase* in the scalability of the resulting search space.

2   Related Work

Finite Isometric Embeddings are excellently summarised by Blumenthal [1]. He uses the phrase *four-point property* to mean a space that is 4-embeddable in 3-dimensional Euclidean space: that is, that for any four objects in the original space it is possible to construct a distance-preserving tetrahedon. Wilson [17] shows various properties of such spaces, and Blumenthal points out that results given by Wilson, when combined with work by Menger [15], generalise to show that some spaces with the four-point property also have the n-point property: any n points can be isometrically embedded in an $(n-1)$-dimensional Euclidean space (ℓ_2^{n-1}). In a later work, Blumenthal [2] shows that any space which is isometrically embeddable in a Hilbert space has the n-point property. This single result applies to many metrics, including Euclidean, Cosine, Jensen-Shannon and Triangular [7], and is sufficient for our purposes here.

Dimensionality Reduction aims to produce low-dimensional encodings of high-dimensional data, preserving the local structure of some input data. See [11,18] for comprehensive surveys on this topic.

The *Principal Component Analysis* (PCA) [12] is the most popular of the techniques for unsupervised dimensionality reduction. The idea is to find a linear transformation of n-dimensional to k-dimensional vectors $(k \leq n)$ that best preserves the *variance* of the input data. Specifically, PCA projects the data along the direction of its first k principal components, which are the eigenvectors of the covariance matrix of the (centered) input data.

According to the *Johnson-Lindenstrauss Flattening Lemma* (JL) (see e.g. [14, pag. 358]), a projection can also be used to embed a finite set of n euclidean vectors into a k-dimensional euclidean space space $(k < n)$ with a "small" distortion. Specifically the Lemma asserts that for any n-points of ℓ_2 and every $0 < \epsilon < 1$ there is a mapping into ℓ_2^k that preserves all the interpoint distances within factor $1 + \epsilon$, where $k = O(\epsilon^{-2} \log n)$. The low dimensional embedding given by the Johnson Lindenstrauss lemma is particularly simple to implement.

General Metric Spaces do not allow either PCA or JL as these require access to a coordinate space. Mao et al. [13] pointed out that multidimensional-methods can be indirectly applied to metric space by using the *pivot space* model. In that case each metric object is represented by its distance to a finite set of pivots.

In the general metric space context, perhaps the best known technique is *metric Multidimensional Scaling* (MDS) [8]. MDS aims to preserve *inter-point distances* using spectral analysis. However, when the number m of data points is large the classical MDS is too expensive in practice due to a requirement for $O(m^2)$ distance computations and spectral decomposition of a $m \times m$ matrix.

The *Landmark MDS* (LMDS) [9] is a fast approximation of MDS. LMDS uses a set of k *landmark* points to compute $k \times m$ distances of the data points from the pivots. It applies classical MSD to these points and uses a distance-based triangulation procedure to project the remaining data points.

LAESA. [16] is a more tractable mechanism which has been used for metric filtering, rather than approximate search. n reference objects are somehow selected. For each element of the data, the distances to these points are recorded in a table. At query time, the distances between the query and each reference point are calculated. The table can then be scanned row at a time, and each distance compared; if, for any reference object p_i and data object s_j the absolute difference $|d(q, p_i) - d(s_j, p_i)| > t$, then from triangle inequality it is impossible for s_j to be within distance t of the query, and the distance calculation can be avoided. LAESA can be used as an efficient pre-filter for exact search when memory size is limited, and we make an experimental comparison with the new lower-bound mechanism we describe in this paper.

3   Upper and Lower Bounds from Simplexes

For any $(n + 1)$ objects u_i in a supermetric space (U, d), there exists a simplex in ℓ_2^n where each vertex v_i corresponds to one object u_i and whose edge lengths correspond to distances in the original space, i.e. $\ell_2(v_i, v_j) = d(u_i, u_j)$.

We show now how this property can be used to give bounds on distances between two elements of U whose distance cannot be directly measured. This is useful in many different search paradigms where the bounds are required between an arbitrary elements $s_i \in S \subset U$ which has been pre-processed before a search, and an element $q \in U$ which is not known when the pre-processing occurs.

Our strategy is to choose a set of n reference points $P \subset U$, from which an isometric $(n - 1)$-dimensional simplex σ is created. Now, given a further point $u \in U$, and all the distances $d(p_i, u)$, an n-dimensional simplex σ_u can be created by the addition of a single vertex to σ.

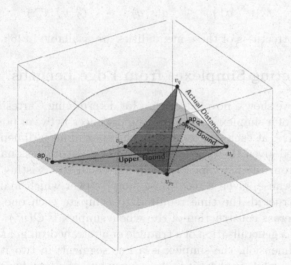

Fig. 1. Tetrahedral embedding of four points into 3D Euclidean space.

For simplicity, Fig. 1 shows an ℓ_2^3 space into which four objects have been projected. Here we have only two reference points, p_1 and p_2. For each element a the notation v_a is used to denote a corresponding point in the ℓ_2^3 space. The distance $d(s, q)$ is not known; however the 4-point property means that the corresponding distance $\ell_2(v_s, v_q)$ must be able to form the final edge of a tetrahedron. From this Figure, the intuition of the upper and lower bounds on $d(s, q)$ is clear, through rotation of the triangle $v_{p_1} v_{p_2} v_q$ around the line $v_{p_1} v_{p_2}$ until it is coincident with the plane in which $v_{p_1} v_{p_2} v_s$ lies. The two possible orientations give the upper and lower bounds, corresponding to the distances between v_s and the two apexes ap_{q_-} and ap_{q_+} of the two possible planar tetrahedra.

The same intuition generalises into many dimensions. The inter-object distances within a set $\{p_i\}$ of n reference objects are used to form a *base* simplex σ_0, with vertices $v_{p_1}, \ldots, v_{p_n}$, in $(n-1)$ dimensions. This corresponds to the line segment $v_{p_1} v_{p_2}$ in the figure, which gives a two-vertex simplex in ℓ_2^1. The simplex σ_0 is contained within a hyperplane of the ℓ_2^n space, and the distances from object s to each p_i are used to calculate a new simplex σ_s, in ℓ_2^n, consisting of a new apex point v_s set above the base simplex σ_0. There are two possible positions in ℓ_2^n for v_s, one on either side of the hyperplane containing σ_0; we denote these as v_s^+, and v_s^- respectively. Now, given the distances between object q and all p_i, there also exist two possible simplexes for σ_q, with two possible positions for v_q denoted by v_q^+ and v_q^-.

The process of rotating a triangle around its base generalises to that of rotating the apex point of any simplex around the hyperplane containing its base simplex. Furthermore, the n-point property guarantees the existence of a simplex σ_1 in ℓ_2^{n+1} which preserves the distance $d(s,q)$ as $\ell_2(v_s, v_q)$. From these observations we immediately have the following inequalities:

$$\ell_2^n(v_s^+, v_q^+) \quad \leq \quad d(s,q) \quad \leq \quad \ell_2^n(v_s^+, v_q^-)$$

Proofs of the correctness of these inequalities are available in [6].

4   Constructing Simplexes from Edge Lengths

In this section, we show a novel algorithm for determining Cartesian coordinates for the vertices of a simplex, given only the distances between points. The algorithm is inductive, at each stage allowing the apex of an n-dimensional simplex to be determined given the coordinates of an $(n-1)$-dimensional simplex, and the distances from the new apex to each vertex in the existing simplex. This is important because, given a fixed base simplex over which many new apexes are to be constructed, the time required to compute each one is $\mathcal{O}(n)$ for n dimensions, whereas construction of the whole simplex is $\mathcal{O}(n^2)$.

A simplex is a generalisation of a triangle or a tetrahedron in arbitrary dimensions. In one dimension, the simplex is a line segment; in two it is the convex hull of a triangle, while in three it is the convex hull of a tetrahedron. In general, the n-simplex of vertices $p_1, \ldots, p_{n+1}$ equals the union of all the line segments joining p_{n+1} to points of the $(n-1)$-simplex of vertices $p_1, \ldots, p_n$.

The structure of a simplex in n-dimensional space is given as an $n+1$ by n matrix representing the cartesian coordinates of each vertex. For example, the following matrix represents four coordinates which are the vertices of a tetrahedron in 3D space:

$$\begin{bmatrix} 0 & 0 & 0 \\ v_{2,1} & 0 & 0 \\ v_{3,1} & v_{3,2} & 0 \\ v_{4,1} & v_{4,2} & v_{4,3} \end{bmatrix}$$

Algorithm 1. nSimplexBuild

Input: $n + 1$ points $p_1, \ldots, p_{n+1} \in (U, d)$
Output: n-dimensional simplex in ℓ_2^n represented by the matrix $\Sigma \in \mathbb{R}^{(n+1) \times n}$

1   $\Sigma = 0 \in \mathbb{R}^{(n+1) \times n}$;
2   **if** $n = 1$ **then**
3     $\delta = d(p_1, p_2)$;
4     $\Sigma = \begin{bmatrix} 0 \\ \delta \end{bmatrix}$;
5   **return** Σ;
6   **end**
7   Σ_{Base} = nSimplexBuild($p_1, \ldots, p_n$);
8   $Distances = 0 \in \mathbb{R}^n$;
9   **for** $1 \le i \le n$ set $Distances[i] = d(p_i, p_{n+1})$;
10   $newApex$ = ApexAddition($\Sigma_{Base}, Distances$);
11   **for** $1 \le i \le n$ and $1 \le j \le i - 1$ set $\Sigma[i][j]$ to $\Sigma_{Base}[i][j]$;
12   **for** $1 \le j \le n$ set $\Sigma[n + 1][j]$ to $newApex[j]$;
13   **return** Σ;

For all such matrices Σ, the invariant that $v_{i,j} = 0$ whenever $j \ge i$ can be maintained without loss of generality; for any simplex, this can be achieved by rotation and translation within the Euclidean space while maintaining the distances among all the vertices. Furthermore, if we restrict $v_{i,j} \ge 0$ whenever $j = i - 1$ then in each row this component represents the *altitude* of the ith point with respect to a base face represented by the matrix cut down from Σ by selecting elements above and to the left of that entry.

4.1   Simplex Construction

This section gives an inductive algorithm (Algorithm 1) to construct a simplex in n dimensions based only on the distances measured among $n + 1$ points.

For the base case of a one-dimensional simplex (i.e. two points with a single distance δ) the construction is simply $\Sigma = \begin{bmatrix} 0 \\ \delta \end{bmatrix}$. For an n-dimensional simplex, where $n > 1$, an $(n - 1)$-dimensional simplex is first constructed using the distances among the first n points. This simplex is used as a simplex base to which a new apex, the $(n + 1)^{th}$ point, is added by the *ApexAddition* algorithm (Algorithm 2).

For an arbitrary set of objects $s_i \in S$, the apex $\phi_n(s_i)$ can be pre-calculated. When a query is performed, only n distances in the metric space require to be calculated to discover the new apex $\phi_n(q)$ in ℓ_2^n.

In essence, the *ApexAddition* algorithm is derived from exactly the same intuition as the lower-bound property explained earlier, at each stage lifting the final dimension out of the same hyperplane into a new dimension to capture the measured distances. Proofs of correctness for both the construction and the lower-bound property are available in [6].

4.2   Bounds

Because of the method we use to build simplexes, the final coordinate always represents the altitude of the apex above the hyperplane containing the base simplex. Given this, two apexes exist, according to whether a positive or negative real number is inserted at the final step of the algorithm.

Algorithm 2. ApexAddition

Input: A $(n-1)$-dimensional base simplex and the distances between a new (unknown) apex point and the vertices of the base simplex:

$$\Sigma_{Base} = \begin{bmatrix} 0 & & & \\ v_{2,1} & 0 & & \Large 0 \\ v_{3,1} & v_{3,2} & \ddots & \\ \vdots & & \ddots & 0 \\ v_{n,1} & & \cdots & v_{n,n-1} \end{bmatrix} \in \mathbb{R}^{n \times n-1}$$

$$Distances = \begin{bmatrix} \delta_1 \cdots \delta_n \end{bmatrix} \in \mathbb{R}^n$$

Output: The cartesian coordinates of the new apex point

1 $Output = \begin{bmatrix} \delta_1 \; 0 \cdots 0 \end{bmatrix} \in \mathbb{R}^n$;
2 **for** $i = 2$ **to** n **do**
3      $l = \ell_2(\Sigma_{Base}[i], Output)$;
4      $\delta = Distances[i]$;
5      $x = \Sigma_{Base}[i][i-1]$;
6      $y = Output[i-1]$;
7      $Output[i-1] = y - (\delta^2 - l^2)/2x$;
8      $Output[i] = +\sqrt{y^2 - (Output[i-1])^2}$;
9 **end**
10 **return** $Output$

As a direct result of this observation, and those given in Sect. 3, we have the following bounds for any two objects s_1 and s_2 in the original space:

Let

$$\phi_n(s_1) = (x_1, x_2, \ldots, x_{n-1}, x_n)$$
$$\phi_n(s_2) = (y_1, y_2, \ldots, y_{n-1}, y_n)$$

then

$$\sqrt{\sum_{i=1}^{n}(x_i - y_i)^2} \;\leq\; d(s_1, s_2) \;\leq\; \sqrt{\sum_{i=1}^{n-1}(x_i - y_i)^2 + (x_n + y_n)^2}$$

From the structure of these calculations, it is apparent that they are likely to converge rapidly around the true distance as the number of dimensions used becomes higher, as we will show in Sect. 5. It can also be seen that the cost of calculating both of these values together, especially in higher dimensions, is essentially the same as a simple ℓ_2 calculation.

Finally, we note that the lower-bound function is a proper metric, but the upper-bound function is not even a semi-metric: even although it is a Euclidean distance in the apex space, one of the domain points is constructed by reflection across a hyperplane and thus the distance between a pair of identical points is in general non-zero.

5 Measuring Distortion

We define distortion for an approximation (U', d') of a space (U, d) mapped by a function $f : U \to U'$ as as the smallest D such that, for some scaling factor r

$$r \cdot d'(f(u_i), f(u_j)) \quad \leq \quad d(u_i, u_j) \quad \leq \quad D \cdot r \cdot d'(f(u_i), f(u_j))$$

We have measured this for a number of different spaces, and present results over the SISAP *colors* benchmark set which are typical and easily reproducible. Summary results are shown in Fig. 2.

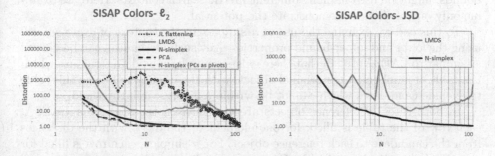

Fig. 2. Distortion measurements for various dimensionality reduction strategies for the *colors* data set. The left figure gives measurements for Euclidean distance, the right for Jensen-Shannon distance where only LMDS and n-simplex are applicable. The *colors* data set has 112 physical dimensions.

In each case, the X-axis represents the number of dimensions used for the representation, with the distortion plotted against this. For Euclidean distance, there are two entries for n-simplex: one for randomly-selected reference points, and the other where the choice of reference points is guided by the use of PCA. In the latter case we select the first n principal components (eigenvectors of the covariance matrix) as pivots.

It can be seen that n-simplex outperforms all other strategies except for PCA, which is not applicable to non-Euclidean spaces. LMDS is the only other mechanism applicable to general metric spaces[1]; this is a little more expensive than n-simplex to evaluate, and performs relatively badly. The comparison with JL is a slightly unfair, as the JL lemma applies only for very high dimensions in an evenly distributed space; we have also tested such spaces, and JL is still out-performed by n-simplex, especially at lower dimensions.

The distortion we show here is only for the lower-bound function of n-simplex. We have measured the upper-bound function also, which gives similar results. Unlike the lower-bound, the upper-bound is not a proper metric; however for non-metric approximate search it should be noted that the mean of the lower- and upper-bound functions give around half the distortion plotted here.

The implications of these results for exact search should be noted. For Euclidean search, the distortion has dropped to almost zero at between 20 and 30 dimensions, implying the possibility of accurate search using data which is less than one-quarter of the original size. For Jensen-Shannon, more dimensions will be required, but the cost of the ℓ_2 metric required to search the compressed space is around one-hundredth the cost of the original metric. In the next section we present experimental results consistent with these observations.

6 Exact Search: Indexing with n-Simplex

The simplex-building mechanism, along with the observations of upper and lower bounds, might be used in many different metric search contexts. Here, we examine only one of these to demonstrate the potential.

To this end we examine the use of n-simplex in the context of exact search, using the lower and upper-bound properties. Any such mechanism can be viewed as similar to LAESA [16], in that there exists an underlying data structure which is a table of numbers, n per original object, with the intention of using this table to exclude candidates which cannot be within a given search threshold.

In both cases, n reference objects are chosen from the space. For LAESA, each row of the table is filled, for one element of the data, with the distances from the candidate to each reference object. For n-simplex, each row is filled for one element of the data with the Cartesian coordinates of the new apex formed in n dimensions by applying these distances to an $(n-1)$-dimensional simplex formed from the reference objects.

The table having been established, a query notionally proceeds by measuring the distances from the query object to each reference point object. In the case of LAESA, the metric for comparison is Chebyshev: that is, if any pairwise difference is greater than the query threshold, the object from which that row was derived cannot be a solution to the query. For n-simplex, the metric used is ℓ_2: that is, if the apex represented in a row is further than the query threshold from the apex generated from the query, again the object from which that apex was derived cannot be a solution to the query.

---

[1] In [9] the authors note it works better for some metrics than for others; in our understanding, it will work well only for spaces with the n-point property.

In both cases, there are two ways of approaching the table search. It can be performed sequentially over the whole table, in which case either metric can be terminated within a row if the threshold is exceeded, without continuing to the end of the row. Alternatively the table can itself be re-indexed using a metric index. Although this compromises the amount of space available for the table itself, it may avoid many of the individual row comparisons.

In the context of re-indexing we also note that, in the case of n-simplex, the Euclidean metric used over the table rows itself has the four-point property, and so the Hilbert Exclusion property as described in [5] may be used.

In all cases the result is a filtered set of candidate objects which is guaranteed to contain the correct solution set. In general, this set must be re-checked against the original metric, in the original space. For n-simplex however the upper-bound condition is checked first; if this is less than the query threshold, then the object is guaranteed to be an element of the result set with no further check required.

6.1   Experiment - SISAP *colors*

Any such mechanism will perform differently over data sets with different characteristics and we cannot yet provide a full survey. To give useful comparisons with other studies in the literature, we apply the techniques to the SISAP *colors* [10] data set, using three different supermetrics: Euclidean, Cosine, and Jensen-Shannon[2]. We chose this data set because (a) it has only positive values and is therefore indexable by all of the metrics, and (b) it shows an interesting non-uniformity, in that its intrinsic dimensionality [4] for all metrics is much less than its physical dimensionality (112). It should thus give an interesting "real world" context to assess the relative value of the different mechanisms. Although it is a relatively small set, further experiments performed on much larger sets with different properties give quite consistent results, which we do not have space to report here.

For Euclidean distance, we used the three benchmark thresholds; for the other metrics, we chose thresholds that return around 0.01% of the data. In all cases the first 10% of the file is used to query the remaining 90%. Pivots are randomly-selected both for LAESA and n-simplex approach.

For each metric, we tested different mechanisms with different allocations of space: 5 to 50 numbers per data element, thus the space used per object is between 4.5% and 45% of the original. All results reported are for exact search, that is the initial filtering is followed by re-testing within the original space where required. Five different mechanism were tested, as follows:

sequential LAESA (L_{seq}) each row of the table is scanned sequentially, each element of each row is tested against the query and that row is abandoned if the absolute difference is greater than the threshold.
reindexed LAESA (L_{rei}) the data in the table is indexed using a monotone hyperplane tree, searched using the Chebyshev metric.

---

[2] For precise definitions of the non-Euclidean metrics used, see [5].

Table 1. Elapsed Times - SISAP *colors*, Euclidean distance. All times are in seconds, for executing 11268 queries over 101414 data. The *Tree* times are independent of the row as reference points are not used.

Dims	$t_0 = 0.051768$					$t_1 = 0.082514$					$t_2 = 0.131163$				
	L_{seq}	L_{rei}	N_{seq}	N_{rei}	$Tree$	L_{seq}	L_{rei}	N_{seq}	N_{rei}	$Tree$	L_{seq}	L_{rei}	N_{seq}	N_{rei}	$Tree$
5	18.6	28.0	13.8	5.8	5.5	33.4	80.9	22.4	29.0	18.1	56.2	201.6	34.9	70.4	54.4
10	17.7	22.1	15.0	3.3		30.3	67.9	20.3	14.7		58.1	220.3	25.5	50.6	
15	16.3	15.2	14.6	**3.0**		26.7	59.7	20.2	12.1		45.8	159.5	**24.4**	44.7	
20	19.0	16.3	18.9	3.3		28.2	56.6	19.4	**11.5**		46.8	189.3	27.8	48.3	
25	22.5	16.9	20.4	3.4		27.4	56.8	22.3	13.4		45.5	167.5	26.2	40.1	
30	20.9	16.8	20.4	3.5		28.6	57.3	24.5	13.6		45.9	181.2	28.5	45.1	
35	22.0	16.4	21.3	3.9		28.7	65.0	22.5	13.9		43.9	163.0	31.2	44.9	
40	23.1	17.3	22.1	4.0		28.8	55.9	22.8	14.3		49.4	180.5	34.2	46.1	
45	22.5	18.7	22.2	4.4		32.0	61.5	27.7	15.0		48.5	169.8	37.1	44.9	
50	21.3	17.1	18.9	4.5		32.0	59.0	24.0	15.5		55.2	207.6	34.5	45.3	

sequential n-simplex (N_{seq}) each row of the table is scanned sequentially, for each element of each row the square of the absolute difference is added to an accumulator, the row is abandoned if the accumulator exceeds the square of the threshold, and the upper-bound is applied if the end of the row is reached before re-checking in the original space.

reindexed n-simplex (N_{rei}) the data in the table is indexed using a monotone hyperplane tree using the Hilbert Exclusion property, and searched using the Euclidean metric; the upper-bound is applied for all results, before re-checking in the original space.

normal indexing (*Tree*) the space is indexed using a monotone hyperplane tree with the Hilbert Exclusion property, without the use of reference points.

The monotone hyperplane tree is used as, in previous work, this has been found to be the best-performing simple indexing mechanism for use with Hilbert Exclusion.

Measurements different figures are measured for each mechanism: the elapsed time, the number of original-space distance calculations performed and, in the case of the re-indexing mechanisms, the number of re-indexed space calculations. All code is available online for independent testing[3].

The tests were run on a 2.8 GHz Intel Core i7, running on an otherwise bare machine without network interference. The code is written in Java, and all data sets used fit easily into the Java heap without paging or garbage collection occurring.

---

[3] https://richardconnor@bitbucket.org/richardconnor/metric-space-framework.git.

Table 2. Elapsed Times - SISAP *colors* with Cosine and Jensen-Shannon distances, and a 30-dimensional generated Euclidean space. All times are in seconds. The generated Euclidean space is evenly distributed in $[0,1]^{30}$, and gives the elapsed time for executing 1,000 queries against 9,000 data, with a threshold calculated to return one result per million data ($t = 0.7269$)

Dims	SISAP *colors*										30-dim ℓ_2^{30}					
	Cosine (t = 0.042)					Jensen-Shannon (t = 0.135)										
	L_{seq}	L_{rei}	N_{seq}	N_{rei}	Tree	L_{seq}	L_{rei}	N_{seq}	N_{rei}	Tree	Dims	L_{seq}	L_{rei}	N_{seq}	N_{rei}	Tree
5	10.3	4.5	8.8	1.0	3.1	248.4	335.5	61.9	65.5	124.8	3	0.5	2.5	0.5	1.6	1.4
10	9.8	3.4	10.4	0.8		155.3	233.2	29.0	29.3		6	0.5	2.3	0.5	1.8	
15	12.7	2.4	11.7	**0.7**		103.5	163.2	22.3	17.2		9	0.5	2.4	0.4	1.3	
20	16.5	2.8	16.7	**0.7**		95.7	162.8	23.8	**14.7**		12	0.5	2.6	0.3	1.2	
25	17.9	2.8	17.7	0.8		87.2	155.6	25.9	16.1		15	0.5	2.8	0.3	1.0	
30	18.1	2.6	17.4	0.9		67.7	130.4	27.0	16.5		18	0.6	3.4	0.3	1.0	
35	17.7	3.1	17.1	1.1		69.6	136.3	27.9	17.2		21	0.6	3.3	**0.2**	1.1	
40	18.1	3.0	18.1	1.0		62.4	131.2	27.8	17.1		24	0.7	2.9	**0.2**	1.1	
45	17.4	2.7	18.2	1.1		61.1	133.4	29.7	18.4		27	0.7	3.5	0.3	1.2	
50	17.6	3.5	17.3	1.4		58.3	130.4	30.6	18.6		30	0.7	3.5	0.3	1.4	

Results. As can be seen in Table 1, N_{rei} consistently and significantly outperforms the normal index structure at between 15 and 25 dimensions, depending on the query threshold. It is also interesting to see that, as the query threshold increases, and therefore scalability decreases, N_{seq} takes over as the most efficient mechanism, again with a "sweet spot" at 15 dimensions.

Table 2 shows the same experiment performed with Cosine and Jensen-Shannon distances. In these cases, the extra relative cost saving from the more expensive metrics is very clear, with relative speedups of 4.5 and 8.5 times respectively. In the Jensen-Shannon tests, the relatively very high cost of the metric evaluation to some extent masks the difference between N_{seq} and N_{rei}, but we note that the latter maintains scalability while the former does not. Finally, in the essentially intractable Euclidean space, with a relatively much smaller search threshold, N_{seq} takes over as the fastest mechanism.

Scalability. Table 3 shows the actual number of distance measurements made, for Euclidean and Jensen-Shannon searches of the *colors* data. The number of calls required in both the original and re-indexed spaces are given. Note that original-space calls are the same for both table-checked and re-indexed mechanisms; the number of original-space calls include those to the reference points, from which the accuracy of the n-simplex mechanism even in small dimensions can be appreciated. By 50 dimensions almost perfect accuracy is achieved for Euclidean search 50 original-space calculations are made, but in fact even at 10 dimensions almost every apex value can be deterministically determined as either a member or otherwise of the solution set based on its upper and lower bounds. At 20 dimensions, only 10 elements of the 101414-element data set have

Table 3. Distance Calculations Performed in Original and Re-indexed Space (figures given are thousands of calculations per query)

Dims	Euclidean (t = 0.051768)					Jensen-Shannon (t = 0.135)				
	Original space			Re-indexed		Original space			Re-indexed	
	L	N	Tree	L_{rei}	N_{rei}	L	N	Tree	L_{rei}	N_{rei}
5	2.75	0.38	1.48	5.28	1.76	12.77	2.29	5.97	18.40	6.91
10	1.33	0.05	1.48	4.40	1.23	7.81	0.58	5.97	19.66	6.32
15	0.57	0.04	1.48	3.24	1.13	4.62	0.16	5.97	15.46	4.99
20	0.51	0.03	1.48	3.42	1.15	3.89	0.11	5.97	15.85	4.80
25	0.43	0.04	1.48	3.15	1.18	3.65	0.09	5.97	14.88	4.87
30	0.37	0.04	1.48	3.02	1.21	2.53	0.08	5.97	13.83	4.70
35	0.34	0.04	1.48	2.85	1.31	2.59	0.08	5.97	13.56	4.86
40	0.33	0.04	1.48	2.95	1.29	2.14	0.08	5.97	13.48	4.64
45	0.31	0.05	1.48	2.82	1.32	1.95	0.08	5.97	13.74	4.89
50	0.27	0.05	1.48	2.57	1.33	1.83	0.08	5.97	12.63	4.87

bounds which straddle the query threshold. This indeed reflects the results presented in Fig. 2 where it is shown that for $n \geq 20$ the n-simplex lower bound is practically equivalent to the Euclidean distance to search *colors* data.

Equally interesting is the number of re-indexed distance measurements. This requires further investigation: for *n*-simplex, these are generally less than for the original space. This seems to hold for all data other than perfectly evenly-distributed (generated sets), for which the scalability is the same. The implication is that the re-indexed metric has better scalability properties than the original, although we would have expected indexing over the lower-bound function to be less, rather than more, scalable.

7   Conclusions and Further Work

Based on observations made over half a century ago, we have observed that a class of useful metric spaces have the *n*-point property. We have discovered a practical application for this previously abstract knowledge, by showing that irregular simplexes of any dimension can be constructed from only their edge lengths. This then allows upper and lower bounds to be calculated for any two objects, when the only knowledge available is their respective distances to a fixed set of reference objects.

There are a number of ways in which this knowledge can be used towards efficient search for suitable spaces. We have so far examined only one in detail, where a Euclidean space is extracted and used to pre-filter exact search. Over the benchmark SISAP *colors* data set, for some different metrics, this technique gives the best-recorded performance for exact search. However we believe the real power of this technique will emerge with huge data sets and more expensive metrics, and is yet to be experienced.

Acknowledgements. The work was partially funded by Smart News, "Social sensing for breaking news", co-funded by the Tuscany region under the FAR-FAS 2014 program, CUP CIPE D58C15000270008.

References

1. Blumenthal, L.M.: A note on the four-point property. Bull. Amer. Math. Soc. **39**(6), 423–426 (1933)
2. Blumenthal, L.M.: Theory and Applications of Distance Geometry. Clarendon Press, London (1953)
3. Chávez, E., Navarro, G.: Metric databases. In: Rivero, L.C., Doorn, J.H., Ferraggine, V.E. (eds.) Encyclopedia of Database Technologies and Applications, pp. 366–371. Idea Group (2005)
4. Chávez, E., Navarro, G., Baeza-Yates, R., Marroquín, J.L.: Searching in metric spaces. ACM Comput. Surv. **33**(3), 273–321 (2001)
5. Connor, R., Cardillo, F.A., Vadicamo, L., Rabitti, F.: Hilbert exclusion: improved metric search through finite isometric embeddings. ACM Trans. Inform. Syst. **35**(3), 17:1–17:27 (2016)
6. Connor, R., Vadicamo, L., Rabitti, F.: High-Dimensional Simplexes for Supermetric Search. arXiv e-prints, July 2017
7. Connor, R.: A tale of four metrics. In: Amsaleg, L., Houle, M.E., Schubert, E. (eds.) SISAP 2016. LNCS, vol. 9939, pp. 210–217. Springer, Cham (2016). doi:10.1007/978-3-319-46759-7_16
8. Cox, M.A.A., Cox, T.F.: Multidimensional Scaling. Springer, Heidelberg (2008). pp. 315–347
9. De Silva, V., Tenenbaum, J.B.: Sparse multidimensional scaling using landmark points. Technical report (2004)
10. Figueroa, K., Navarro, G., Chávez, E.: Metric spaces library (2007). http://www.sisap.org
11. Fodor, I.K.: A survey of dimension reduction techniques. Technical report, Center for Applied Scientific Computing, Lawrence Livermore National Laboratory (2002)
12. Jolliffe, I.: Principal Component Analysis. Wiley, New York (2014)
13. Mao, R., Miranker, W.L., Miranker, D.P.: Dimension reduction for distance-based indexing. In: SISAP 2010, pp. 25–32. ACM (2010)
14. Matoušek, J.: Lectures on Discrete Geometry, Graduate Texts in Mathematics. Springer, New York (2013)
15. Menger, K.: Untersuchungen ber allgemeine metrik. Math. Ann. **100**, 75–163 (1928)
16. Micó, M.L., Oncina, J., Vidal, E.: A new version of the nearest-neighbour approximating and eliminating search algorithm (AESA) with linear preprocessing time and memory requirements. Patt. Recogn. Lett. **15**(1), 9–17 (1994)
17. Wilson, W.A.: A relation between metric and euclidean spaces. Am. J. Math. **54**(3), 505–517 (1932)
18. Yang, L.: Distance Metric Learning: A Comprehensive Survey (2006)
19. Zezula, P., Amato, G., Dohnal, V., Batko, M.: Similarity Search: The Metric Space Approach. Advances in Database Systems, vol. 32. Springer, Heidelberg (2006)

Improving k-NN Graph Accuracy Using Local Intrinsic Dimensionality

Michael E. Houle[1], Vincent Oria[2], and Arwa M. Wali[2,3]([✉])

[1] National Institute of Informatics, Tokyo 101-8430, Japan
meh@nii.ac.jp
[2] New Jersey Institute of Technology, Newark, NJ 07102, USA
{vincent.oria,amw7}@njit.edu
[3] King Abdulaziz University, Jeddah, Saudi Arabia

Abstract. The k-nearest neighbor (k-NN) graph is an important data structure for many data mining and machine learning applications. The accuracy of k-NN graphs depends on the object feature vectors, which are usually represented in high-dimensional spaces. Selecting the most important features is essential for providing compact object representations and for improving the graph accuracy. Having a compact feature vector can reduce the storage space and the computational complexity of search and learning tasks. In this paper, we propose NNWID-Descent, a similarity graph construction method that utilizes the NNF-Descent framework while integrating a new feature selection criterion, Support-Weighted Intrinsic Dimensionality, that estimates the contribution of each feature to the overall intrinsic dimensionality. Through extensive experiments on various datasets, we show that NNWID-Descent allows a significant amount of local feature vector sparsification while still preserving a reasonable level of graph accuracy.

Keywords: Intrinsic dimensionality · k-nearest neighbor graph · Feature selection · Vector sparsification

1 Introduction

The k-nearest neighbor (k-NN) graph is a key data structure used in many applications, including machine learning, data mining, and information retrieval. Some prominent examples for k-NN graph utilization include object retrieval [21], data clustering [3], outlier detection [8], manifold ranking [9], and content-based filtering methods for recommender systems [22]. In applications such as multimedia and recommender systems where data objects are represented by high-dimensional vectors, the so-called 'curse of dimensionality' poses a significant challenge to k-NN graph construction: as the dimensionality increases, the discriminative ability of similarity measures diminishes to the point where methods such as k-NN graph search that depend on them lose their effectiveness.

The construction of k-NN graphs using brute-force techniques requires quadratic time, and is practical only for small datasets [4]. One recent technique that efficiently constructs an approximate k-NN graph in a generic metric

© Springer International Publishing AG 2017
C. Beecks et al. (Eds.): SISAP 2017, LNCS 10609, pp. 110–124, 2017.
DOI: 10.1007/978-3-319-68474-1_8

space is NN-Descent [4]. NN-Descent is an iterative algorithm that follows a simple transitivity principle: two neighbors of a given data object have a higher chance of being neighbors of each other. When ground truth class information is available, the accuracy of a k-NN graph can be measured in terms of the proportion of edges that connect nodes sharing the same class label. A common approach for maximizing k-NN graph accuracy is to incorporate dimensionality reduction techniques in the graph construction process. This can be done either independently as a preprocessing step using techniques such as Sparse Principal Component Analysis (Sparse PCA) [25], or integrated within the graph construction process itself, such as feature weighting [7] or other supervised feature selection approaches [23]. However, supervised feature selection would depend on ground truth information, which may not be always available.

In [15], an unsupervised method is presented, NNF-Descent, that iteratively and efficiently improves k-NN graph construction using the Local Laplacian Score (LLS) as a feature selection criterion. LLS favors those features that have high global variance among all objects, but less variance among the neighborhood of a given target object. The NNF-Descent method identifies locally noisy features relative to each object in the dataset — that is, those features having larger LLS scores. The noisy features are then gradually modified using a local sparsification process so as to decrease the distances between related objects, and thereby increase k-NN graph accuracy. NNF-Descent has already shown significant improvement in the semantic quality of the graphs produced, and superior performance over its competitors on several image databases [15]. However, NNF-Descent is a conservative method in that only a fixed small number of noisy features are sparsified in each iteration. With greater rates of feature sparsification, the k-NN graph accuracy tends to decrease. This also occurs when increasing the neighborhood size k beyond (roughly) 10. NNF-Descent is designed for datasets with dense feature vectors. In sparse datasets, vectors may contain very few non-zero features, in which case the sparsification process may incorrectly remove valuable features [15].

In this paper, we address the problem of improving the tradeoff between k-NN graph accuracy and the degree of data sparsification. We present the NNWID-Descent similarity graph construction method, which utilizes the NNF-Descent framework with a new feature selection criterion, Support-Weighted Intrinsic Dimensionality (support-weighted ID, or wID) [14]. Support-weighted ID is an extension of the Local Intrinsic Dimensionality (LID) measure introduced in [1,12], and is used within NNWID-Descent to identify and retain relevant features of each object. Unlike LLS, which is a variance-based measure, support-weighted ID penalizes those features that have lower locally discriminative power as well as higher density. In fact, support-weighted ID measures the ability of each feature to locally discriminate between objects in the dataset.

The remainder of this paper is organized as follows. Section 2 provides background on the relevant feature selection research literature, and on unsupervised approaches in particular. An overview of the NNF-Descent framework is presented in Sect. 3. We outline the proposed NNWID-Descent method in Sect. 4.

In Sect. 5, the performance of our method — with experimental results and analysis on several real datasets — is compared to NNF-Descent and other competing methods from the literature. Finally, we conclude in Sect. 6 with a discussion of future research directions.

2 Related Work

A brief review of feature selection techniques is provided in this section, with a particular emphasis on unsupervised methods.

2.1 Supervised Feature Selection and k-NN Graph Construction

Feature selections methods are commonly used in supervised learning methods to maximize their predictive accuracy. For example, Han et al. [7] proposed a Weight Adjusted k-Nearest Neighbor (WAKNN) classification scheme where the weights of the features are learned using an iterative algorithm. In [23], a supervised feature selection method is presented that uses an improved k-NN graph-based text representation model to reduce the number of features and predict the category of the text in the test set.

2.2 Unsupervised Feature Selection

In unsupervised feature selection methods, class information is not available, and thus it is difficult to decide the importance of a feature — especially when many of the features may be redundant or irrelevant [5]. Most existing unsupervised feature selection approaches are customized to a particular search or clustering algorithm.

Unsupervised feature selection methods can be further classified into global feature selection methods and local feature selection methods. In global feature selection methods, the features are selected based on their relevancy that has been computed globally using the entire dataset. The Laplacian Score (LS) [10] is one of the most popular unsupervised filter-based methods for generic data. LS selects the features to be used for all objects in the dataset based on their ability to discriminate among object classes. LS favors features that have high variance on the entire datasets and low variance within local neighborhoods. Local feature selection methods are based on the idea that the discriminative power and the importance of a feature may vary from one neighborhood to another; they aim to select features based on their relevancy to a given neighborhood. For example, Li et al. [18] introduced a localized feature selection algorithm for clustering that is able to reduce noisy features within individual clusters. Their algorithm computes, adjusts, and normalizes the scatter separability for individual clusters before applying a backward search technique to find the optimal (local) feature subsets for each cluster. Mitra et al. [20] introduced an algorithm that partitions the original feature set into clusters based on a k-NN graph principle. To detect and remove redundant features, their algorithm uses a pairwise feature similarity measure, the Maximum Information Compression index, that find the linear correlation between features in the clusters. This algorithm has a low computational complexity, since it does not involve any search for feature subsets [20]. However, their model may be too restrictive for real datasets, since correlations among features within clusters may not exist, or may be non-linear when they do exist [24].

3   Overview of NNF-Descent

As the basis for the work presented in this paper, in this section we provide an overview of the NNF-Descent algorithm [15]. We also describe its feature selection criterion, the Local Laplacian score LLS, and discuss its utilization in feature ranking and sparsification processes.

3.1   Local Laplacian Score, Feature Ranking, and Sparsification

Local Laplacian Score LLS is used for feature ranking with respect to individual data objects. Assume we have a dataset X with n data objects, each represented by a D-dimensional feature vector $\mathbf{f} = (f_1, f_2, \ldots, f_D)$. We further assume that the vectors are normalized. Then, for an object $x_i \in X$, the LLS score for each of its feature f_i can be computed using the following formula:

$$LLS(f_i) = \sum_j \frac{(f_i - f_j)^2 S_{ij}}{var(\mathbf{f})} \tag{1}$$

where $var(\mathbf{f})$ is the variance of feature $\mathbf{f}$, and S_{ij} is the (Gaussian) RBF kernel similarity between two object vectors x_i and x_j defined as:

$$S_{ij} = \begin{cases} exp(\frac{-\|x_i - x_j\|^2}{2\sigma^2}), & \text{if } i \text{ and } j \text{ are connected;} \\ 0, & \text{otherwise.} \end{cases} \tag{2}$$

Here, σ is a bandwidth parameter. S_{ij} favors neighboring objects x_i and x_j that are likely to share the same class label. A smaller value for $LLS(f_i)$ indicates that the feature is stable among the neighbors of object x_i. The features are ranked for each object in decreasing order of their LLS values, and the top-ranked proportion Z of the ranked list is deemed to be noise. In the sparsification process, the impact of noisy features is minimized by changing their values in the feature vectors to the global mean, which is zero due to normalization.

3.2   NNF-Descent

The NNF-Descent framework interleaves k-NN graph construction using NN-Descent [4] with a feature ranking and sparsification process. Algorithm 1 gives the complete algorithm for NNF-Descent. After normalizing the original vectors of the dataset X, the algorithm starts by computing the initial approximate k-NN graph using NN-Descent [4] (lines 1–2). The NN-Descent procedure depends on the so-called *local join* operation. Given a target point p, the local join operation checks whether any neighbor of p's neighbors is closer to p than any points currently in its neighbor list, and also whether pairs of neighbors of p can likewise improve each other's tentative neighbor list. Noisy features are gradually identified using LLS, ranked, and then sparsified.

Algorithm 1. NNF-Descent

Input   : Dataset X, distance function dist, neighborhood size K, sparsification
             rate Z, number of iterations T
Output: k-NN graph G
1 Normalize the original feature vectors of X;
2 Run NN-Descent(X, dist, K) to convergence to obtain an initial k-NN graph G;
3 **repeat**
4 | Generate a list L of all data points of X in random order;
5 | **foreach** *data point* $p \in L$ **do**
6 | | Rank the features of p in descending order of their LLS scores, as
   | | computed over the current k-NN list of p;
7 | | Change the value of the top-ranked Z-proportion of features to 0;
8 | | Recompute the distances from p to its k-NN and RNN points;
9 | | Re-sort the k-NN lists of p and its RNNs;
10 | | For each pair (q, r) of points from the k-NN list and RNN list of p,
   | | compute dist(q, r);
11 | | Use $(q, \text{dist}(q, r))$ to update the k-NN list of r, and use $(r, \text{dist}(q, r))$ to
   | | update the k-NN list of q;
12 | **end**
13 **until** *maximum number of iterations T is reached*;
14 Return G

---

4   Improving NN-Descent Graph with Weighted ID

The NNF-Descent framework, which integrates feature ranking and sparsification with k-NN graph construction, serves as the basis for the method presented in this paper, NNWID-Descent. In NNWID-Descent, instead of feature variance, a measure of the discriminability of features is used for feature ranking. In this section, we first provide a brief overview of this measure of discriminability, the Support-Weighted Local Intrinsic Dimensionality or support-weighted ID (Sect. 4.1). The utilization of support-weighted ID as a feature selection criterion is then presented in Sect. 4.2. Finally, the details of the proposed NNWID-Descent algorithm is given in Sect. 4.3.

4.1 Support-Weighted Local Intrinsic Dimensionality

As an alternative to the Local Laplacian Score, we propose in this paper a new feature evaluation strategy based on the Local Intrinsic Dimension ('Local ID', or 'LID') model originally appearing [12]. Given a distribution of distances with a univariate cumulative distribution function F that is positive and continuously differentiable in the vicinity of distance value x, the indiscriminability of F at x is given by

$$\mathrm{ID}_F(x) \triangleq \frac{x \cdot F'(x)}{F(x)}. \tag{3}$$

The indiscriminability reflects the growth rate of the cumulative distance function at x; it can be regarded as a probability density associated with the neighborhood of radius x (that is, $F'(x)$), normalized by the cumulative density of the neighborhood (that is, $F(x)/x$). The local intrinsic dimension has been shown to be equivalent to a notion of local intrinsic dimensionality, which can be defined as the limit $\mathrm{ID}_F^* = \lim_{x \to 0^+} \mathrm{ID}_F(x)$. However, the notion of local ID as proposed in [13,14] is considerably more general, in that the original model of [12] has been extended to handle multivariate real-valued functions that are not necessarily the cumulative distribution functions of distance distributions.

When considering a distance distribution on a space of many features, it is natural to ask which variables or features are contributing most to the overall discriminability of the function or cumulative distribution function (as the case may be). Two variables or features with the same local ID value may not necessarily have the same impact on the overall ID value. To illustrate this, let Φ and Ψ be the respective cumulative distribution functions of two univariate distance distributions on distance variable x.

The indiscriminability $\mathrm{ID}_\Phi(x)$ can be thought of as having a 'support' equal to the probability measure associated with distance x — namely, $\Phi(x)$; similarly, the support for $\mathrm{ID}_\Psi(x)$ would be $\Psi(x)$. Even when the indiscriminabilities $\mathrm{ID}_\Phi(x)$ and $\mathrm{ID}_\Psi(x)$ are equal, if (say) the support $\Phi(x)$ greatly exceeded $\Psi(x)$, one would be forced to conclude that the features associated with ID_Φ are more significant than those of ID_Ψ, at least within the neighborhood of radius x.

For the comparison of the discriminabilities of different features in our proposed adaptation of NNF-Descent, we will adopt the following Support-Weighted ID complexity measure. This measure has the highly desirable theoretical advantage of being additive across features (for more details we refer the reader to [14]).

Definition 1 (Support-Weighted ID [14]). *Let F be a real-valued multivariate function over a normed vector space $(\mathbb{R}^m, \|\cdot\|)$, and let $\mathbf{x} \neq \mathbf{0} \in \mathbb{R}^m$ be a vector of positive norm. The support-weighted indiscriminability of F at $\mathbf{x}$ is defined as*

$$\mathrm{wID}_F(\mathbf{x}) \triangleq F(\mathbf{x})\,\mathrm{ID}_F(\mathbf{x}) = \mathbf{x} \cdot \nabla F(\mathbf{x}). \tag{4}$$

Estimating support-weighted ID for the purpose of assessing indiscriminability can be complicated by the need to standardize the distance within which the indiscriminabilities are measured — in a k-NN graph, each neighborhood is associated with its own potentially-unique k-NN distance. If each feature were to assessed at widely-varying distances, there would be no basis for the fair comparison of feature performance.

In practice, however, estimation of ID requires samples that are the result of a k-nearest neighbor query on the underlying dataset. Across such samples, standardization can be achieved using the local ID representation theorem:

Theorem 1 (Local ID Representation Theorem [13]). *Let $\Phi : \mathbb{R} \to \mathbb{R}$ be a real-valued function, and let $v \in \mathbb{R}$ be a value for which $\mathrm{ID}_\Phi(v)$ exists. Let x and w be values for which x/w and $\Phi(x)/\Phi(w)$ are both positive. If Φ is non-zero and continuously differentiable everywhere in the interval $[\min\{x, w\}, \max\{x, w\}]$, then*

$$\frac{\Phi(x)}{\Phi(w)} = \left(\frac{x}{w}\right)^{\mathrm{ID}_\Phi(v)} \cdot G_{\Phi,v,w}(x), \text{ where} \tag{5}$$

$$G_{\Phi,v,w}(x) \triangleq \exp\left(\int_x^w \frac{\mathrm{ID}_\Phi(v) - \mathrm{ID}_\Phi(t)}{t} \, \mathrm{d}t\right), \tag{6}$$

whenever the integral exists.

For a univariate cumulative distribution function Φ at distance x, we can use Theorem 1 with $v = 0$ to relate the support $\Phi(x)$ with the support at another desired distance w. If n is the size of the dataset that we are given, we choose the distance at which over n selection trials one would expect k samples to fall within the neighborhood — that is, w would satisfy $\Phi(w) = k/n$. The support-weighted ID would thus be:

$$\mathrm{wID}_\Phi(x) = \Phi(x)\,\mathrm{ID}_\Phi(x) = \frac{k\,\mathrm{ID}_\Phi(x)}{n} \cdot \left(\frac{x}{w}\right)^{\mathrm{ID}_\Phi^*} \cdot G_{\Phi,0,w}(x). \tag{7}$$

In [13] it is shown that (under certain mild assumptions) the function $G_{\Phi,0,w}(x)$ tends to 1 as $x, w \to 0$ (or equivalently, as $n \to \infty$); also, $\mathrm{ID}_\Phi(x)$ would tend to ID_Φ^*, for which reliable estimators are known [1,11]. Thus, for reasonably large dataset sizes, we could use the following approximation:

$$\mathrm{wID}_\Phi(x) \approx \frac{k\,\mathrm{ID}_\Phi^*}{n} \cdot \left(\frac{x}{w}\right)^{\mathrm{ID}_\Phi^*}. \tag{8}$$

4.2   Defining Support-weighted ID (wID) for each Feature

Let $X = \{x_1, x_2, x_3, \dots, x_n\}$ be a dataset consisting of n objects such that each object x_i is represented as a feature vector in $\mathbb{R}^D$. The set of features is denoted as $F = \{1, 2, \dots, D\}$ such that $j \in F$ is the j-th feature in the vector representation. Since the factor k/n in Eq. 8 can be regarded as constant, the support-weighted ID criterion for feature f_j of object x_i can be simplified:

$$\mathrm{wID}_i(f_j) = \mathrm{ID}_{f_j} \cdot \left(\frac{a_f}{w_{f_j}}\right)^{\mathrm{ID}_{f_j}} \tag{9}$$

where ID_{f_j} is the local intrinsic dimensional estimate for the neighborhood, and w_{f_j} is the distance to the k-th nearest neighbor with respect to feature f_j, respectively. a_f is any positive constant representing the distance value x. For simplicity, a_f can be set as an average of a sample of k-NN distances across many objects for feature f_j. Equation (9) helps to find the most discriminative features by considering both the density of neighborhood around each object and the complexity of local ID with respect to a particular feature f_j.

For feature ranking, a straightforward method is used for selecting the most local discriminative features for each object using wID_i, in which the D features are ranked in descending order of $\mathrm{wID}_i(f_j)$, and a proportion Z of the top-ranked features are determined as candidates for sparsification. Assuming that the feature vectors have been normalized, the sparsification process (described in Sect. 3.1) will set the values of the least important features to 0.

4.3 NNWID-Descent

Algorithm 2 shows how NNWID-Descent proceeds. The input parameters are K, Z, and T, where $K \geq k$ is the working neighborhood size during the construction of the output k-NN graph, Z is a fixed proportion of features that are sparsified in each iteration, and T is the total number of desired iterations. The feature sparsification rate Z should be relatively small.

The algorithm has two phases: an initialization phase, and a sparsification and refinement phase. In the initialization phase, the algorithm computes a K-NN graph using NN-Descent after normalizing the original vectors of the dataset X (lines 2–4). This step is crucial, since a neighborhood of reasonably high quality is needed for the subsequent refinement phase to be effective.

In line 4, the value of a_f for each feature is precomputed for use in calculating wID values, during the sparsification and refinement phase. As will be described in Sect. 5.4, the value a_f can be computed as the average of the K-NN distances using the feature f alone, over a sample of the data objects. The K-NN graph entries are then improved using the sparsification and refinement phase (Lines 6–16). This phase includes three steps: feature ranking, sparsfication, and graph updating. In lines 9–10, the features are ranked in decreasing order according to the wID values obtained from the set of K-NN distances determined by each feature alone. For each object p, the top Z-proportion of features are then sparsified (line 11). As will be described in Sect. 5.4, the value Z is chosen depending on the density of the dataset X. As in [15], only non-zero features are candidates for sparsification, since features with value 0 do not provide discriminative information in the vicinity of p, and thus do not affect the quality of the K-NN graph. Ignoring zero features will ensure that once sparsified, a feature will not be evaluated again in subsequent iterations. Sparsifying a feature vector for p in one iteration will more likely change the nearest neighbors for each feature of p; for this reason, to determine the correct wID value in subsequent iterations, recomputation of the K-NN distances is required for each feature.

Lines 12–14 correspond to Lines 8–11 in NNF-Descent (Algorithm 1) which identify the local join operation and graph update step to improve the graph

---

Algorithm 2. NNWID-Descent

    Input   : Dataset X; distance function dist, neighborhood size K, sparsification
               rate Z, number of iterations T

    Output: k-NN graph G

1 {Initialization Phase}

2 Normalize the original feature vectors of X;

3 Run NN-Descent(X, dist, K) to convergence to obtain an initial K-NN graph G;

4 For each feature f, set the value of a_f to the average of K-NN distances
    computed for the feature over a sample of objects.

5 {Sparsification and Refinement Phase}

6 **repeat**

7   Generate a list L of all data points of X in random order;

8   **foreach** *data point $p \in L$* **do**

9   For each feature, compute the K-NN distances from p with respect to X;

10   Rank the features of p in descending order of their wID scores (Eq. 9),
            as computed over the current K-NN list of p;

11   Change the value of the top-ranked Z-proportion of features to 0;

12   Recompute the distances from p to its K-NN and RNN points;

13   Re-sort the K-NN lists of p and its RNNs;

14   For each pair (q, r) of points from the K-NN list and RNN list of p,
            compute $\text{dist}(q, r)$;

15   Use $(q, \text{dist}(q, r))$ to update the K-NN list of r, and use $(r, \text{dist}(q, r))$ to
            update the K-NN list of q;

16   **end**

17 **until** *maximum number of iterations T is reached*;

18 Return G

---

accuracy. In the implementation, we set $K \geq k$ to be the length for both RNN and NN lists used in computing wID.

The time complexity of NNWID-Descent can be divided according to its phases as follows: For the initialization phase, data normalization and NN-Descent —in terms of distance computation until convergence— take $O(Dn)$ and $O(K^2Dn)$ time, respectively. Computing the values of a_f for all features using average k-NN distances takes $O(Dn^2)$. For each iteration of the sparsification and refinement phase, feature ranking and selection using wID takes $O(KDn + D \log D)$ time per object, with total time in $O(KDn^2 + Dn \log D)$ over all objects. As with NN-Descent, assuming that the lengths of the RNN lists are in $O(K)$, each iteration of NNWID-Descent takes $O(K^2Dn)$ time for the neighbor list update step. However, the optimizations that have been defined for NN-Descent in [4] can also applied for NNWID-Descent to speed up the local join operation and update steps.

5   Experiments

For the comparison of NNWID-Descent with competing methods, we conducted experiments to study the influence on performance of varying the feature sparsification rate Z and the working neighbor list size K.

5.1   Datasets

Six real datasets of varying sizes and densities were considered, of which five are image sets:

- The Amsterdam Library of Object Images (ALOI) [6] contains 110,250 images of 1000 small objects. Each image is described by a 641-dimensional feature vector based on color and texture histograms.
- The MNIST dataset [17] contains 70,000 images of handwritten digits. Each image is represented by 784 gray-scale texture values.
- Google-23 [16] contains 6,686 faces extracted from images of 23 celebrities. The dimension of the face descriptors is 1,937.
- The Isolated Letter Speech Recognition dataset (ISOLET) [19] contains 7797 objects generated by having 150 subjects speak the name of each letter of the alphabet twice. Each object is described by 617 features, and were scaled so that all values lie in the interval $[-1.0, 1.0]$.
- The Human Activity Recognition Using Smartphones dataset (HAR) [2] contains 10,299 instances of accelerometer and gyroscope data from 30 subjects performing 6 different activities. Each instance is represented by a feature vector of 561 time and frequency domain variables.
- The Relative Location of CT dataset (RLCT) [19] contains 53500 axial CT slice images from 74 different patients. Each CT slice is described by two histograms in polar space. The feature vectors of the images are of 385 dimensions.

5.2   Competing Methods

The performance of NNWID-Descent is contrasted with that of 3 competitors:

- NNF-Descent: uses LLS criterion for feature ranking and sparsification (as described in Sect. 3).
- Random: as per NNF-Descent, except that for each object the features to be sparsified are selected randomly. The rationale for the comparison with this method is to establish a baseline for the performance of the feature ranking and sparsification criterion.
- Sparse PCA: is similar to wID in such that it takes into account the dataset sparsity. In this method, the feature extraction and graph construction are conducted as two separate processes. To allow a fair comparison with other methods, after choosing the highest principal components, an exact k-NN graph is computed (at a computation cost of $O(Dn^2)$).

5.3 Performance Measure

We use the graph accuracy as a performance measure. The class labels of data objects were used to measure the quality of the resulting k-NN graph at every iteration. The accuracy of the resulting k-NN graph is evaluated, as in [15], using the following formula:

$$graph\ accuracy = \frac{\#correct\ neighbors}{\#data \times K}, \qquad (10)$$

where the 'correct' neighbors share the same label as the query object.

5.4 Default Parameters

Except for the case of Sparse PCA, the feature vectors were normalized within each dataset in each experiment performed, and the Euclidean (L_2) distance was employed. In NNWID-Descent, for the datasets Google-23, HAR, and ISOLET, the value of a_f in the weight parameter of Equation (9) is set to be the average of distances (using feature f) computed from the neighbors of all objects in the dataset; for ALOI, MNIST, and RLCT, the average was computed over a random sample of 100 objects. Furthermore, for all features, the value a_f is precomputed in advance using the original feature vectors without sparsification. For simplicity, the number of neighbors used for computing wID and LLS is set to be equal to the input parameter K.

5.5 Effects of Varying the Sparsification Rate Z

Parameter Setting. In this experiment, we tested the effect on performance of varying Z while keeping K fixed. The choices of Z is varied with different datasets as it depends heavily on the density of the feature vectors. For example, in each iteration, smaller choice of Z ($= 0.0025\%$) for the sparse datasets (MNIST, ALOI, ISOLET, and RLCT) was required to produce gradual changes in graph accuracy with acceptable performance. On the other hand, the dense datasets (Google-23 and HAR) require a larger starting point for Z ($= 0.1\%$)to produce perceptible changes in performance from iteration to iteration. For Sparse PCA, the parameter controlling sparsity was set to Z, and the number of principle components selected were set to ZD. The total number of iterations T is set to 70 for all datasets except ALOI, for which T is set to 40. For all methods in the comparison, the value of K is fixed at 100. Figure 1 shows plots of the graph accuracy in each iteration for all the methods, across a range of Z values.

Results and Analysis. On five of the six datasets, compared with its competitors, NNWID-Descent achieves consistent improvements for graph accuracy and resistance to performance degradation as sparsification increases — for ISOLET, it is outperformed only by Random. For the MNIST dataset, Sparse PCA has

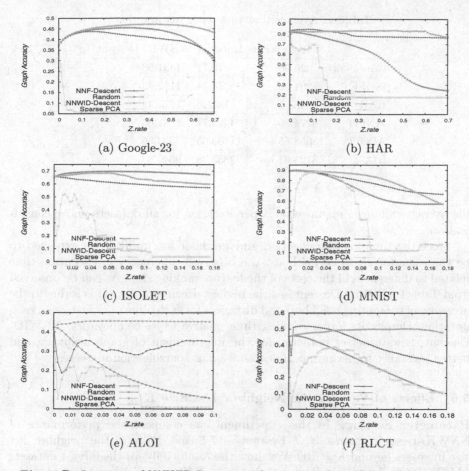

Fig. 1. Performance of NNWID-Descent with varying values of Z, and $K = 100$.

a performance comparable to that of NNWID-Descent for small sparsification rates.

It is important to realize that obtaining accurate estimates of wID requires that the neighborhood be of generally good quality. In NNWID-Descent, the recomputation of neighborhoods after sparsification at each iteration is essential to the quality of wID estimation. However, using distance values computed from the current k-NN graph may lead to less accurate ID estimation, if the initial graph is of low quality.

Execution Time. The cost of sparsification and refinement dominates the overall computational performance of the three methods that employ this strategy. For these methods, the execution time for the sparsification and refinement phase is displayed in Table 1. The displayed times account for the operations of feature ranking, sparsification, and updating of neighbor lists. The table shows

Table 1. Average time in seconds per iteration.

	NNF-Descent	Random	NNWID-Descent
Google-23	320.96	70.77	1431.56
ISOLET	204.92	73.34	1152.43
HAR	248.75	141.46	1275.44
MNIST	5274.55	4429.77	8281.03
ALOI	13053.55	11183.65	55363.56
RLCT	3125.64	2853.78	9549.33

the average running time in seconds per iteration for all datasets under consideration.

Since the time for sparsification and neighbor list updating is expected to be the same for all three methods, the observed differences in execution time related to differences in the costs of the feature ranking step. As can be observed from Table 1, NNWID-Descent has the highest execution cost. This is due to the necessity of computing neighborhood distances for each object per feature in each iteration. Despite its larger running time relative to its competitors, NNWID-Descent shows a better potential for the improvement of graph accuracy, and better resistance to performance degradation as sparsification increases.

5.6 Effects of Varying the Neighbor List Size K

Parameter Setting. In this experiment, we compare the performance of NNWID-Descent against NNF-Descent and Sparse PCA as the neighbor list size increases beyond $K = 100$. We show the results only for the largest datasets (ALOI, MNIST and RLCT), as the values of K are too large relative to the size of the other datasets. Concretely, K is set to 100, 200, 400, and 800, and Z is fixed at 4% for MNIST and RLCT, and at 2% for ALOI. These Z values represent approximately the peak graph accuracy achieved in Fig. 1. The performances across these choices of K are plotted in Fig. 2.

Results and Analysis. We note that for ALOI and RLCT, NNWID-Descent still provides better accuracy than other methods as the neighborhood list size K is increased. With MNIST, Sparse PCA outperforms other methods as K increases, which indicates that this method can lead to a reasonable graph accuracy for a sparse dataset when Z is small. For all methods, the performance degrades as K increases. In addition, we observe that the relative performances of all methods shown when varying K (Fig. 2) is still consistent with the performances observed when varying Z (Fig. 1).

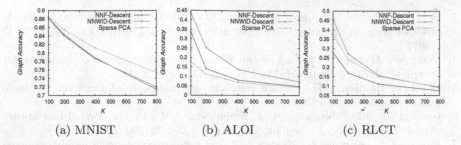

(a) MNIST   (b) ALOI   (c) RLCT

Fig. 2. Performance of NNWID-Descent with different values of K and fixed Z ($Z =$ 4% for RLCT, MNIST, and 2% for ALOI).

6   Conclusion and Future Work

In this paper, we presented the NNWID-Descent similarity graph construction method, which utilizes the NNF-Descent framework with a new unsupervised feature selection criterion. This method aimed to improve or maintain k-NN graph accuracy while achieving a significant amount of sparsification of object feature vectors. We proposed the use of support-weighted ID (wID) to identify relevant features with higher discriminative power local to each object. NNWID-Descent ranks the features according to their wID values, then sparsifies those features achieving the smallest values.

With respect to the correctness of k-NN graph produced using six real datasets, NNWID-Descent has been shown to generally outperform its closest competitors, NNF-Descent and Sparse PCA. NNWID-Descent can be applied to obtain more compact representations for high-dimensional features vectors, which is important to reduce the storage and computational complexity for many applications. However, the ID estimator used in NNWID-Descent generally requires relatively large dataset sizes to provide a reasonable accuracy. Of the six datasets used in our experiments, three are considered too small for the extreme-value-theoretic LID model to be applicable. Further improvement of NNWID-Descent could be achieved through the development of ID estimators that can more accurately handle smaller dataset sizes and smaller neighborhood sample sizes.

Acknowledgments. M.E. Houle acknowledges the financial support of JSPS Kakenhi Kiban (A) Research Grant 25240036 and JSPS Kakenhi Kiban (B) Research Grant 15H02753. V. Oria acknowledges the financial support of NSF Research Grant DGE 1565478.

References

1. Amsaleg, L., Chelly, O., Furon, T., Girard, S., Houle, M.E., Kawarabayashi, K., Nett, M.: Estimating local intrinsic dimensionality. In: KDD, pp. 29–38 (2015)
2. Anguita, D., Ghio, A., Oneto, L., Parra, X., Reyes-Ortiz, J.L.: A public domain dataset for human activity recognition using smartphones. In: ESANN (2013)

3. Brito, M., Chávez, E., Quiroz, A., Yukich, J.: Connectivity of the mutual k-nearest-neighbor graph in clustering and outlier detection. Stat. Probab. Lett. **35**(1), 33–42 (1997)
4. Dong, W., Moses, C., Li, K.: Efficient K-nearest neighbor graph construction for generic similarity measures. In: WWW, pp. 577–586 (2011)
5. Dy, J.G., Brodley, C.E.: Feature selection for unsupervised learning. J. Mach. Learn. Res. **5**, 845–889 (2004)
6. Geusebroek, J.M., Burghouts, G.J., Smeulders, A.W.M.: The Amsterdam library of object images. Int. J. Comput. Vis. **61**(1), 103–112 (2005)
7. Han, E.-H.S., Karypis, G., Kumar, V.: Text categorization using weight adjusted k-nearest neighbor classification. In: Cheung, D., Williams, G.J., Li, Q. (eds.) PAKDD 2001. LNCS (LNAI), vol. 2035, pp. 53–65. Springer, Heidelberg (2001). doi:10.1007/3-540-45357-1_9
8. Hautamaki, V., Karkkainen, I., Franti, P.: Outlier detection using k-nearest neighbour graph. In: ICPR, vol. 3, pp. 430–433, August 2004
9. He, J., Li, M., Zhang, H.J., Tong, H., Zhang, C.: Manifold-ranking based image retrieval. In: ACM MM, pp. 9–16 (2004)
10. He, X., Cai, D., Niyogi, P.: Laplacian score for feature selection. In: NIPS, vol. 186, p. 189 (2005)
11. Hill, B.M.: A simple general approach to inference about the tail of a distribution. Ann. Stat. **3**(5), 1163–1174 (1975)
12. Houle, M.E.: Dimensionality, discriminability, density & distance distributions. In: ICDMW, pp. 468–473 (2013)
13. Houle, M.E.: Local intrinsic dimensionality I: an extreme-value-theoretic foundation for similarity applications. In: SISAP, pp. 1–16 (2017)
14. Houle, M.E.: Local intrinsic dimensionality II: multivariate analysis and distributional support. In: SISAP, pp. 1–16 (2017)
15. Houle, M.E., Ma, X., Oria, V., Sun, J.: Improving the quality of K-NN graphs through vector sparsification: application to image databases. Int. J. Multimedia Inf. Retrieval **3**(4), 259–274 (2014)
16. Houle, M.E., Oria, V., Satoh, S., Sun, J.: Knowledge propagation in large image databases using neighborhood information. In: ACM MM, pp. 1033–1036 (2011)
17. Lecun, Y., Bottou, L., Bengio, Y., Haffner, P.: Gradient-based learning applied to document recognition. Proc. IEEE **86**(11), 2278–2324 (1998)
18. Li, Y., Dong, M., Hua, J.: Localized feature selection for clustering. Pattern Recogn. Lett. **29**(1), 10–18 (2008)
19. Lichman, M.: UCI Machine Learning Repository (2013). http://archive.ics.uci.edu/ml
20. Mitra, P., Murthy, C., Pal, S.K.: Unsupervised feature selection using feature similarity. IEEE TPAMI **24**(3), 301–312 (2002)
21. Qin, D., Gammeter, S., Bossard, L., Quack, T., van Gool, L.: Hello neighbor: accurate object retrieval with k-reciprocal nearest neighbors. In: CVPR 2011, pp. 777–784, June 2011
22. Sarwar, B., Karypis, G., Konstan, J., Riedl, J.: Application of dimensionality reduction in recommender systems – a case study. Technical report, DTIC Document (2000)
23. Wang, Z., Liu, Z.: Graph-based KNN text classification. In: FSKD, vol. 5, pp. 2363–2366, August 2010
24. Yu, L., Liu, H.: Efficient feature selection via analysis of relevance and redundancy. J. Mach. Learn. Res. **5**, 1205–1224 (2004)
25. Zou, H., Hastie, T., Tibshirani, R.: Sparse principal component analysis. J. Comput. Graph. Stat. **15**(2), 265–286 (2006)

Distances for Complex Objects

Dynamic Time Warping and the (Windowed) Dog-Keeper Distance

Jörg P. Bachmann[✉] and Johann-Christoph Freytag

Humboldt-Universität zu Berlin, 10099 Berlin, Germany
joerg.bachmann@informatik.hu-berin.de,
freytag@dbis.informatik.hu-berlin.de

Abstract. Finding similar time series is an important task in multimedia retrieval, including motion gesture recognition, speech recognition, or classification of hand-written letters. These applications typically require the similarity (or distance) measure to be robust against outliers and time warps. Time warps occur if two time series follow the same path in space, but need specific time adjustments. A common distance measure respecting time warps is the dynamic time warping (DTW) function. The edit distance with real penalties (ERP) and the dog-keeper distance (DK) are variations of DTW satisfying the triangle inequality. In this paper we propose a novel extension of the DK distance called windowed dog-keeper distance (WDK). It operates on sliding windows, which makes it robust against outliers. It also satisfies the triangle inequality from the DK distance. We experimentally compare our measure to the existing ones and discuss the conditions under which it shows an optimal classification accuracy. Our evaluation also contributes a comparison of DK and DTW. For our experiments, we use well-known data sets such as the cylinder-bell-funnel data set and data sets from the UCI Machine Learning Repository.

Keywords: Time series · Metric · Multimedia retrieval

1 Introduction

Many applications require to find similar time series to a given pattern. One common application of finding similar time series is multimedia retrieval, including motion gesture recognition, speech recognition, and classification of handwritten letters. All these tasks have in common that the time series of same classes (e.g., same spoken words or same gestures) follow the same path in space, but have some temporal displacements. Another example is tracking the GPS coordinates of two cars driving the same route from A to B. Although we want these time series to be recognized as being similar, driving style, traffic lights, and traffic jams might result in large temporal differences. Distance functions such as dynamic time warping (DTW) [11], edit distance with real penalties (ERP) [4], and the dog-keeper distance (DK) [6] respect this semantic requirement.

© Springer International Publishing AG 2017
C. Beecks et al. (Eds.): SISAP 2017, LNCS 10609, pp. 127–140, 2017.
DOI: 10.1007/978-3-319-68474-1_9

Another requirement for similarity functions is their computational performance since it is common to compare a sample time series to a large set of time series. To improve performance we might improve the computation time of one time series comparison or we might reduce the number of comparisons. Assuming that SETH [3] holds, Bringmann and Künnemann proved that there is no algorithm computing the exact value of DTW in less than quadratic time [3]. Similar results were proven for the DK distance [2] and the edit distance [1]. However, we are usually not interested in the exact distance values, but in the set of the nearest neighbours. A common approach for pruning elements as possible candidates are lower bounds to the distance function.

Keogh and Ratanamahatana exhaustively compared nine different time series distance functions including DTW and ERP on 38 time series data sets coming from different domains [5]. They also compared eight different time series representations including Discrete Fourier Transformation (DFT) and Symbolic Aggregate approXimation (SAX) [10]. In their work, they investigated contradictory claims about the effectiveness of different time series distance functions and representations. Their first major insight is that there is little difference in the effectiveness between different time series representations excluding some rare cases. They say there is no clear winner for the choice of the time series distance function, although elastic distance functions, such as DTW, ERP, LCSS, or EDR are more accurate, especially on small data sets.

To the best of our knowledge, DTW has not been compared to the DK distance. If DTW is the time warping equivalent to the L_1-norm, then the DK distance is the equivalent to the L_∞-norm and thereby more sensitive to noise or outliers within time series. On the other hand, we could observe a speed-up by an order of magnitude in our experimental evaluation. Why does the DK distance perform much better although the algorithm is quite similar to that of DTW? Can we improve the robustness of the DK distance?

The first contribution of our paper is the windowed DK distance (WDK), which is a modification of the DK distance to satisfy the triangle inequality. We evaluate the performance of the four time warping distance functions DTW, ERP, DK, and WDK by comparing the results of k-nearest neighbour classifiers on four different multimedia time series data sets coming from different domains. The second contribution is that we also investigate the reason for the low computation time of the DK and WDK distance functions.

The rest of this paper is structured as follows. Section 2 introduces basic terms and notations and reviews the time series distance functions DTW, ERP, and DK. Section 3 defines the WDK distance function and provides an algorithm for its computation. Section 4 evaluates these four distance functions on four multimedia time series data sets. Section 5 concludes the paper.

2   Preliminaries and Concepts

This section introduces basic notations and concepts used in this paper.

It is hard to find an open access proof for the triangle inequality of the DK distance in modern mathematical language. Therefore we provide a new proof in this section that shows this well known fact again. This also proves the triangle inequality for the WDK distance proposed in this paper.

Basic Notation: With $\mathbb{N}$, $\mathbb{R}$, $\mathbb{R}_{\geqslant c}$ we denote the set of non-negative integers, the set of reals, and the set of all reals $\geqslant c$, for some $c \in \mathbb{R}$, respectively. An $m \times n$ matrix is denoted by $A = (a_{i,j})$. Given a matrix A, $A_{i,j}$ denotes the element in the i-th row and j-th column.

By $\mathbb{R}^k$, for $k \in \mathbb{N}$, we denote the set of all vectors of length k. For a vector $v \in \mathbb{R}^k$ we write v_i for the entry at position i.

For mappings $f : A \longrightarrow B$ and $g : B \longrightarrow C$, we denote the image of f as $f(A) := \{f(x) \mid x \in A\}$ and $g \circ f : x \mapsto g(f(x))$ the concatenation of g and f. Furthermore, $\inf f$ and $\sup f$ are the infimum and the supremum of $f(A)$ respectively.

Norms and Metric Spaces: By $\|\cdot\|_p$, for $p \in \mathbb{R}_{\geqslant 1}$, we denote the well known L_p-*norm* on $\mathbb{R}^k$; i.e., $\|v\|_p = \left(\sum_{i=1}^k |v_i|^p\right)^{1/p}$ for all $v \in \mathbb{R}^k$.

Recall that a *pseudo metric space* $(\mathbb{M}, d)$ consists of a set $\mathbb{M}$ and a distance function $d : \mathbb{M} \times \mathbb{M} \longrightarrow \mathbb{R}_{\geqslant 0}$ satisfying the following axioms:

$$\forall\, x, y \in \mathbb{M} : d(x, y) = d(y, x).$$
$$\forall\, x, y, z \in \mathbb{M} : d(x, z) \leqslant d(x, y) + d(y, z).$$

A *metric space* is a pseudo metric space which also satisfies $\forall\, x, y \in \mathbb{M} : d(x, y) = 0 \iff x = y$. Note that if $\|\cdot\|$ is an arbitrary vector norm and $d(\cdot, \cdot)$ is defined as $d(u, v) := \|u - v\|$, then $(\mathbb{R}^k, d)$ is a metric space. By d_p, for $p \in \mathbb{R}_{\geqslant 1}$, we denote the usual L_p-distance, i.e., the particular distance function with $\mathsf{d}_p(x, y) = \|x - y\|_p$.

Time Series: A time series T of length ℓ over a metric space $\mathbb{M}$ is a sequence $T = (t_1, \cdots, t_\ell)$ with $t_i \in \mathbb{M}$ for $1 \leqslant i \leqslant \ell$. We denote $\mathtt{Tail}(T) := (t_2, \cdots, t_n)$ as the time series when removing first element. In the rest of the paper, we consider $\mathbb{M} = \mathbb{R}^k$ for some $k \in \mathbb{N}$. We denote time series with the letters S, T, and R.

Time Series Distances: The algorithms for the computation of DTW, ERP, and DK are very similar. They differ in how they handle a time warping step and whether they take the maximum along a warping path or sum up these values. DTW and ERP sums the values up while the DK distance takes the maximum.

For a formal definition, let $S = (s_1, \cdots, s_m)$ and $T = (t_1, \cdots, t_n)$ be two time series, gap a globally constant element (0 as proposed by [4]), and $\mathsf{d}(s, t)$ a distance function for the elements of the time series. The well known distance function DTW is defined as follows.

$$\mathtt{DTW}(S, ()) = \infty \quad \mathtt{DTW}((), T) = \infty \quad \mathtt{DTW}((s), (t)) = \mathsf{d}(s, t)$$

$$\mathtt{DTW}(S, T) = \min \begin{cases} \mathsf{d}(s_1, t_1) + \mathtt{DTW}(\mathtt{Tail}(S), \mathtt{Tail}(T)) \\ \mathsf{d}(s_1, t_1) + \mathtt{DTW}(S, \mathtt{Tail}(T)) \\ \mathsf{d}(s_1, t_1) + \mathtt{DTW}(\mathtt{Tail}(S), T) \end{cases}$$

ERP differs from DTW by including gap elements to the time series on warping steps.

$$\text{ERP}(S,()) = \infty \quad \text{ERP}((),T) = \infty \quad \text{ERP}((s),(t)) = d(s,t)$$

$$\text{ERP}(S,T) = \min \begin{cases} d(s_1,t_1) + \text{DTW}(\text{Tail}(S),\text{Tail}(T)) \\ d(s_1,\text{gap}) + \text{DTW}(S,\text{Tail}(T)) \\ d(\text{gap},t_1) + \text{DTW}(\text{Tail}(S),T) \end{cases}$$

The DK distance is similar to DTW and differs by taking the maximum distance along a warping path instead of the sum.

$$\text{DK}(S,()) = \infty \quad \text{DK}((),T) = \infty \quad \text{DK}((s),(t)) = d(s,t)$$

$$\text{DK}(S,T) = \min \begin{cases} \max\{d(s_1,t_1),\text{DK}(\text{Tail}(S),\text{Tail}(T))\} \\ \max\{d(s_1,t_1),\text{DK}(S,\text{Tail}(T))\} \\ \max\{d(s_1,t_1),\text{DK}(\text{Tail}(S),T)\} \end{cases}$$

Note that ERP and DK satisfy the triangle inequality and therefore are metric distance functions [4, 7]. See Fig. 1 for sketches of the behaviour of these distance functions.

Fig. 1. Example time series with example warping paths sketching the behaviour of DTW (left), DK (center), and ERP (right). Distances between states are marked with solid lines while the circled and squared time series are connected using dashed lines. DTW sums up the distances along the warping path (all solid lines). DK is the largest distance along the warping path (longest solid line). ERP sums up the distances along the warping path (all solid lines). However, when warping (second circle from the left and third square from the right), states are compared to the gap element (empty square).

Algorithm 1 shows a pseudo code for computing the DK distance between two time series similar to the algorithm proposed by Eiter and Mannila [6]. We extended the algorithm considering a threshold as third parameter for early abandoning. The idea of early abandoning works the same in all algorithms for the mentioned time series distance functions. After computing the next row (or column) of the matrix D, the minimum value in that row is a lower bound for the final distance value. If that value already exceeds the threshold, the algorithm stops and returns the lower bound.

Dog-Keeper is a Metric: Satisfying the triangle inequality might be an opportunity for indexing the data using metric index structures. In the following we

Algorithm 1. Pseudo Code for the Dog-Keeper Distance with Early Abandoning

```
1   Input:  S = (s₁,···,s_ℓ),  T = (t₁,···,t_m),  τ
2   Output: Lower bound for the dog-keeper distance
3
4   D_{1,1} = d(s₁,t₁)
5   if D_{1,1} ⩾ τ
6      return D_{1,1}
7   for i in 2,···,ℓ
8      D_{i,1} = max {D_{i-1,1}, d(s_i,t₁)}
9   for j in 2,···,m
10     ε = ∞
11     for i in 1,···,ℓ
12        pred = min{D_{i-1,j}, D_{i,j-1}, D_{i-1,j-1}}
13        D_{i,j} = max {d(s_i,t_j), pred}
14        ε = min {ε, D_{i,j}}
15     if ε ⩾ τ
16        return ε
17  return D_{ℓ,m}
```

want to provide a new proof in modern mathematical language that shows that the dog-keeper distance satisfies the triangle inequality. Therefore, we prove the triangle inequality for th Fréchet distance. Since the dog-keeper distance is the discrete special case of the Fréchet distance, the proof also holds for the dog-keeper distance.

Let $\mathbb{M} := \mathbb{R}^k$ be the space of states, $\mathsf{d} : \mathbb{M} \times \mathbb{M} \longrightarrow \mathbb{R}_{\geqslant 0}$ be a metric on all states. We denote the set of all (piecewise continuous) curves over $[0,1] \subset \mathbb{R}$ by

$$\mathcal{T} := \{f \colon [0,1] \longrightarrow \mathbb{M}\}$$

and the set of all time warps over [0,1] by

$$\Sigma := \{\sigma \colon [0,1] \longrightarrow [0,1]\},$$

where all $\tau \in \Sigma$ are continuous, strictly monotonically increasing, and $\inf \tau = 0$, $\sup \tau = 1$. For $f, g \in \mathcal{T}$, let $\delta_\infty(f,g) := \max_{x \in [0,1]} \mathsf{d}(f(x), g(x))$ be the maximum distance of f and g.

Definition 1 (Fréchet Distance). *Let $f, g \in \mathcal{T}$ be two curves over $[0,1]$. The Fréchet distance DK of f and g is defined as*

$$DK(f,g) := \inf_{\sigma,\tau \in \Sigma} \delta_\infty(f \circ \sigma, g \circ \tau)$$

Using this notation we prove the following theorem.

Theorem 1. *The Fréchet distance DK satisfies the triangle inequality, i.e.,*

$$\forall f,g,h \in \mathcal{T} : DK(f,h) \leqslant DK(f,g) + DK(g,h).$$

To prove the triangle inequality, we first prove the following lemma showing that the δ_∞ distance does not change when applying the same temporal adjustment to both curves. The second lemma then reduces the search to all warping functions applied to one time series only.

Lemma 1. *Let $f, g \in \mathcal{T}$ be two arbitrary curves and $\sigma \in \Sigma$ be an arbitrary time warp. Then, the following equation holds:*

$$\delta_\infty(f, g) = \delta_\infty(f \circ \sigma, g \circ \sigma)$$

Proof. Consider the mapping

$$\theta \colon [0, 1] \longrightarrow \mathbb{R}_{\geqslant 0}$$
$$x \longmapsto \mathsf{d}(f(x), g(x)).$$

Then,

$$\delta_\infty(f, g) = \sup\left(\theta([0, 1])\right), \text{and}$$
$$\delta_\infty(f \circ \sigma, g \circ \sigma) = \sup\left(\theta \circ \sigma([0, 1])\right)$$

Since $\theta([0,1]) = \theta(\sigma([0,1]))$, the desired equation $\delta_\infty(f, g) = \delta_\infty(f \circ \sigma, g \circ \sigma)$ follows. □

Lemma 2. *Let $f, g \in \mathcal{T}$ be two arbitrary curves. Then the following equation holds:*

$$DK(f, g) = \inf_{\sigma \in \Sigma} \delta_\infty(f, g \circ \sigma)$$

Proof. Consider two sequences $(\sigma_i)_{i \in \mathbb{N}}$ and $(\tau_i)_{i \in \mathbb{N}}$ with $\sigma_i, \tau_i \in \Sigma$ for $i \in \mathbb{N}$, such that

$$\delta_\infty(f \circ \sigma_i, g \circ \tau_i) \xrightarrow{\;i \to \infty\;} DK(f, g).$$

Since each σ_i is invertable, Lemma 1 can be applied on $\delta_\infty(f \circ \sigma_i, g \circ \tau_i)$ with σ_i^{-1}, i.e. we obtain

$$\delta_\infty(f, g \circ \tau_i \circ \sigma_i^{-1}) = \delta_\infty(f \circ \sigma_i \circ \sigma_i^{-1}, g \circ \tau_i \circ \sigma_i^{-1})$$
$$= \delta_\infty(f \circ \sigma_i, g \circ \tau_i) \xrightarrow{\;i \to \infty\;} DK(f, g).$$

Thus, we have a sequence $(\theta_i)_{i \in \mathbb{N}} := (\tau_i \circ \sigma_i^{-1})_{i \in \mathbb{N}}$ with $\theta_i \in \Sigma$ for $i \in \mathbb{N}$, such that $\delta_\infty(f, g \circ \theta_i) \xrightarrow{\;i \to \infty\;} DK(f, g)$.

On the other hand,

$$\inf_{\sigma, \tau \in \Sigma} \delta_\infty(f \circ \sigma, g \circ \tau) \leqslant \inf_{\theta \in \Sigma} \delta_\infty(f, g \circ \theta).$$

Hence, $DK(f, g) = \inf_{\theta \in \Sigma} \delta_\infty(f, g \circ \theta)$. □

Proof (Proof of Theorem 1). Consider some arbitrary but fixed $f, g, h \in \mathcal{T}$. Since $\mathrm{DK}(f,g) = \inf_{\sigma \in \Sigma} \delta(f, g \circ \sigma)$ (Lemma 2), an infinite sequence $(\sigma_i)_{i \in \mathbb{N}}$ exists with $\sigma_i \in \Sigma$ for all $i \in \mathbb{N}$, such that

$$\delta_\infty(f, g \circ \sigma_i) \xrightarrow{\ i \to \infty\ } \mathrm{DK}(f,g).$$

Analogously, a sequence $(\tau_i')_{i \in \mathbb{N}}$ with $\tau_i' \in \Sigma$ for all $i \in \mathbb{N}$ exists, such that

$$\delta_\infty(g, h \circ \tau_i') \xrightarrow{\ i \to \infty\ } \mathrm{DK}(g,h).$$

Considering the sequence $(\tau_i)_{i \in \mathbb{N}}$ with $\tau_i = \tau_i' \circ \sigma_i \in \Sigma$ and using Lemma 1, we obtain

$$\delta_\infty(g \circ \sigma_i, h \circ \tau_i) = \delta_\infty(g, h \circ \tau_i') \xrightarrow{\ i \to \infty\ } \mathrm{DK}(g,h).$$

Recall that $(\mathcal{T}, \delta_\infty)$ is a metric space, thus the triangle inequality holds for each $i \in \mathbb{N}$:

$$\delta_\infty(f, h \circ \tau_i) \leqslant \delta_\infty(f, g \circ \sigma_i) + \delta_\infty(g \circ \sigma_i, h \circ \tau_i)$$

Since $\mathrm{DK}(f,h) = \inf_{\tau \in \Sigma} \delta_\infty(f, h \circ \tau)$, we obtain the triangle inequality:

$$\mathrm{DK}(f,h) \leqslant \lim_{i \to \infty} \delta_\infty(f, h \circ \tau_i) \leqslant \mathrm{DK}(f,g) + \mathrm{DK}(g,h) \qquad \square$$

3 Windowed Dog-Keeper Distance

If there is one outlier in a time series, then this outlier dominates the DK distance, i.e. it dominates the maximum along a path through the matrix in Algorithm 1. Hence, the DK distance is not robust against outliers. In the case of DTW or ERP, the error of the outlier is relatively small compared to the sum of all small errors. One of our contributions is the windowed dog-keeper distance described below.

By comparing sliding windows with the L_1-norm instead of single elements, the same behaviour is possible for the DK distance. If there is an outlier within one time series, the error will not dominate the sum of distances within two sliding windows. For a formal definition, consider the sequence of sliding windows as a new time series.

Definition 2 (Windowed Time Series). *Let $n \in \mathbb{N}$ be an arbitrary window size and $T = (t_1, \ldots, t_\ell)$ be an arbitrary time series. The k-th n-window of T is the subsequence*

$$T_k^n = (t_k, \ldots, t_{k+n-1})$$

The n-windowed time series of T is the sequence

$$T^n = (T_1^n, \ldots, T_{k+n-1}^n)$$

Comparing two time series now is based on comparing windows. Here we might use the advantage of the L_1-metric to improve the robustness against outliers.

Definition 3 (Window Distance). *Consider two n-windows $P = (p_1, \cdots, p_n)$ and $Q = (q_1, \cdots, q_n)$. Then*

$$d(P, Q) = \sum_{i=1}^{n} d(p_i, q_i)$$

We now define the windowed dog-keeper distance (WDK).

Definition 4 (Windowed Dog-Keeper Distance). *Let S and T be two time series and n be an arbitrary window size. The n-windowed dog-keeper distance (n-WDK) of S and T is the dog-keeper distance of their n-windowed time series, i.e.*

$$WDK_n(S, T) := DK(S^n, T^n)$$

If it is clear from the context, we omit the parameter n.

Corollary 1. *The windowed distance is a metric.*

Note that the 1-WDK distance is equivalent to the DK distance, thus n-WDK can be seen a generalization of the DK distance.

The WDK distance is more robust against outliers as the experiments show. However, it comes with a price. The distance measure is less robust against local time warping, since the time series can drift apart within one window. Hence, the window size is a tuning parameter to choose between robustness against outliers and robustness against time warps. The larger the window size is, the more we gain robustness against outliers. With shrinking window size we increase the robustness against strong time warps.

Computation: When computing the WDK distance the naive way, there is a lot of redundancy. For example, computing the 2-window distances $d(S^1, T^1) = d(s_1, t_1) + d(s_2, t_2)$ and $d(S^2, T^2) = d(s_2, t_2) + d(s_3, t_3)$ each includes computing $d(s_2, t_2)$. The first improvement of the WDK algorithm caches these values.

The second improvement optimizes the computation of the sum of the distance values along an n-window by first computing integral matrices. For time series S and T of length m and n respectively, the integral matrix $\int(S, T)$ is defined as

$$\int(S, T)_{i,j} = \begin{cases} \sum_{k=0}^{i-1} d\left(s_{i-k}, t_{j-k}\right) & \text{if } i \leqslant j \\ \sum_{k=0}^{j-1} d\left(s_{i-k}, t_{j-k}\right) & \text{else} \end{cases} \tag{1}$$

where $\int(S, T)_{i,j}$ is the entry in the i-th row and the j-th column. Less formally, we sum up the values of the matrix $(d(s_i, t_j))$ along diagonals. The n-window distance $\mathsf{d}(T_i^n, S_j^n)$ is computed as a difference of two matrices:

$$\mathsf{d}(T_{i-n+1}^n, S_{j-n+1}^n) = \begin{cases} \int(S,T)_{i,j} - \int(S,T)_{i-n,j-n} & \text{if } i,j \geqslant n \\ \int(S,T)_{i,j} & \text{else} \end{cases} \quad (2)$$

Finally, the Fréchet distance is computed based on the window distance. Algorithm 2 represents the algorithm in pseudo code. Line 5 to 11 compute the integral matrix, line 14 to 16 compute the window distances. For time series of length ℓ and m, these sections have complexity $\mathcal{O}(\ell \cdot n)$. The rest of the code computes the Fréchet Distance similarly to Algorithm 1 but on the window distances thus the overall complexity is $\mathcal{O}(\ell \cdot n)$. Furthermore, the complexity does not depend on the window size.

Example 1. Consider the following example: $S = (1,2,1,5,6)$, $T = (2,1,6,5,6)$. When computing the 3-WDK distance function, the matrices in Algorithm will contain the following elements if they did not stop because of early abandoning:

$$I = \begin{pmatrix} 0 & 0 & 0 & 0 & 0 & 0 \\ 0 & 1 & 0 & 5 & 4 & 5 \\ 0 & 0 & 2 & 4 & 8 & 8 \\ 0 & 1 & 0 & 7 & 8 & 13 \\ 0 & 3 & 5 & 1 & 7 & 9 \\ 0 & 4 & 8 & 5 & 2 & 7 \end{pmatrix} \quad W = \begin{pmatrix} 7 & 8 & 13 \\ 1 & 6 & 9 \\ 5 & 2 & 5 \end{pmatrix} \quad D = \begin{pmatrix} 7 & 8 & 13 \\ 7 & 7 & 9 \\ 7 & 7 & 7 \end{pmatrix}$$

Thus, the 3-WDK distance of S and T is 7.

4   Experimental Evaluation

We evaluate the performance of the time series distance functions DTW, ERP, DK, and WDK on four data sets. Our first choice is the well-known cylinder-bell-funnel data set (CBF) as an example of noisy data. The other three data sets come from the UCI Machine Learning Repository [9]. We chose the following labeled multidimensional multimedia data sets: the Character Trajectories sata set (CT), the Spoken Arabic Digit data set (SAD), and the Australian Sign Language signs (High Quality) data set (ASL) [8].

Data Preparation: We prepared the data sets by normalizing them individually. The CT data set consists of three-dimensional time series holding the derivative of the trajectory and the pressure of the pen. We first integrated the derivative to retrieve the actual pen coordinates. The resulting time series have been normalized using the L_2-norm.

The Spoken Arabic Digits data set has been normalized using the L_1-norm. Furthermore, we removed the 23 shortest time series to assure that each time series has a length of at least 20 elements, such that we can evaluate the WDK distance for window sizes up to 20.

The ASL data set consists of 22-dimensional time series, 11 dimensions for each hand holding position, rotation and five finger bend information. We normalized the position information of the hands using the L_2 norm.

Algorithm 2. Pseudo Code for the n-Windowed Dog-Keeper Distance

```
1   Input: S = (s₁,···,sₗ),  T = (t₁,···,tₘ),  τ
2   Output: Lower bound for the dog-keeper distance
3
4   // compute integral matrix I_{i,j} = ∫(S,T)_{i,j} as in Equation (1)
5   for i in 0,···,ℓ
6       I_{i,0} = 0
7   for j in 1,···,m
8       I_{0,j} = 0
9   for i in 1,···,ℓ
10      for j in 1,···,m
11          I_{i,j} = I_{i-1,j-1} + d(sᵢ,tⱼ)
12
13  // compute the n−window distances W_{i,j} = d(Tᵢⁿ,Sⱼⁿ) as in Equation (2)
14  for i in n+1,···,ℓ+1
15      for j in n+1,···,m+1
16          W_{i-n,j-n} = I_{i-1,j-1} − I_{i-n-1,j-n-1}
17
18  // compute the DK distance as in Algorithm 1.
19  D_{1,1} = W_{1,1}
20  if D_{1,1} ⩾ τ
21      return D_{1,1}
22  for i in 2,···,ℓ−n+1
23      D_{i,1} = max {D_{i-1,1},W_{i,1}}
24  for j in 2,···,m−n+1
25      ε = ∞
26      for i in 1,···,ℓ−n+1
27          pred = min {D_{i-1,j},D_{i,j-1},D_{i-1,j-1}}
28          D_{i,j} = max {W_{i,i},pred}
29          ε = min {ε,D_{i,j}}
30      if ε ⩾ τ
31          return ε
32  return D_{ℓ-n+1,m-n+1}
```

Retrieval Correctness: We use the data sets to evaluate the quality of the distance functions experimentally. Since we have chosen labeled time series, we can evaluate the correctness using a k-nearest neighbour classifier. We specifically ran a Leave-One-Out cross-validation on each data set. In order to evaluate the discriminability of the distance functions we ran the tests for different k from 1 to values larger than the class size.

Figure 2 shows that DTW has almost best retrieval results on the noisy CBF data set. ERP decreases in quality with increasing k. For small k, all distance

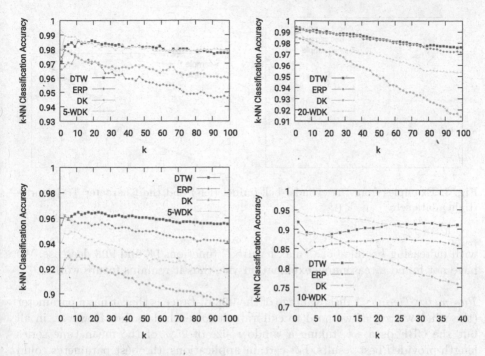

Fig. 2. Classification Accuracy on the CBF (top left), CT (top right), SAD (bottom left) and ASL (bottom right) data sets.

functions provide similar quality. We did not expect the WDK distance to perform well on that data set since it is very noisy (cf. Fig. 3).

In contrast to the CBF data set, Fig. 2 shows that the retrieval results decrease linearly with increasing k on the CT data set. Although there is nearly no difference in retrieval quality for small k, there is a clear tendency for large k. DTW and ERP have identical behaviour, while the DK is way behind. This experiment also shows that WDK improves the DK distance. Figure 3 shows two representational examples from the data set. We could not find any outliers in the data set and there is little need for warping. On the other hand, the distance between two points along the characters differ on long parts of the path, thus there are windows with a large distance to each other. This could be the reason for the good performance of DTW and ERP but the bad performance of DK and WDK.

Figure 2 shows the results for the SAD data set. The results are similar to those on the CT data set. The WDK distance improves the DK distance but loses against DTW and ERP.

Most interesting results for the WDK distance can be found in the ASL data set, shown in Fig. 2. Since both distance functions DTW and ERP drop to around 90% correctness, 10-WDK is nearly 95%. Another interesting observation here is that both "sum natured" functions DTW and ERP are increasing in correctness

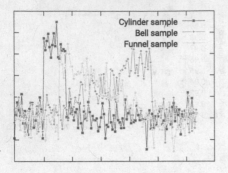 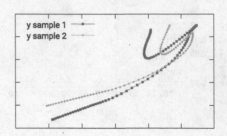

Fig. 3. Examples from the cylinder-bell-funnel (left) and the Character Trajectories (right) data sets.

with increasing k while both "max natured" functions DK and WDK decrease. We have not found a reasonable explanation yet, thus it remains future work.

Parameter Tuning: A disadvantage of the WDK distance is that it has a parameter (the window size) as we need to calibrate it for each data set. However, in all but the CBF data set, taking a window size of 25% of the mean time series length provided best results. For certain applications, the best parameter could be evaluated on a sample of the data set beforehand.

Figure 4 shows that the window size adjusts a trade-off as we expected. There is an optimal value and the classification Accuracy decreases monotone with diverging window size.

Computation Time: Table 1 shows the relative computation times with DTW as the base line for the 1-nearest neighbour classifier. Although the algorithms of all distance functions are quite similar the DK and WDK distance functions ran faster

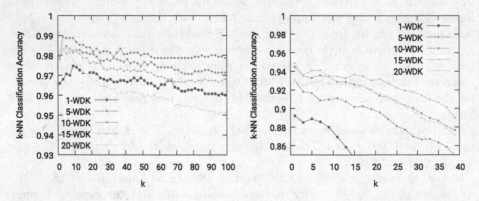

Fig. 4. Classification Accuracy on the CBF (left) and the Australian Sign Language (right) data sets for WDK with different window sizes.

by more than an order of magnitude. These differences can not be explained by implementation details. The only plausible explanation is the early abandoning.

Since DTW and ERP sum up the errors along the warping path, the probability for later abandoning increases. On the other hand, the DK distance takes the maximum value along the best warping path and therefore aborts computation most likely during the first step. We call the number of columns we need to compute before the computation can be aborted the point of early abandoning. The only exception is a value of 0 which means that the first elements of the time series are compared only.

Table 1. Computation time in relation to the computation time of DTW

	DTW	ERP	DK	WDK
CBF	1	0.89	0.23	0.13
Spoken digits	1	1.2	0.06	0.04
Signs	1	1.24	0.05	0.08
Character	1	1.39	0.13	0.06

Table 2 shows measurements of the number of comparisons which are aborted immediately after comparing the first elements of the time series on the ASL data set. It shows that 94.9% and 99.6% of the computations of the DK and WDK distance abort immediately, resp. The mean point of early abandoning for DTW and ERP is more than 10, which means that in most cases more than 10 columns of the matrix are filled.

Table 2. Point of early abandoning.

	DTW	ERP	DK	WDK
Immediate	0%	0%	94.9%	99.6%
Mean	10.8	13.3	0.39	0.21

5   Conclusion and Future Work

In this paper we compared the performance of different time warping distance functions on multimedia time series data sets. We have chosen data sets for motion gesture recognition, speech recognition, and classification of handwritten letters. This work extends existing evaluations by comparing the dog-keeper distance against DTW and ERP on these data sets. Although DTW has the best classification results on most data sets, we could show that the dog-keeper distance has nearly same results. For 1-nearest neighbour classification, the error rate of the dog-keeper distance was no more than 3% worse.

We also observed a significant difference in computation time. Our investigation showed that the reason is the very early abandoning.

We also improved the dog-keeper distance by comparing sliding windows instead of single elements. Our experimental evaluation shows that this modification did increase the classification correctness of the dog-keeper distance. On the Australian Sign Language data set (ASL), it even outperforms the other distance functions in retrieval quality. Furthermore, it inherited the property of early abandoning from the dog-keeper distance and even improved these values. On the Australian Sign Language data set, 99.6% of the comparisons already stopped after comparing the first elements of the time series. Hence, it seems there is no need for any further optimization using lower bounds.

It remains future work to investigate and compare to these functions with the Sakoe Chiba band [11] applied. We expect nearly the same behaviour from the dog-keeper and windowed dog-keeper distances. However, there are lower bounds to DTW with a Sakoe Chiba band applied which drastically improve retrieval times.

References

1. Backurs, A., Indyk, P.: Edit distance cannot be computed in strongly subquadratic time (unless SETH is false). CoRR, abs/1412.0348 (2014)
2. Bringmann, K.: Why walking the dog takes time: Frechet distance has no strongly subquadratic algorithms unless SETH fails. CoRR, abs/1404.1448 (2014)
3. Bringmann, K., Künnemann, M.: Quadratic conditional lower bounds for string problems and dynamic time warping. CoRR, abs/1502.01063 (2015)
4. Chen, L., Ng, R.: On the marriage of Lp-norms and edit distance. In: Proceedings of the Thirtieth International Conference on Very Large Data Bases, VLDB 2004, vol. 30, pp. 792–803. VLDB Endowment (2004)
5. Ding, H., Trajcevski, G., Scheuermann, P., Wang, X., Keogh, E.: Querying and mining of time series data: experimental comparison of representations and distance measures. Proc. VLDB Endow. 1(2), 1542–1552 (2008)
6. Eiter, T., Mannila, H.: Computing discrete Fréchet distance. Technical report, Technische Universität Wien (1994)
7. René Fréchet, M.: Sur quelques points du calcul fonctionnel. 22. Rendiconti del Circolo Mathematico di Palermo (1906)
8. Kadous, M.W.: Temporal classification: extending the classification paradigm to multivariate time series. Ph.D. thesis, School of Computer Science and Engineering, University of New South Wales (2002)
9. Lichman, M.: UCI machine learning repository (2013)
10. Lin, J., Keogh, E., Wei, L., Lonardi, S.: Experiencing SAX: a novel symbolic representation of time series. Data Min. Knowl. Discov. 15(2), 107–144 (2007)
11. Sakoe, H., Chiba, S.: Dynamic programming algorithm optimization for spoken word recognition. In: Waibel, A., Lee, K. (eds.) Readings in Speech Recognition, pp. 159–165. Morgan Kaufmann Publishers Inc., San Francisco (1990)

Fast Similarity Search with the Earth Mover's Distance via Feasible Initialization and Pruning

Merih Seran Uysal[1]($\boxtimes$), Kai Driessen[1], Tobias Brockhoff[1], and Thomas Seidl[2]

[1] Data Management and Exploration Group, RWTH Aachen University,
Aachen, Germany
{uysal,driessen,brockhoff}@cs.rwth-aachen.de
[2] Database Systems Group, LMU Munich, Munich, Germany
seidl@dbs.ifi.lmu.de

Abstract. The Earth Mover's Distance (EMD) is a similarity measure successfully applied to multidimensional distributions in numerous domains. Although the EMD yields very effective results, its high computational time complexity still remains a real bottleneck. Existing approaches used within a filter-and-refine framework aim at reducing the number of exact distance computations to alleviate query time cost. However, the refinement phase in which the exact EMD is computed dominates the overall query processing time. To this end, we propose to speed up the refinement phase by applying a novel feasible initialization technique (INIT) for the EMD computation which reutilizes the state-of-the-art lower bound IM-Sig. Our experimental evaluation over three real-world datasets points out the efficiency of our approach (This work is partially based on [12]).

Keywords: Earth Mover's Distance · Similarity search · Lower bound · Filter distance · Initialization · Refinement phase

1 Introduction

The Earth Mover's Distance (EMD) [9] is a prominent effective distance-based similarity measure which computes the minimum amount of work required to transform one data representation into the other. Since the EMD is robust to outlier noise, it has been used and investigated in numerous fields, such as computer vision [8] and multimedia [1,13]. The EMD is applied to feature-based data representations, such as a histogram denoting a predefined shared set of features in a feature space, or a signature which comprises an object-specific set of features. Signatures denoting individual binning have been utilized to represent various kinds of data, such as biological [5] and multimedia data [13].

The EMD-based similarity search and query processing yield very effective results, however, at the cost of high computational query time. Since the EMD is a transportation problem, it can be solved by efficient linear programming techniques, such as interior point [3] and transportation simplex algorithms [4].

© Springer International Publishing AG 2017
C. Beecks et al. (Eds.): SISAP 2017, LNCS 10609, pp. 141–155, 2017.
DOI: 10.1007/978-3-319-68474-1_10

The empirical computational time complexity of the EMD is super-cubic in the feature dimensionality [9]. Hence, there has been much research devoted to accelerate the EMD-based similarity search [1,2,6,9,13,16,17]. These existing works are based on a filter-and-refinement framework which aims at reducing the number of the exact EMD computations. However, with increasing feature dimensionality and data cardinality, it is becoming a real challenge to attain high efficiency for the EMD-based query processing. The main reason here lies in the expensive EMD computation performed in the refinement step which dominates the whole query processing time cost.

In this paper, our aim is to improve the efficiency of the refinement phase of the EMD-based query processing within a filter-and-refine framework which utilizes the state-of-the-art IM-Sig as a filter distance function. As the refinement phase requires much time due to the expensive EMD computation, we focus on the acceleration of the refinement phase by the explicit *reutilization* of the filter distance which has already been computed in the filter step. Unlike most of the existing filters, IM-Sig is an optimization problem and constraint relaxation of the EMD. This property of the IM-Sig allows for the derivation of a particular initialization for the computation of the EMD. In this way, the EMD computation in the refinement phase does not start with an arbitrary standard initialization, instead, it makes use of an *elaborate* initialization which *reutilizes* the IM-Sig filter distance information already obtained in the previous phase. Furthermore, we use an early pruning algorithm which adapts the optimal multi-step algorithm [11] and generates lower bounds to the EMD in the intermediate steps of the EMD computation. This is carried out by making use of the advantageous dual feasibility property of the interior point methods in operations research. Any dual feasible objective function value in each step of the EMD computation is a lower bound to the EMD. In this way, it is possible to safely prune any non-promising data object in the intermediate steps of a single EMD computation, which overall leads to high efficiency improvement.

In summary, our key contributions are listed as follows: We introduce an efficient feasible **init**ialization technique INIT *reusing* IM-Sig filter to efficiently compute EMD in the refinement phase (Sect. 4.1). We use an **e**arly **p**runing optimal multi-step algorithm EP which further boosts the refinement phase (Sect. 4.2). Our extensive experimental evaluation over three real-world datasets indicates the high efficiency improvement of our proposed technique INIT and algorithm EP (Sect. 5). Results report that the application of the state-of-the-art filter IM-Sig within a filter-and-refine framework is remarkably outperformed by the application of our proposals.

2 Preliminaries

2.1 Earth Mover's Distance (EMD)

The EMD [9] is a similarity measure which determines the dissimilarity between any two signatures by transforming one signature into the other. A signature $Q = \{\langle q_i, h_i \rangle\} = \{\langle q_1, h_1 \rangle, \cdots, \langle q_n, h_n \rangle\}$ is a set of tuples each of which is denoted by a feature $q_i \in \mathbb{F}$ and a weight $h_i \in \mathbb{R}^+$ associated with that feature, where $\mathbb{F}$ denotes a feature space. Signatures may have different number of features, and features of a signature may differ from those of another signature. To compute the EMD between signatures Q and P, features of Q and P are considered as earth hills and holes, respectively, where the EMD denotes the minimum amount of work performed to transfer the earth from hills to holes. In our paper, for the sake of simplicity, we comply with the presentation of the linear optimization problems and use the following notations: The earth movement between the features q_i and p_j is indicated by the flow x_{ij}, and the ground distance between those features is given by δ_{ij}. The definition of the EMD is given below.

Definition 1 (EMD). *Given signatures* $Q = \{\langle q_i, h_i \rangle\}$ *and* $P = \{\langle p_j, t_j \rangle\}$ *of normalized total weight* $m = \sum_i h_i = \sum_j t_j$ *satisfying for index sets* I_Q, I_P $\forall i \in I_Q = \{1, \cdots, n_q\}$ $h_i \in \mathbb{R}^+$ *and* $\forall j \in I_P = \{1, \cdots, n_p\} t_j \in \mathbb{R}^+$, *EMD*$(Q, P)$ *is computed by solving the following linear program:*

$$\text{EMD}(Q, P) = \min \sum_{i \in I_Q} \sum_{j \in I_P} x_{ij} \cdot \frac{\delta_{ij}}{m}, \text{subject to:}$$

$$
\begin{array}{lll}
\forall i \in I_Q & \sum_{j \in I_P} x_{ij} = h_i & \text{(Source constraint)} \\
\forall j \in I_P & \sum_{i \in I_Q} x_{ij} = t_j & \text{(Target constraint)} \\
\forall i \in I_Q \; \forall j \in I_P & x_{ij} \geq 0 & \text{(Non-negativity constraint)}
\end{array}
$$

For the *primal* minimization problem above, the source constraints guarantee that for any feature q_i the sum of all outgoing flows x_{ij} may not exceed its capacity. Similarly, the target constraints state that for any target feature p_j the sum of all incoming flows x_{ij} may not exceed the capacity of p_j.

Since in the upcoming sections we will deal with dual linear programs to derive our proposed initialization technique, we present the *dual* maximization problem of the EMD which has the same optimum value as calculated by the minimization problem given in Definition 1:

$$\text{EMD}(Q, P) = \max \sum_{i \in I_Q} u_i \cdot h_i + \sum_{j \in I_P} v_j \cdot t_j, \text{subject to:}$$

$$
\begin{array}{lll}
\forall i \in I_Q \; \forall j \in I_P & u_i + v_j + z_{ij} = \dfrac{\delta_{ij}}{m} \\
\forall i \in I_Q & u_i \in \mathbb{R} \\
\forall j \in I_P & v_j \in \mathbb{R} \\
\forall i \in I_Q \; \forall j \in I_P & z_{ij} \in \mathbb{R}^{\geq 0}.
\end{array}
\tag{1}
$$

In the dual problem above, each variable u_i and v_j corresponds to a source constraint and a target constraint in the primal problem, respectively. The utilization of the slack variables $z_{ij} \in \mathbb{R}$ makes it possible to define constraints as equalities. Note that solving the dual EMD problem with the maximization objective function yields the same optimum result as the primal EMD.

2.2   Filter-and-refine Framework

While theoretical time complexity of the EMD computed by simplex algorithms is exponential in the feature dimensionality, the empirical time complexity is super-cubic. We focus on the most prominent and often utilized k-nearest-neighbor (k-nn) query in databases, since it does not require the determination of any parameter. The optimal multi-step algorithm in [11] has been proven to be optimal in the number of exact distance computations. Given a query signature Q and a database $\mathcal{D}$ of signatures, the key idea is to apply two steps in order to process the k-nn query efficiently. In the filter step, a lower-bounding filter distance function is applied to $\mathcal{D}$ in order to generate a ranking of lower bounds in ascending order. In each iteration, an unprocessed object is retrieved from the ranking if its lower bound does not exceed the current k-nn distance threshold. In the refinement phase, such an object is considered as a candidate, i.e. the exact distance to the query is computed. If the exact distance does not exceed the current k-nn distance threshold, the object is inserted into the result set and the threshold is updated to indicate the distance of the k-th object in the result set. The algorithm terminates when an unprocessed object is retrieved from the ranking with a lower bound exceeding the threshold distance. Since the applied filter distance function is a lower bound to the exact distance function, the generated result set preserves *completeness*.

3   Related Work

Approaches in [1,10,16,17] focus on effective filtering by introducing lower bounds on histogram distributions. Hence, they are bounded by fixed predefined features in the underlying feature space, i.e. histograms share the same features in a feature space. However, individual object-specific distributions (a.k.a. signatures) come up in numerous applications and domains, and the techniques given above cannot be applied to signatures. Fortunately, researchers have proposed lower bounds applicable to signatures reflecting individual distributions per data object [2,6,9,13]. The *projected emd (Pemd)* [2] lower-bounds the EMD on signatures by computing the EMD for projected signatures where each of them comprises features projected on an individual dimension of the feature space. The *simple relaxation (sRelax)* lower bound [6] moves the entire earth (capacity) of each feature i of a source signature to the nearest-neighbor feature in a target signature. The *centroid-based (Rubner)* lower bound [9] is another filter which computes the distance between mean signatures. Although the filter time is considerably low, the efficiency of the query processing is hindered by the high number of exact EMD computations, as analyzed in [13]. The *IM-Sig* lower bound [13] is another lower bound which replaces the target constraint of the EMD by a new one stating that each flow may not exceed the target capacity. The drawback of all these techniques is that they attempt to derive lower bounds for the filter phase of the filter-and-refine framework to reduce the number of candidates. However, as mentioned before, the expensive EMD computation in the refinement phase is a bottleneck, causing high query time cost [13].

While the state-of-the-art IM-Sig successfully and remarkably reduces the number of exact EMD computations, the refinement phase still results in efficiency degradation with increasing feature dimensionality (Fig. 8 in [13]) and data cardinality. Thus, the existing results have been encouraging enough to merit further investigation into accelerating the refinement phase, which we will immediately present in the upcoming section.

4 Speeding up Refinement Phase

In this section, we first introduce our efficient feasible **init**ialization technique INIT *reusing* the IM-Sig filter to efficiently compute the EMD in the refinement phase (Sect. 4.1). Then, we use a modified algorithm **e**arly **p**runing optimal multi-step algorithm EP which safely discards non-promising objects and further boosts the refinement phase (Sect. 4.2).

For the remainder of this paper, we use the following notations: Regarding the computation of the IM-Sig or EMD, the earth is transferred from a signature $Q = \{\langle q_i, h_i \rangle\}$ to a signature $P = \{\langle p_j, t_j \rangle\}$ of normalized total weight $m = \sum_i h_i = \sum_j t_j = 1$. Furthermore, $I_Q = \{1, \cdots, n_q\}$ and $I_P = \{1, \cdots, n_p\}$ represents the set of indices of features in Q and P, respectively. In addition, feature capacities take positive real numbers: $\forall i \in I_Q \; h_i \in \mathbb{R}^+$ and $\forall j \in I_P \; t_j \in \mathbb{R}^+$. The ground distance between two features q_i and p_j is denoted by $\delta(q_i, p_j)$ while the linear programs of the EMD and IM-Sig require it to be a constant δ_{ij}, hence, we use $\delta(q_i, p_j)$ and δ_{ij} interchangeably, where necessary. In the linear problems of the EMD, x_{ij} is a primal variable representing the flow between features q_i and p_j. Dual variables and slack variables are denoted by u_i, v_j and z_{ij}, respectively. Similarly, for the IM-Sig, x_{ij}^I is a primal variable, and dual variables are represented by u_i^I, v_{ij}^I. Below, we present the IM-Sig lower bound [13] as a linear program which we will use later on.

Definition 2 (IM-Sig). *Given signatures $Q = \{\langle q_i, h_i \rangle\}$ and $P = \{\langle p_j, t_j \rangle\}$, the Independent Minimization for Signatures IM-Sig(Q, P) is computed by solving the following linear program:*

IM-Sig$(Q, P) = \min \sum_{i \in I_Q} \sum_{j \in I_P} x_{ij}^I \cdot \frac{\delta_{ij}}{m}$, *subject to:*

$$\forall i \in I_Q \qquad \sum_{j \in I_P} x_{ij}^I = h_i \qquad \text{(Source constraint)}$$
$$\forall i \in I_Q \forall j \in I_P \qquad x_{ij}^I \leq t_j \qquad \text{(IM-Sig target constr.)}$$
$$\forall i \in I_Q \; \forall j \in I_P \qquad x_{ij}^I \geq 0 \qquad \text{(Non-negativity constraint)}$$

The dual feasible solution to the IM-Sig is used to initialize the EMD computation. To this end, we will use the *dual* maximization problem of the IM-Sig which has the same optimum value as calculated by the minimization problem given in Definition 2:

$$\text{IM-Sig}(Q, P) = \max \sum_{i \in I_Q} u_i^I \cdot h_i + \sum_{i \in I_Q} \sum_{j \in I_P} v_{ij}^I \cdot t_j, \text{ s.t.:}$$

$$\forall i \in I_Q \, \forall j \in I_P \qquad u_i^I + v_{ij}^I \leq \frac{\delta_{ij}}{m} \qquad (2)$$

$$\forall i \in I_Q \qquad u_i^I \in \mathbb{R}$$

$$\forall i \in I_Q \, \forall j \in I_P \qquad v_{ij}^I \leq 0. \qquad (3)$$

In order to introduce the feasible initialization of the EMD computation reusing the IM-Sig information, we first present some basic definitions and notations for the IM-Sig computation (c.f. Algorithm 1 in [13]). Given two signatures $Q = \{\langle q_i, h_i \rangle\}$ and $P = \{\langle p_j, t_j \rangle\}$, for any feature q_i, a permutation $\pi_i = (\pi_i(1), \cdots, \pi_i(n_p))$ of target feature indices $\{1, \cdots, n_p\}$ is built satisfying $\delta(q_i, p_{\pi_i(k)}) \leq \delta(q_i, p_{\pi_i(k+1)})$ with $1 \leq i \leq n_q$ and $1 \leq k < n_p$. Hence, entries in the permutation are determined according to the distances between the feature q_i and target features in signature P in ascending order. Let $b_i \in \{1, \cdots, n_p\}$ be the number of target features which receive earth from q_i, i.e. the following statements $\sum_{r=1}^{b_i - 1} t_{\pi_i(r)} < h_i$ and $\sum_{r=1}^{b_i} t_{\pi_i(r)} \geq h_i$ hold. We call $p_{\pi_i(b_i)}$ the *boundary feature* of q_i to indicate that it is the last feature in the permutation order to receive earth from the source feature q_i. By utilizing this notion, we can now define *target feature sets* which we will use in the remainder of this section:

Definition 3 (target feature sets). *Given a feature q_i and its permutation $\pi_i = (\pi_i(1), \cdots, \pi_i(n_p))$ of target feature indices, F_i and B_i are sets of target features receiving earth from q_i, and Z_i is the set of target features which do not receive any earth from q_i:*
$F_i := \{p_{\pi_i(j)} | 1 \leq j < b_i\}$
$B_i := \{p_{\pi_i(b_i)}\}$
$Z_i := \{p_{\pi_i(j)} | b_i < j \leq n_p\}.$

The definition above states that we can partition the target features into three sets with respect to earth transfer from a source feature q_i: boundary set B_i, flow set F_i and zero-flow set Z_i. B_i consists of the *boundary feature* which denotes the last feature receiving earth from the source feature q_i in the permutation order. In addition, the IM-Sig flow [13] is presented as follows:

Definition 4 (IM-Sig flow). *IM-Sig flow between two given features q_i and p_j is given by:*

$$x_{ij}^I = \begin{cases} t_j & \text{if } p_j \in F_i \\ h_i - \sum_{r=1}^{b_i-1} t_{\pi_i(r)} & \text{if } p_j \in B_i \\ 0 & \text{else } (p_j \in Z_i) \end{cases}$$

After giving the fundamental notations, below we present the details about how to initialize the EMD computation which reutilizes the IM-Sig filter which has already been computed in the filter phase.

4.1   Feasible Initialization Technique (INIT)

We aim at using the IM-Sig filter as initialization of the interior point method
(IPM) used to compute the EMD. Moreover, we guarantee that this derived initialization starts with a dual feasible solution to the EMD so that each iteration
of the IPM preserves dual feasibility, i.e. each iteration yields a lower bound to
the EMD. As will be presented later, this valuable property of INIT requires a
considerably smaller number of IPM iterations to compute the EMD.

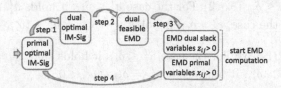

Fig. 1. Illustration of the feasible initialization technique INIT in four steps.

We present in 4 steps how to boost the refinement phase by the explicit
reutilization of the filter distance information (Fig. 1). For sake of simplicity, we
assume that signatures are 1-normalized signatures, i.e. the total weight of any
signature is equal to 1 (m in Definition 1). In the first step, the dual optimal
IM-Sig solution is derived from a given primal optimal IM-Sig solution computed
by the algorithm in [13]. Then, in the second step, a feasible solution to the dual
EMD problem is derived from the dual optimal IM-Sig solution. In the third
step, dual slack variables z_{ij} of the EMD problem are guaranteed to be strictly
positive, which is one of the two requirements of the IPM. In the fourth step, the
second requirement of the IPM is considered: a valid primal initialization for the
interior point method (IPM) is generated by ensuring strict positiveness of the
primal variables x_{ij} of the EMD problem. Overall, instead of starting with an
arbitrary initialization, the IPM starts with the targeted initialization reusing
the IM-Sig. Four steps to derive this initialization are presented below.

Step 1: Given a primal optimal IM-Sig solution computed by the algorithm
in [13], the dual optimal IM-Sig solution is generated in the first step. The
corresponding theorem is given as follows:

Theorem 1. (optimal dual im-sig). *Given an optimal solution x^I to the
primal IM-Sig problem, an optimal solution of the corresponding dual problem is
obtained by setting the dual variables of the IM-Sig as follows:*

$$\forall i \in I_Q \qquad u_i^I := \delta_{ij}, \; j = b_i \tag{4}$$

$$\forall i \in I_Q \; \forall j \in I_P \quad v_{ij}^I := \begin{cases} 0 & if \, u_i^I \le \delta_{ij} \\ \delta_{ij} - u_i^I & else \end{cases} \tag{5}$$

Proof. First we show the feasibility of the dual solution of the IM-Sig, and then prove the complementary slackness [14] which states that primal feasible and dual feasible solutions are optimal.

Feasibility. To show the feasibility of the dual solution, we analyze the constraints in Eqs. 2 and 3, and show that they are fulfilled by using the dual variables of IM-Sig (Eqs. 4 and 5) :

- Show $u_i^I + v_{ij}^I \leq \delta_{ij}$ (Eq. 2): For the case $u_i^I \leq \delta_{ij}$, it holds $u_i^I + v_{ij}^I = u_i^I + 0 = u_i^I \leq \delta_{ij}$. For the case $u_i^I > \delta_{ij}$, it holds $u_i^I + v_{ij}^I = u_i^I + (\delta_{ij} - u_i^I) = \delta_{ij}$.

- Show $v_{ij}^I \leq 0$ (Eq. 3): For the case $u_i^I \leq \delta_{ij}$, it holds that $v_{ij}^I = 0$, and for the other case $u_i^I > \delta_{ij}$, it holds that $v_{ij}^I = \delta_{ij} - u_i^I < 0$.

So far, we have seen that we obtain a dual feasible solution, if we are given a primal optimal feasible solution and apply the variable assignment given in Eqs. 4 and 5. Now, we prove the complementary slackness of the primal and dual solution to the IM-Sig, which is required to show the optimality of the solutions.

Optimality. According to the complementary slackness theorem [14], the primal feasible solution x^I and the dual feasible solution (u^I, v^I) to the IM-Sig are optimal if and only if the following equations are fulfilled:

$$\forall i \in I_Q \; \forall j \in I_P \quad x_{ij}^I \cdot (\delta_{ij} - u_i^I - v_{ij}^I) = 0 \tag{6}$$

$$\forall i \in I_Q \qquad u_i^I \cdot \left(h_i - \sum_{j \in I_P} x_{ij}^I \right) = 0 \tag{7}$$

$$\forall i \in I_Q \; \forall j \in I_P \quad v_{ij}^I \cdot (t_j - x_{ij}^I) = 0 \tag{8}$$

For the sake of simplicity, for any $i \in I_Q$ we denote $p_{\pi_i(j)}$ and $p_{\pi_i(b_i)}$ by the notation p_j and p_b, respectively. In this way, p_b is the last target feature which receives earth from the feature q_i regarding the permutation considering distance order (c.f. Definition 3).

- Show Eq. 6: By Eq. 4 and Definition 3, u_i^I is equal to the distance between the features q_i and p_b, i.e. $u_i^I = \delta_{ib}$.
  If $p_j \in B_i$, then p_j is exactly the boundary element p_b which leads to $u_i^I = \delta_{ib}$, thus it holds $x_{ij}^I \cdot (\delta_{ib} - \delta_{ib} - 0) = 0$.
  If $p_j \in Z_i$, then the feature p_j does not receive any earth from q_i, i.e. $x_{ij}^I = 0$, thus it holds $0 \cdot (\delta_{ij} - u_i^I - v_{ij}^I) = 0$.
  If $p_j \in F_i$, the amount of the earth transferred from q_i to p_j is equal to the capacity of p_j which is t_j (Definition 4), and we consider three cases. Case $u_i^I < \delta_{ij}$: This case does not apply because $u_i^I = \delta_{ib}$ is greater than or equal to δ_{ij}, since $\forall p_j \in F_i \; \delta_{ij} \leq \delta_{ib}$ by Definition 3. Case $u_i^I = \delta_{ij}$: It holds $v_{ij}^I = 0$, hence $x_{ij}^I \cdot (\delta_{ij} - \delta_{ij} - 0) = 0$. Case $u_i^I > \delta_{ij}$: It holds $x_{ij}^I \cdot (\delta_{ij} - u_i^I - (\delta_{ij} - u_i^I)) = 0$.

- Show Eq. 7: Since x^I is a primal feasible solution, the constraint $\forall i \in I_Q \ \sum_{j \in I_P} x_{ij} = h_i$ is fulfilled, hence it holds $u_i^I \cdot \left(h_i - \sum_{j \in I_P} x_{ij}^I \right) = 0$.

- Show Eq. 8: There are three cases we need to consider:
  If $p_j \in B_i$, then $u_i^I = \delta_{ij}$ holds. If $p_j \in Z_i$, then $u_i^I = \delta_{ib} \leq \delta_{ij}$ holds.
  If $p_j \in F_i$, p_j receives an amount of earth which is equal to its capacity t_j. Hence, $v_{ij}^I \cdot (t_j - x_{ij}^I) = v_{ij}^I \cdot (t_j - t_j) = 0$. Consequently, we proved the optimality of the dual solution to the IM-Sig problem by using complementary slackness.

Step 2: After obtaining a dual optimal IM-Sig solution computed by Theorem 1 a dual feasible EMD solution is generated which is formulated as below:

Theorem 2 (feasible dual emd). *Given an optimal solution* (u^I, v^I) *to the dual IM-Sig, a feasible solution* (u, v) *to the dual EMD is obtained as follows:*

$$\forall i \in I_Q \quad u_i := u_i^I \tag{9}$$

$$\forall j \in I_P \quad v_j := \min_{i \in I_P} v_{ij}^I \tag{10}$$

Proof. By using the equations in the theorem, we obtain $u_i + v_j = u_i^I + \min_{i \in I_P} v_{ij}^I \leq u_i^I + v_{ij}^I \overset{Eq.\,2}{\leq} \delta_{ij}$ (recall that $m = 1$ for the sake of simplicity in this section).

Step 3: After generating a dual feasible solution (u, v) to the EMD in the second step, now we will provide a valid dual initialization for the interior point method (IPM). To this end, the dual slack variables z_{ij} of the EMD problem will be guaranteed to be strictly positive, i.e. $z_{ij} > 0$. Since the given tuple (u, v) is a feasible solution of the dual EMD problem, the constraint $u_i + v_j + z_{ij} = \delta_{ij}$ (Eq. 1) is fulfilled. We propose to modify the slack variables z_{ij} and dual variables v_j of the EMD as follows: We replace v_j and z_{ij} by $v_j - \epsilon$ and $z_{ij} + \epsilon$, respectively ($\epsilon \in \mathbb{R}^{>0}$). By substituting the variables in the constraint (Eq. 1), we obtain $u_i + v_j + z_{ij} = u_i + (v_j - \epsilon) + (z_{ij} + \epsilon) = \delta_{ij}$. Thus, our modification preserves the dual feasibility of the EMD solution, and since $z_{ij} > 0$ is fulfilled, a valid dual initialization for the IPM is provided successfully.

Step 4: In the last step, a valid primal initialization for the interior point method (IPM) is generated by ensuring strict positiveness of the primal variables x_{ij} of the EMD problem. We propose to define the primal EMD variables as follows: $x_{ij} := \max\{x_{ij}^I, \zeta\}$ where $\zeta \in \mathbb{R}^{>0}$ is a small real number used to provide the strict positiveness of x_{ij}, i.e. $x_{ij} > 0$.

So far, we have seen that the application of four steps presented above leads to a valid initialization of the interior point method used to compute the EMD by utilizing the IM-Sig solution. Now, let us examine our proposed feasible initialization technique INIT on a single EMD computation between two signatures of dimensionality 64, as depicted in Fig. 2. When the EMD computation starts with a standard initialization, it first takes a value of -0.19 after which the

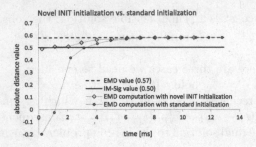

Fig. 2. An example EMD computation between two signatures (of dimensionality 64) with a standard initialization and our proposed technique INIT.

intermediate steps yield higher values. After 12.8 ms, the final EMD value is reached (0.57). When, however, the EMD computation starts with our proposed INIT technique starting with the IM-Sig filter distance which is computed as 0.5, the intermediate steps yield considerably higher values (i.e. lower bounds) than those for the standard initialization. After 10.8 ms, the final EMD value is reached and the EMD computation is finalized. As presented in this example, making use of the IM-Sig filter distance by our proposed feasible initialization technique INIT leads to higher values in the intermediate steps, and reaches the EMD value in a shorter time than a standard initialization.

4.2 Early Pruning (EP)

The prerequisite of the early pruning algorithm is the dual feasibility of the computation of the EMD. As stated in [3], if the dual constraints are fulfilled at the current iteration of the interior point method algorithm, then the next iteration's intermediate solution preserves the dual feasibility, as well.

Recall that according to the *weak duality theorem* in the linear programming field [14], we know that given a primal (minimization) problem, any feasible solution to the dual (maximization) problem yields an objective function value which is smaller than or equal to the optimum value of the dual objective function. Applying this notion to our proposed early pruning approach, the key idea is to start with a dual feasible initialization which leads to dual feasible solutions obtained in the intermediate steps of the algorithm. Since each dual feasible solution is a lower bound to the exact distance (in our case EMD), it is directly possible to verify whether each intermediate solution has already exceeded the current k-nn distance or not. If the intermediate objective function value exceeds the current k-nn distance, then there is no need to continue the distance computation any more, as further iterations will not lead to smaller values than the current k-nn distance. In this case, the object can be safely pruned due to the lower-bounding property, which contributes to the alleviation of the computational time cost.

5 Experimental Evaluation

5.1 Experimental Setup

Datasets. We use three real-world datasets: The first one is the prominent medical image dataset IRMA [7] which includes 12,000 anonymous radiographs. PUBVID [15] is the second dataset which comprises 250,000 public videos from the internet. We generate the third dataset by downloading 500,000 videos from the web site *vine.co*, and we refer to this dataset as SOCIAL. For signature generation, 10,000 pixels of each image in IRMA are extracted and represented in a 7-dimensional feature space comprising dimensions of the relative spatial information (x, y values), color information (CIELAB values), and texture information (contrast, coarseness values). Then, features are clustered by using a k-means algorithm which outputs k clusters whose centroids represent representatives (features) of the generated signature. Hence, each feature of a signature is denoted by a 7-dimensional feature vector. Any signature in PUBVID and SOCIAL is generated by extracting 10,000 pixels from the associated video and applying a k-means clustering algorithm in an 8-dimensional feature space.

Algorithms and Methods. We consider existing lower bounds to the EMD on signatures mentioned in Sect. 3: projected EMD *(Pemd)* lower bound [2], simple relaxation *(sRelax)* lower bound [6], centroid-based *(Rubner)* lower bound [9], and the Independent Minimization for Signatures *(IM-Sig)* lower bound [13]. In order to study the multi-step k-nearest-neighbor (k-nn) filter-and-refine framework, we implemented two algorithms: The baseline is the optimal multistep algorithm [11] for filter and refinement which is referred to as *FAR*, while the second algorithm is its modified version, early pruning algorithm, represented by *EP*. The aforementioned lower bounds are investigated associated with the algorithms FAR and EP. Furthermore, since we are interested in boosting the refinement phase, we investigate the performance of our proposed feasible initialization technique *INIT* and the standard initialization of the EMD computed by the interior-point method [3].

System Setup. Results are averaged out of 100 queries randomly chosen from the associated datasets. Table 1 gives the parameters regarding signature dimensionality and data cardinality of each dataset (default values are shown in bold and K represents 1,000).

Table 1. Parameter setting for datasets.

	Signature size	Data cardinality
IRMA	4 8 16 **32** 64	4K 6K 8K 10K **12K**
PUBVID	4 8 16 **32** 64	50K 100K 150K 200K **250K**
SOCIAL	4 8 16 **32** 64	100K 200K 300K 400K **500K**

The implementation of programs is performed in JAVA and experiments were conducted on a (single-core) 2.2 GHz computer with Windows Server 2008 OS and 6 GB of main memory. To determine the ground distance among features, Manhattan (L_1) distance is used. However, proposed technique and algorithm are not restricted to any ground distance function. In addition, due to space limitation we set the query parameter k to 100 for k-nn filter-and-refine algorithms.

5.2   Experimental Results

Figure 3(a)-(c) plots query time for lower bounds (Rubner, Pemd, sRelax, and IM-Sig) associated with algorithms (FAR, EP) across signature dimensionality over the datasets IRMA, PUBVID, and SOCIAL. Any chosen lower bound, the number of EMD computations in the refinement phase of the FAR algorithm is the same as for the EP algorithms which we omit due to space limitations. IM-Sig exhibits the lowest number of exact distance computations with increasing signature dimensionality. At signature size 64, when IM-Sig is applied, EMD computations are performed for only 2.6% and 4.1% of the datasets PUBVID and SOCIAL, respectively. Recall that the difference between FAR and EP is that the latter applies early pruning within a single EMD computation in the refinement phase thanks to provided dual feasibility of the interior point method (IPM). This impacts on the efficiency of overall query time which can be noticed in Fig. 3(a)-(c). For each dataset, it is observed that any lower bound utilized within EP algorithm results in much smaller query time than for that lower bound within FAR algorithm. At signature size 64, IM-Sig applied within EP attains the best efficiency for all datasets, while IM-Sig applied within FAR outperforms all lower bounds applied within FAR. For PUBVID at signature size 64, EP with IM-Sig performs 1.72 times faster than FAR with IM-Sig, while EP with sRelax, Pemd, and Rubner runs 2.4, 3.0, and 3.2 times faster than FAR with these lower bounds, respectively.

Now we compare the query performance with the most successful filter IM-Sig within filter-and-refine algorithms (FAR, EP) by utilizing two initializations of the EMD computation. Recall that for any single EMD computation in the refinement phase the standard initialization is used, while the EMD computation using INIT is carried out by explicitly reutilizing the IM-Sig filter information.

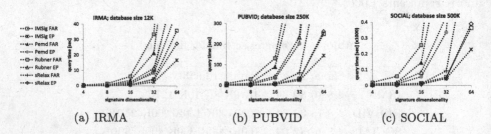

(a) IRMA   (b) PUBVID   (c) SOCIAL

Fig. 3. Experimental evaluation of lower bounds with the filter-and-refine (FAR) algorithm and early pruning (EP) algorithm regarding query time.

Figure 4(a)-(b) plots number of iterations of EMD computation and query time for IM-Sig with the standard initialization and our initialization technique INIT associated with FAR and EP algorithms across signature dimensionality over IRMA. The application of INIT within EP algorithm (IM-Sig EP INIT in the plot) leads to 3.3 times smaller number of iterations of IPM than those for the standard initialization of EMD computations within FAR algorithm (IM-Sig FAR in the plot) at signature size 64. The query response time of IM-Sig EP INIT is 2.65 times faster than that for IM-Sig FAR. This can be elucidated by the fact that INIT using the IM-Sig flow information for the initialization of the EMD computation attains a reduction in IPM iterations. Thus, it leads to a speed up in the refinement phase. Moreover, the dual feasibility of IPM enables the EP algorithm to terminate earlier, contributing to the query time cost reduction. In addition, the number of iterations of IPM and query time across data cardinality are depicted in Fig. 4(c)-(d). All curves show a constant behavior where IM-Sig EP INIT outperforms the other ones with increasing data size.

Figure 5 depicts results over PUBVID dataset where IM-Sig EP INIT outperforms other combinations of methods with both increasing signature size and data cardinality. IM-Sig EP INIT shows the lowest number of IPM iterations for the EMD computation and the highest query efficiency improvement. It requires 3.8 times smaller number of IPM iterations than for IM-Sig FAR and is 2.11 times faster (signature size 64). Moreover, IM-Sig EP INIT remarkably shows the best efficiency improvement by being 40.16 times faster than Rubner FAR. Similarly, IM-Sig EP INIT indicates up to 3.81 times smaller number of IPM iterations than for IM-Sig FAR with increasing data size. These results rely on two facts: INIT attains tight lower bounds in the intermediate steps of the EMD computation (c.f. Fig. 2). Moreover, EP checks each iteration of the EMD computation in the refinement phase and interrupts the EMD computation if the intermediate value of the objective function exceeds the current k-nn distance.

In order to determine the scalability of INIT and EP, we conduct experiments on SOCIAL dataset. As Fig. 6 presents, the application of IM-Sig within EP algorithm using INIT initialization (IM-Sig EP INIT in the plots) outperforms all other combinations of methods, exhibiting certainly similar behavior to the previous plots in Figs. 4 and 5. This indicates the high stability of our feasible initialization technique INIT and EP. In particular, IM-Sig EP INIT indicates

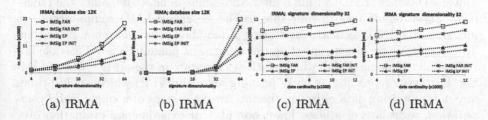

| (a) IRMA | (b) IRMA | (c) IRMA | (d) IRMA |

Fig. 4. Experimental evaluation of IM-Sig with standard and our initialization (INIT) technique with FAR algorithm and EP algorithm on IRMA dataset.

154   M.S. Uysal et al.

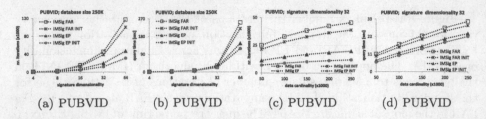

(a) PUBVID   (b) PUBVID   (c) PUBVID   (d) PUBVID

Fig. 5. Experimental evaluation of IM-Sig with standard and our proposed initialization (INIT) technique within FAR and EP algorithm on PUBVID dataset.

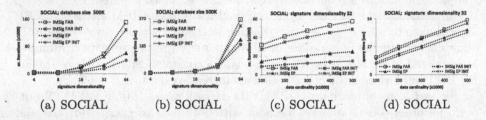

(a) SOCIAL   (b) SOCIAL   (c) SOCIAL   (d) SOCIAL

Fig. 6. Experimental evaluation of IM-Sig with standard and our initialization (INIT) technique within FAR our proposed EP algorithm on SOCIAL dataset.

up to 3.9 times smaller number of IPM iterations than for IM-Sig FAR, being 1.9 times faster than the latter. IM-Sig EP INIT performs considerably well, when compared to Rubner FAR which is beaten at a factor of 50.7 in terms of efficiency. In addition, with increasing data cardinality, IM-Sig FAR with different initializations exhibits a higher number of iterations, while IM-Sig applied within EP shows considerably smaller slope in the plots. IM-Sig EP INIT shows up to 3.9 times smaller number of IPM iterations than for IM-Sig FAR with increasing data size, and exhibits the best efficiency improvement for SOCIAL dataset.

6 Conclusion

In this paper, we study how to speed up the similarity search based on the Earth Mover's Distance (EMD) within a filter-and-refine architecture. Since the refinement phase requires much time due to the expensive EMD computation, we introduce the INIT technique and using a modified filter-and-refine algorithm. To this end, we propose to speed up the refinement phase by a novel feasible initialization technique (INIT) which reutilizes the state-of-the-art lower bound IM-Sig. Furthermore, we use an early pruning algorithm (EP) which safely prunes any non-promising data object in the intermediate steps of a single EMD computation, regardless of utilized lower bound. Our experimental evaluation over three real-world datasets points out the efficiency of our approach.

References

1. Assent, I., Wenning, A., Seidl, T.: Approximation techniques for indexing the earth mover's distance in multimedia databases. In: ICDE, p. 11 (2006)
2. Cohen, S.D., Guibas, L.J.: The earth mover's distance: lower bounds and invariance under translation, Technical report. Stanford University (1997)
3. Gondzio, J.: Interior point methods 25 years later. EJOR **218**(3), 587–601 (2012)
4. Hillier, F., Lieberman, G.: Introduction to Linear Programming. McGraw-Hill, New York (1990)
5. Hinneburg, A., Lehner, W.: Database support for 3D-protein data set analysis. In: SSDBM, pp. 161–170 (2003)
6. Kusner, M.J., Sun, Y., Kolkin, N.I., Weinberger, K.Q.: From word embeddings to document distances. In: ICML, pp. 957–966 (2015)
7. Lehmann, T., et al.: Content-based image retrieval in medical applications. Methods Inf. Med. **43**(4), 354–361 (2004)
8. Pele, O., Werman, M.: A linear time histogram metric for improved SIFT matching. In: Forsyth, D., Torr, P., Zisserman, A. (eds.) ECCV 2008. LNCS, vol. 5304, pp. 495–508. Springer, Heidelberg (2008). doi:10.1007/978-3-540-88690-7_37
9. Rubner, Y., Tomasi, C., Guibas, L.: A metric for distributions with applications to image databases. In: ICCV, pp. 59–66 (1998)
10. Ruttenberg, B.E., Singh, A.K.: Indexing the earth mover's distance using normal distributions. PVLDB **5**(3), 205–216 (2011)
11. Seidl, T., Kriegel, H.: Optimal multi-step k-nearest neighbor search. In: SIGMOD, pp. 154–165 (1998)
12. Uysal, M.S.: Efficient Similarity Search in Large Multimedia Databases. Apprimus Verlag (2017)
13. Uysal, M.S., et al.: Efficient filter approximation using the EMD in very large multimedia databases with feature signatures. In: CIKM, pp. 979–988 (2014)
14. Vanderbei, R.J., Progr, L.: Foundations and Extensions. Springer, US (2014)
15. Vandersmissen, B., et al.: The rise of mobile and social short-form video: an in-depth measurement study of vine. In: SoMuS, vol. 1198, pp. 1–10 (2014)
16. Wichterich, M., et al.: Efficient emd-based similarity search in multimedia databases via flexible dimensionality reduction. In: SIGMOD, pp. 199–212 (2008)
17. Xu, J., Zhang, Z., et al.: Efficient and effective similarity search over probabilistic data based on earth mover's distance. PVLDB **3**(1), 758–769 (2010)

A New Perspective on the Tree Edit Distance

Stefan Schwarz, Mateusz Pawlik[✉], and Nikolaus Augsten

University of Salzburg, Salzburg, Austria
mateusz.pawlik@sbg.ac.at

Abstract. The tree edit distance (TED), defined as the minimum-cost sequence of node operations that transform one tree into another, is a well-known distance measure for hierarchical data. Thanks to its intuitive definition, TED has found a wide range of diverse applications like software engineering, natural language processing, and bioinformatics. The state-of-the-art algorithms for TED recursively decompose the input trees into smaller subproblems and use dynamic programming to build the result in a bottom-up fashion. The main line of research deals with efficient implementations of a recursive solution introduced by Zhang in the late 1980s. Another more recent recursive solution by Chen found little attention. Its relation to the other TED solutions has never been studied and it has never been empirically tested against its competitors. In this paper we fill the gap and revisit Chen's TED algorithm. We analyse the recursion by Chen and compare it to Zhang's recursion. We show that all subproblems generated by Chen can also origin from Zhang's decomposition. This is interesting since new algorithms that combine the features of both recursive solutions could be developed. Moreover, we revise the runtime complexity of Chen's algorithm and develop a new traversal strategy to reduce its memory complexity. Finally, we provide the first experimental evaluation of Chen's algorithm and identify tree shapes for which Chen's solution is a promising competitor.

1 Introduction

Data featuring hierarchical dependencies are often modelled as trees. Trees appear in many applications, for example, the JSON or XML data formats; human resource hierarchies, enterprise assets, and bills of material in enterprise resource planning; natural language syntax trees; abstract syntax trees of source code; carbohydrates, neuronal cells, RNA secondary structures, and merger trees of galaxies in natural sciences; gestures; shapes; music notes.

When querying tree data, the evaluation of tree similarities is of great interest. A standard measure for the tree similarity, successfully used in numerous applications, is the *tree edit distance* (TED). TED is defined as the minimum-cost sequence of node edit operations that transform one tree into another. In the classical setting [16, 18], the edit operations are node deletion, node insertion, and label renaming. In this paper we consider ordered trees in which the sibling order matters. For ordered trees TED can be solved in cubic time, whereas the problem is NP-complete for unordered trees.

C. Beecks et al. (Eds.): SISAP 2017, LNCS 10609, pp. 156–170, 2017.
DOI: 10.1007/978-3-319-68474-1_11

In 1989, Zhang and Shasha proposed a recursive solution for TED [18]. The recursion decomposes trees into smaller subforests. New subforests are generated by either deleting the leftmost or the rightmost root node of a given subforest. A good choice (left or right) is essential for the runtime efficiency of the resulting algorithm. We call *Zhang decomposition* an algorithm that implements Zhang and Shasha's recursive formula.

Most TED algorithms, including the following, are dynamic programming implementations of the Zhang decomposition and differ in the strategy of left vs. right root deletion. Zhang and Shasha's own algorithm [18] runs in $O(n^4)$ time and $O(n^2)$ space for trees with n nodes. Klein [11] proposes an algorithm with $O(n^3 \log n)$ time and space complexity. Demaine et al. [7] further reduce the runtime complexity to $O(n^3)$ ($O(n^2)$ space), which is currently the best known asymptotic bound for TED. The same bounds are achieved by Pawlik and Augsten in their RTED [13] and AP-TED$^+$ [14] algorithms. According to a recent result [2] it is unlikely that a truly subcubic TED solution exists.

Although TED is cubic in the worst case, for many practical instances the runtime is much faster. For example, Zhang and Shasha's algorithm [18] runs in $O(n^2 \log^2 n)$ time for trees with logarithmic depth. Pawlik and Augsten [13] dynamically adapt their decomposition strategy to the tree shape and show that their choice is optimal. They substantially improve the performance for many practically relevant tree shapes. AP-TED$^+$ [14] is a memory and runtime optimized version of RTED and is the state of the art in computing TED.

In this paper we study an algorithm that does not fall into the mainstream category of Zhang decompositions, namely the TED algorithm introduced by Chen in 2001 [6]. Chen proposes an alternative recursive solution for TED and provides a dynamic programming implementation of his recursion. In terms of asymptotic runtime complexity, Chen's algorithm is known to be more efficient than all other algorithms for deep trees with a small number of leaves. Unfortunately, this algorithm has received little attention in literature. In particular, its relation to Zhang decompositions has never been studied. Further, we are not aware of any implementation or empirical evaluation of the algorithm. We revisit Chen's algorithm and make the following contributions:

- We perform the first analytical comparison of the decompositions by Chen and Zhang. Although the decompositions seem very different at the first glance, we show that all subproblems resulting from Chen's recursion can also be generated in a Zhang decomposition. Chen mainly differs in the way solutions for larger subproblems are generated from smaller subproblems. This is an important insight and opens the path to future research that unifies both decompositions into a single, more powerful decomposition.
- We revise the runtime complexity of Chen's algorithm. In the original paper, a significant reduction of the runtime complexity is based on the assumption of a truly subcubic algorithm for the (min,+)-product of quadratic-size matrices. Unfortunately, there is no such algorithm and even its existence remains an open problem. We adjust the asymptotic bounds accordingly and discuss the impact of the change.

- Memory is a major bottleneck in TED computations. We propose a new technique to reduce the memory complexity of Chen's algorithm from $O((n+l^2)\min\{l,d\})$ to $O((n+l^2)\log(n))$ for trees with l leaves and depth d. This is achieved by a smart traversal of the input trees that reduces the size of the intermediate result. Our technique is of practical relevance and is used in our implementation of Chen's algorithm.
- We implement and empirically compare Chen's algorithm to the state-of-the-art TED solutions. We identify tree shapes for which Chen outperforms all Zhang decomposition algorithms both in runtime and the number of intermediate subproblems. To the best of our knowledge, we are the first to implement Chen's algorithm and experimentally evaluate it.

The remaining paper is organised as follows. Section 2 analyses the relationship between Chen's algorithm and Zhang decompositions. In Sects. 3 and 4 we revise the runtime complexity and improve the memory complexity of Chen's algorithm, respectively. We experimentally evaluate Chen's algorithm in Sect. 5. Section 6 draws conclusions and points to future research directions.

2   Chen's Algorithm and Zhang Decompositions

In this section we analyse the relation of Chen's algorithm to the mainstream solutions for TED, namely *Zhang decompositions*. At the first glance, Chen's and Zhang's approaches seem very different and hard to compare. We tackle this problem in three steps: (1) We represent all subforests resulting from Chen's decomposition in the so-called *root encoding*, which was developed by Pawlik and Augsten [13] to index the subforests of Zhang decompositions. (2) We rewrite Chen's recursive formulas using the root encoding and compare them to Zhang's formulas. (3) We develop a Zhang decomposition strategy that always generates a superset of the subproblems resulting from Chen decomposition. These results lead to the important conclusion that Chen and Zhang decompositions can be combined into a single new decomposition strategy. This is a new insight that may lead to new, more powerful algorithms in the future. We refer to the end of this section for more details.

Trees, forests and nodes. A *tree* F is a directed, acyclic, connected graph with *labeled* nodes $N(F)$ and edges $E(F) \subseteq N(F) \times N(F)$, where each node has at most one incoming edge. A *forest* F is a graph in which each connected component is a tree; each tree is also a forest. We write $v \in F$ for $v \in N(F)$. In an edge (v, w), node v is the *parent* and w is the *child*, $p(w) = v$. A node with no parent is a *root* node, a node without children is a *leaf*. Children of the same node are *siblings*. A node x is an ancestor of node v iff $x = p(v)$ or x is an ancestor of $p(v)$; x is a *descendant* of v iff v is an ancestor of x. A *subforest* of a tree F is a forest with nodes $N' \subseteq N(F)$ and edges $E' = \{(v, w) : (v, w) \in E(F), v \in N', w \in N'\}$. F_v is the *subtree rooted in node* v of F iff F_v is a subforest of F and $N(F_v) = \{x : x = v \text{ or } x \text{ is a descendant of } v \text{ in } F\}$.

Node traversals. The nodes of a forest F are strictly and totally ordered such that (a) $v < w$ for any edge $(v, w) \in E(F)$, and (b) for any two nodes f, g, if $f < g$ and f is not an ancestor of g, then $f' < g$ for all descendants f' of f. The tree traversal that visits all nodes in ascending order is the *left-to-right preorder*. The *right-to-left preorder* visits the root node first and recursively traverses the subtrees rooted in the children of the root node in descending node order.

Example 1. In tree F in Fig. 1, the left (right) subscript of a node is its left-to-right (right-to-left) preorder number.

2.1   Representing Relevant Subproblems

All TED algorithms are based on some recursive solution that decomposes the input trees into smaller subtrees and subforests. Distances for larger tree parts are computed from the distances between smaller ones. A pair of subtrees or subforests that appears in a recursive decomposition is called a *relevant subproblem*. To store and retrieve the distance results for relevant subproblems they must be uniquely identified. Pawlik and Augsten [13] developed the *root encoding* to index all relevant subproblems that can appear in a Zhang decomposition.

Definition 1 (Root Encoding). *[13] Let the leftmost root node l_F and the rightmost root node r_F be two nodes of tree F, $l_F \leq r_F$. The root encoding F_{l_F, r_F} defines a subforest of F with nodes $N(F_{l_F, r_F}) = \{l_F, r_F\} \cup \{x : x \in F, x$ succeeds l_F in left-to-right preorder and x succeeds r_F in right-to-left preorder$\}$ and edges $E(F_{l_F, r_F}) = \{(v, w) \in E(F) : v \in F_{l_F, r_F} \wedge w \in F_{l_F, r_F}\}$.*

Example 2. In tree F in Fig. 1(a), subforest $F_{b, j}$ in root encoding (black nodes) is obtained from F by removing all predecessors of b in left-to-right preorder and all predecessors of j in right-to-left preorder.

Chen [6] also uses a recursive formula, but the decomposition rules are different from Zhang's rules. The result of Chen's decomposition are subtrees and subforests. The subforests can be of two different types (using the original notation). $\mathscr{G}_F(l', l'')$ is a subforest of tree F composed of all maximum-size subtrees having their leaf nodes between leaves l' and l'' in left-to-right preorder. $\mathscr{F}_F(v[1..p])$ is a subforest of tree F composed of the subtrees rooted in the first p children of node v (left-to-right preorder). Interestingly, all subtrees and subforests in Chen's decomposition are expressible in the root encoding.

Example 3. Subforest $\mathscr{G}_F(c, j)$ in Fig. 1(a) (root encoding $F_{b, j}$) consists of the largest subtrees having all leaves between c and j. $\mathscr{F}_F(a[1..2])$ in Fig. 1(b) (root encoding $F_{b, f}$) consists of the subtrees rooted at the first two children of a.

Theorem 1. *Every subtree and subforest that results from Chen's recursive decomposition can be represented in root encoding.*

Proof. A subtree F_v is represented as $F_{v, v}$ in root encoding. We show that both subforest types, (a) $\mathscr{G}_F(l', l'')$ and (b) $\mathscr{F}_F(v[1..p])$, also have a root encoding.

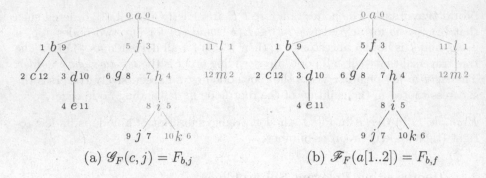

(a) $\mathcal{G}_F(c,j) = F_{b,j}$ (b) $\mathscr{F}_F(a[1..2]) = F_{b,f}$

Fig. 1. Subforests of an example tree F in Chen's and root encodings.

(a): Let a and b be the leftmost and rightmost root nodes of the forest $\mathcal{G}_F(l', l'')$. Then, the leftmost leaf of F_a is l' and the rightmost leaf of F_b is l''. We show that $\mathcal{G}_F(l', l'') = F_{a,b}$. The proof is by contradiction. (i) Assume a node $x \in F_{a,b}$ such that $x \notin \mathcal{G}_F(l', l'')$. Since $x \notin \mathcal{G}_F(l', l'')$, the subtree F_x rooted in x must have a leaf l outside the range l' to l'' (by the definition of $\mathcal{G}_F(l', l'')$), i.e., $l < l'$ or $l > l''$. This, however, is not possible since l' is the leftmost leaf node of F_a and l'' is the rightmost leaf node of F_b. (ii) Assume a node $y \in \mathcal{G}_F(l', l'')$ such that $y \notin F_{a,b}$. Then, by Definition 1, y precedes a in left-to-right preorder or y precedes b in right-to-left preorder. Consider $y < a$ in left-to-right preorder: all nodes that precede a in left-to-right preorder are either to the left of a or are ancestors of a. However, the nodes to the left of a are not in $\mathcal{G}_F(l', l'')$ since they have leaf descendants to the left of l', and ancestors of a are not in $\mathcal{G}_F(l', l'')$ since a is the leftmost root node in $\mathcal{G}_F(l', l'')$. Similar reasoning holds for y and b in right-to-left preorder. Thus y must be in $F_{a,b}$, which contradicts our assumption.

(b): $\mathscr{F}_F(v[1..p])$ is a subforest composed of the subtrees rooted in the first p children of node v. Let $c_1, \ldots, c_p$ be the first p children of node v. Then, according to the definition of the root encoding, $\mathscr{F}_F(v[1..p]) = F_{c_1, c_p}$. c_1 is the leftmost root node and c_p is the rightmost root node of $\mathscr{F}_F(v[1..p])$. Let l_1 be the leftmost leaf of c_1 and l_p be the rightmost leaf of c_p. All nodes in the subtrees rooted at nodes $c_1, \ldots, c_p$ have their left-to-right preorder ids between these of c_1 and l_p, and their right-to-left preorder ids between these of c_p and l_1. Thus, by Definition 1, $\mathscr{F}_F(v[1..p]) = F_{c_1, c_p}$.   □

2.2   Comparing Recursions

Thanks to Theorem 1, which allows us to express all subforests of Chen's decomposition in root encoding, we are able to rewrite Chen's recursive formulas with root encoding. This makes them comparable to Zhang's recursion, which also has a root encoding representation.

The tree edit distance between two forests is denoted $\delta(F, G)$. The trivial cases of the recursion are the same for both Chen and Zhang: $\delta(\emptyset, \emptyset) = 0$,

$\delta(F,\emptyset) = \delta(F - v,\emptyset) + c_d(v)$, $\delta(\emptyset, G) = \delta(\emptyset, G - w) + c_i(w)$, where F and G may be forests or trees, and $\emptyset$ denotes an empty forest. $c_d(v)$, $c_i(w)$, $c_r(v,w)$ are the costs of deleting node v, inserting node w, and renaming the label of v to the label of w, respectively. $F - v$ is the forest obtained from F by removing node v and all edges at v. By $F - F_v$ (v is a root node in forest F) we denote the forest obtained from F by removing subtree F_v. Given forest F and its subforest F', $F - F'$ is a forest obtained from F by removing subforest F'.

Zhang. The recursion by Zhang and Shasha [18] distinguishes two cases.

(a) Both F_v and G_w are trees.

$$\delta(F_v, G_w) = \min \begin{cases} \delta(F_v - v, G_w) + c_d(v) \\ \delta(F_v, G_w - w) + c_i(w) \\ \delta(F_v - v, G_w - w) + c_r(v, w) \end{cases} \tag{1}$$

(b) F_{l_F, r_F} is a forest or G_{l_G, r_G} is a forest.

$$\delta(F_{l_F, r_F}, G_{l_G, r_G}) = \min \begin{cases} \delta(F_{l_F, r_F} - l_F, G_{l_G, r_G}) + c_d(l_F) \\ \delta(F_{l_F, r_F}, G_{l_G, r_G} - l_G) + c_i(l_G) \\ \delta(F_{l_F}, G_{l_G}) + \delta(F_{l_F, r_F} - F_{l_F}, G_{l_G, r_G} - G_{l_G}) \end{cases} \tag{2}$$

In Eq. 2, instead of removing the leftmost root nodes and their subtrees (l_F, l_G, F_{l_G}, G_{l_G}) we can also remove their rightmost root node counterparts (r_F, r_G, F_{r_G}, G_{r_G}), respectively. The choice of left vs. right in each recursive step has an impact on the total number of subproblems that must be computed.

Chen. The recursion by Chen [6] distinguishes four cases. $roots(F_{l_F, r_F})$ and $leaves(F_{l_F, r_F})$ denote the set of all root resp. leaf nodes in forest F_{l_F, r_F}.

(a) Both F_v and G_w are trees. In this case, Chen's recursion is identical to Eq. 1.
(b) F_{l_F, r_F} is a forest and G_w is a tree.

$$\delta(F_{l_F, r_F}, G_w) = \min \begin{cases} \delta(F_{l_F, r_F}, G_w - w) + c_i(w) \\ \min_{s \in roots(F_{l_F, r_F})} \{\delta(F_s, G_w) + \delta(F_{l_F, r_F} - F_s, \emptyset)\} \end{cases} \tag{3}$$

(c) F_v is a tree and G_{l_G, r_G} is a forest.

$$\delta(F_v, G_{l_G, r_G}) = \min \begin{cases} \delta(F_v - v, G_{l_G, r_G}) + c_d(v) \\ \delta(F_v, G_{l_G, r_G} - G_{r_G}) + \delta(\emptyset, G_{r_G}) \\ \delta(F_v, G_{r_G}) + \delta(\emptyset, G_{l_G, r_G} - G_{r_G}) \end{cases} \tag{4}$$

(d) Both F_{l_F,r_F} and G_{l_G,r_G} are forests.

$$\delta(F_{l_F,r_F}, G_{l_G,r_G}) =$$

$$\min \begin{cases} \delta(F_{l_F,r_F}, G_{l_G,r_G} - G_{r_G}) + \delta(\emptyset, G_{r_G}) \\ \delta(F_{l_F,r_F}, G_{r_G}) + \delta(\emptyset, G_{l_G,r_G} - G_{r_G}) \\ \min_{l' \in leaves(F_{l_F,r_F})} \{\delta(F_{l_F,r'_F}, G_{l_G,r_G} - G_{r_G}) + \delta(F_{l''_F,r_F}, G_{r_G}) \quad (5) \\ \qquad\qquad + \delta(F_{l_F,r_F} - F_{l_F,r'_F} - F_{l''_F,r_F}, \emptyset)\} \end{cases}$$

The nodes r'_F and l''_F in Eq. 5 are defined as follows. Let l'' be the next leaf node after l' in F_{l_F,r_F} and $lca(l', l'') \in F$ the lowest common ancestor of the two leaves l' and l'' (not necessarily $lca(l', l'') \in F_{l_F,r_F}$). Then, r'_F (l''_F) is the first descendant of $lca(l', l'')$ in F_{l_F,r_F} that is on the path to l' (l'').

$F_{l_F,r_F} - F_{l_F,r'_F} - F_{l''_F,r_F}$ is a path from $lca(l', l'')$ to a root node (node without a parent) in F_{l_F,r_F} if $lca(l', l'') \in F_{l_F,r_F}$, or it is an empty forest $\emptyset$ otherwise. While this term cannot be expressed in root encoding, the distance in Eq. 5 can be rewritten as follows: $\delta(F_{l_F,r_F} - F_{l_F,r'_F} - F_{l''_F,r_F}, \emptyset) = \delta(F_{l_F,r_F}, \emptyset) - \delta(F_{l_F,r'_F}, \emptyset) - \delta(F_{l''_F,r_F}, \emptyset)$. Similarly, $\delta(F_{l_F,r_F} - F_s, \emptyset) = \delta(F_{l_F,r_F}, \emptyset) - \delta(F_s, \emptyset)$ in Eq. 3.

The correctness of Chen's recursion has only been shown for forests F_{l_F,r_F} that are expressible in the form $\mathscr{G}_F(l', l'')$, where l' (l'') is the leftmost (rightmost) leaf descendant of l_F (r_F); and forests G_{l_G,r_G} that are expressible in the form $\mathscr{F}_G(v[1..p])$, where v is the parent of l_G, and r_G is the p-th child of v [6]. Other forests shapes, although they may have root encoding, are not allowed.

Satisfying this restriction and thanks to the unified notation, we can observe that the recursions by Zhang and Chen can be alternated. Since Chen's decomposition is more efficient for some tree shapes, combining the two formulas may lead to better strategies and new, more efficient algorithms.

2.3 Comparing Relevant Subproblems

The choice of left vs. right in Zhang's decomposition has an impact on the number of relevant subproblems that must be computed (cf. Sect. 2.2). This has first been discussed by Dulucq and Touzet [8]. The RTED algorithm by Pawlik and Augsten [12] computes the optimal strategy and guarantees to minimize the number of subproblems in the class of *path decompositions*. Path decompositions constitute a subclass of Zhang decompositions that includes all currently known Zhang decomposition algorithms. We design a path decomposition algorithm *ChenPaths* that mostly resembles that of Chen and show that the subproblems resulting from ChenPaths are a superset of Chen's subproblems. We evaluate the difference in the subproblems count (ChenPaths vs. Chen) in Sect. 5.

A path decomposition algorithm requires two ingredients [13]: a *path strategy* that assigns a root-leaf path to each subtree pair (F_v, G_w) $(v \in F, w \in G)$ and a *single-path function* that is used to reassemble the results for larger subforests from smaller ones (using dynamic programming). A path decomposition algorithm works as follows: (*step1*) For the input trees (F, G), a root-leaf path is looked up in the path strategy. (*step2*) The algorithm is called recursively for each subtree pair resulting from removal of the root-leaf path from the corresponding input tree. (*step3*) A single path function is executed for the input trees. The single-path function decomposes a forest F_{l_F,r_F} such that, if the rightmost root r_F is on the root-leaf path assigned to F, then the leftmost root nodes are used in Eq. 2, otherwise the rightmost root nodes are used. The path choice affects the relevant subproblems resulting from (*step2*) and (*step3*).

We design *ChenPaths* with a path strategy that maps each subtree pair (F_v, G_w) to the left path in F_v and Δ^A single-path function [13]. Note that for left paths we could apply Δ^L single-path function that results in less subproblems but possibly a subset of Chen's subproblems.

Theorem 2. *The subproblems resulting from ChenPaths algorithm are a superset of the subproblems resulting from Chen's algorithm.*

Proof. As discussed in [13], the subproblems of a path decomposition algorithm are those encountered by all single-path functions executed for subtree pairs resulting from (*step2*). For ChenPaths the subproblems are $\mathcal{F}(F, \Gamma^L(F)) \times \mathcal{A}(G)$, where $\mathcal{F}(F, \Gamma^L(F))$ $(\mathcal{A}(G))$ is the set of all subtrees of F (G) and their subforests obtained by a sequence of rightmost (leftmost and rightmost) root node deletions. The subproblems of Chen's algorithm are $\mathscr{F}(F) \times \mathscr{G}(G)$, where $\mathscr{F}(F)$ is the set of all subtrees of F and their subforests of the form $\mathscr{F}_F(v[1..p])$, and $\mathscr{G}(G)$ is the set of all subtrees and subforests of the form $\mathscr{G}_G(l', l'')$. To show the inclusion of the subproblems it is enough to show the following:

(a) $\mathscr{F}(F) \subseteq \mathcal{F}(F, \Gamma^L(F))$. Every subtree F_v, $v \in F$, is in both sets. Every subforest of the form $\mathscr{F}(v[1..p])$ can be obtained from the subtree F_v by a sequence of rightmost root node deletions that delete root node v and v's children (and all their descendants) from the last child to $p+1$-st. Every subforests obtained this way is in $\mathcal{F}(F, \Gamma^L(F))$.

(b) $\mathscr{G}(G) \subseteq \mathcal{A}(G)$. Every subtree G_w, $w \in G$, is in both sets. Due to Theorem 1, every subforest of the form $\mathscr{G}_G(l', l'')$ can be represented in root encoding. $\mathcal{A}(G)$ is the set of all subforest of G that can be represented in root encoding. ☐

In this section we showed that there are path decompositions – a subclass of the more general class of Zhang decompositions – that can generate all subproblems of Chen's algorithms. This brings Chen's algorithm even closer to the mainstream TED algorithms. It seems likely that the results of Chen may be used to develop a new single-path function that, together with the results of [14], can be used to reduce the number of subproblems needed for TED algorithms. Furthermore, the cost of such a function can be used to compute the optimal-cost path strategy for a given input instance. See [13] for the input/output requirements of a single-path function and [14] for a discussion on how to leverage costs of new single-path functions for optimal strategies.

3   Revisiting the Runtime Complexity

Chen [6] derives for his algorithm a runtime of $O(n^2 + l^2 n + l^{3.5})$ for two trees with n nodes and l leaves. In his analysis, Chen uses a so called *(min,+)-product* of two $l \times l$ matrices, which has a trivial $O(l^3)$-time solution. In order to achieve the term $l^{3.5}$ in the runtime complexity, the (min,+)-product must be solved in time $O(l^{2.5})$. Without that improvement, the respective term becomes l^4, and the overall runtime complexity of Chen's algorithm is $O(n^2 + l^2 n + l^4)$.

Chen interpreted a result by Fredman [10] towards the existence of an efficient (min,+)-product algorithm that runs in $O(l^{2.5})$. Unfortunately, as recent works point out [3,9], it is still a major open problem whether a truly subcubic algorithm (an $O(n^{3-\epsilon})$-time algorithm for some constant $\epsilon > 0$) exists for the (min,+)-product. Fong et al. [9] analyse the related difficulties, Zwick [19] summarizes (in line with Fredman's discussion [10]) that for every n, a separate program can be constructed that solves the (min,+)-product in $O(n^{2.5})$ time, but the size of that program may be exponential in n. As Fredman points out, these results are primarily of theoretical interest and may be of no practical use.

Summarizing, with the current knowledge on (min,+)-product algorithms, the runtime complexity of Chen's algorithm is $O(n^2 + l^2 n + l^4)$. This is also the complexity of our implementation, which does not use the (min,+)-product improvement. Interestingly, even without that improvement, Chen's algorithm is an important competitor for some tree shapes. We discuss the details in Sect. 5.

4   Reducing the Memory Complexity

In this section we reduce the worst-case space complexity of Chen's algorithm. This is an important contribution for making the algorithm practically relevant.

Chen's algorithm uses dynamic programming, i.e., intermediate results are stored for later reuse. The space complexity of Chen's algorithm is $O((l^2 + n) \min\{l, d\})$ for two trees with n nodes, l leaves, and depth d. The complexity is a product of two terms. The first term, $(l^2 + n)$, is the size of arrays used to store intermediate results. The second term, $\min\{l, d\}$, is the maximum number of such arrays that have to be stored in memory concurrently throughout the algorithm's execution. We observe, that there are tree shapes for which Chen's algorithm requires $\lfloor \frac{n}{2} \rfloor$ arrays, for example, a right branch tree (a vertically mirrored version of the left branch tree in Fig. 2(a)). Then, the space complexity has a tight bound of $O((l^2 + n)n)$, which is worse than $O(n^2)$ achieved by other TED algorithms. In this section, we reduce the number of arrays that must be stored concurrently from $\min\{l, d\}$ to $log_2(n)$.

By thoroughly analysing Chen's algorithm we make a few observations. (a) The algorithm traverses the nodes in one of the input trees, say F, and executes one of two functions. These functions (called by Chen *combine* and *upward*) take arrays with intermediate results as an input and return arrays as an output. (b) Due to internals of the functions combine and upward, the traversal of nodes in F must obey the following rules: children must be traversed before parents, and siblings must be traversed from left to right. These rules resemble the so-called *postorder traversal*. (c) After a node $v \in F$ is traversed, exactly one array has to be kept in memory as a result of executing the necessary functions for v and all its descendants. This array must be kept in memory until the algorithm traverses the right sibling of node v.

Observations (b) and (c) suggest that the number of nodes that cause multiple arrays to be kept in memory concurrently, i.e., the nodes waiting for their right siblings to be traversed, strongly depends on the tree shape. For example, in left branch trees at most one node at a time is waiting for its right sibling, whereas in right branch trees all leaf nodes are waiting for their right siblings until the rightmost leaf node is traversed. Our goal is to minimise the number of such nodes. Our solution is based on the so-called *heavy-light decomposition* [15] which introduces a modification to the postorder traversal in observation (b).

We divide the nodes of a tree F into two disjoint sets: *heavy nodes* and *light nodes*. The root of F is light. For each non-leaf node $v \in F$, the child of v that roots the largest (in the number of descendants) subtree is heavy, and all other children are light. In case of ties, we choose the leftmost child with the largest number of descendants to be heavy. The *heavy-light traversal* is similar to the postorder traversal with one exception: the heavy child is traversed before all other children. The remaining children are traversed from left to right.

Theorem 3. *Using the heavy-light traversal for tree F, the maximum number of nodes that cause an additional array to be kept in memory concurrently is at most $\lceil \log_2 n \rceil$.*

Proof. We modify observation (c) for the heavy-light traversal. An array has to be kept in memory for a heavy node until its immediate left and right light siblings (if any) are traversed. For a light node an array has to be kept in memory until its right light sibling is traversed. Nodes never wait for their heavy siblings because the heavy sibling is traversed first.

Consider a path γ in tree F. The number of arrays that have to be kept in memory concurrently is proportional to the number of light nodes on γ. Let $L(\gamma)$ be all light nodes on path γ, and $W(\gamma)$ be all immediate siblings waiting for nodes in $L(\gamma)$. The array for a node in $W(\gamma)$ must be kept in memory until its sibling in $L(\gamma)$ is traversed. That brings us to the conclusion that the maximum number of arrays that have to be kept in memory concurrently equals the maximum number of light nodes on any path in F.

Let $|F| = |N(F)|$ denote the size of tree F. For any light node v, its heavy sibling has more nodes than v. It holds that $|F_{p(v)}| > 2|F_v|$, and $|F_v| < \frac{|F_{p(v)}|}{2}$. Then, each light node v on a path γ decreases the number of consecutive nodes on γ to be at most $\frac{|F_{p(v)}|}{2}$. Hence, the maximum number of light nodes on any path in F is at most $\lceil \log_2 |F| \rceil$. $\qquad\square$

For example, consider left and right branch trees. The heavy-light traversal causes at most one node at a time to wait for its sibling to be traversed. Thus, at most one additional array has to be stored in memory at any time.

With Theorem 3 we reduce the space complexity of Chen's algorithm to $O((l^2+n)\log n)$. For trees with $O(\sqrt{n})$ leaves the complexity becomes $O(n\log n)$. This is remarkable since all other TED algorithms require $O(n^2)$ space independently of the tree shape. So far, space complexities better than $O(n^2)$ were achieved only by approximations (for example, $O(n)$-space pq-gram distance by Augsten et al. [1]), algorithms computing an upper bound for TED (for example, $O(n\log n)$-space constrained tree edit distance by Wang et al. [17]), and algorithms computing the lower bound for TED (for example, $O(n)$-space string edit distance by Chan [4]). Trees with the number of leaves in $O(\sqrt{n})$ are characterised by long node chains, for example, tree representations of RNA secondary structures [5].

5   Experimental Evaluation

In this section we experimentally evaluate Chen's algorithm and compare it to the classical algorithm by Zhang and Shasha (ZS) [18] and the state-of-the-art algorithm AP-TED+ by Pawlik and Augsten [14]. All algorithms were implemented as single-thread applications in Java 1.7. and executed on a single core of a server machine with 8 cores Intel Xeon 2.40 GHz CPUs and 96GB of RAM. The runtime results are averages over three runs.

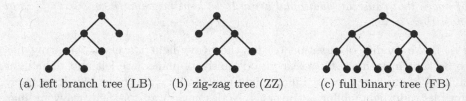

(a) left branch tree (LB)   (b) zig-zag tree (ZZ)   (c) full binary tree (FB)

Fig. 2. Shapes of the synthetic trees

Implementation. We implemented the original algorithm by Chen without the matrix multiplication extension (cf. Section 3). During the implementation process we discovered some minor bugs in Chen's algorithm that we fixed in our implementation. We further extended the implementation with our new traversal strategy to reduce the memory complexity (cf. Section 4). Our tests (not presented due to space limitations) show that the memory usage reduction is significant, for example, in the case of zig-zag trees we reduce the number of arrays concurrently stored in memory from linear to constant. That translates to a reduction of the memory footprint by one order of magnitude already for small trees with 200 nodes. The improvement ratio grows with the tree size.

Datasets. Similar to Pawlik and Augsten [13], we generated trees of five different shapes and varying sizes. Left branch (LB), zig-zag (ZZ), and full binary trees (FB) are shown in Fig. 2. In addition, we created thin and deep trees which favor Chen's algorithm. Thin and deep left branch trees (TDLB) are obtained from LB trees by inserting node chains (of equal length) to the left child of every node. Thin and deep zig-zag trees (TDZZ) are obtained from long node chains by attaching leaf nodes at random positions (alternating between left and right such that the resulting tree resembles a zig-zag tree). For thin and deep trees, we vary the ratio of leaf nodes from 5% to 20%. It is worth mentioning that LB/TDLB trees are the best-case input for ZS, while the performance of AP-TED$^+$ does not depend on the tree shape.

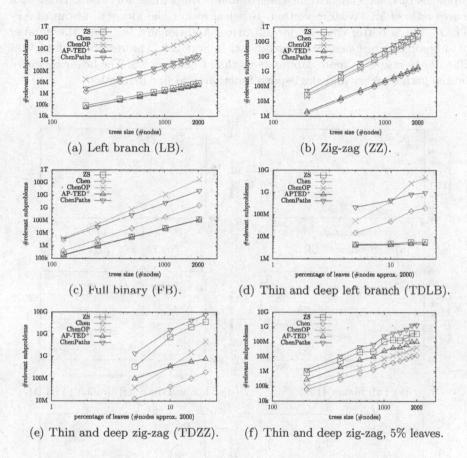

(a) Left branch (LB).

(b) Zig-zag (ZZ).

(c) Full binary (FB).

(d) Thin and deep left branch (TDLB).

(e) Thin and deep zig-zag (TDZZ).

(f) Thin and deep zig-zag, 5% leaves.

Fig. 3. Number of relevant subproblems for different tree shapes.

Number of relevant subproblems. The complexity of TED algorithms is proportional to the number of subproblems that an algorithm has to compute. Figure 3 shows the number of subproblems for different tree shapes. For the LB, FB, and TDLB shapes the leaders are AP-TED$^+$ and ZS, while Chen must compute many more subproblems. For the ZZ shape, the winners are Chen and AP-TED$^+$, ZS performs poorly. For TDZZ trees Chen outperforms its competitors. For TDZZ trees with the leaves ratio of 5% the difference is one order of

168   S. Schwarz et al.

magnitude (Fig. 3(f)). We vary the leaves ratio and observe that Chen results in
the smallest number of subproblems for all tested leave ratios between 5% and
20% (Fig. 3(e)). ZS and AP-TED$^+$ require only a constant number of operations
for each relevant subproblem, while Chen must evaluate the minimum over a
linear number of options (see Eqs. 3 and 5). We count the overall number of
elements in the minima and report the result as ChenOP in Fig. 3. Although
the number of constant time operations is much larger then the number of subproblems in Chen's algorithm, Chen remains the winner for TDZZ trees with
leaves ratio of 5%. With more than 10% leaf ratio Chen looses in favour of AP-
TED$^+$, but is better than ZS for all ratios. Additionally, we mark the number
of subproblems of ChenPaths introduced in Sect. 2.3. The results confirm that
ChenPaths results in more subproblems than Chen and ZS. The latter is caused
by the path strategy and single-path function used in ChenPaths.

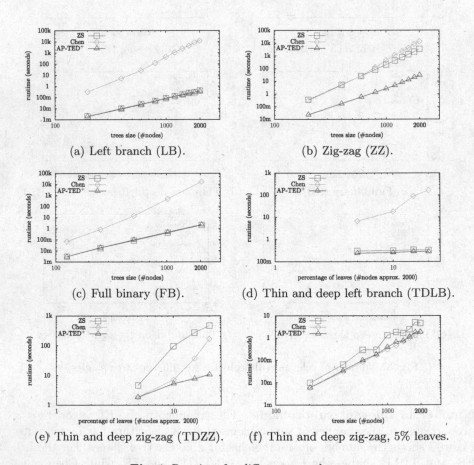

Fig. 4. Runtime for different tree shapes.

Runtime. We compare the runtime of the algorithms for different tree shapes (Fig. 4). The trend is consistent with the results for the number of subproblems. Chen wins only for TDZZ trees with 5% leaf ratio (Fig. 4(f)); the runtime difference to the runner-up AP-TED$^+$ is marginal. Chen's runtime quickly increases with the leaf ratio.

6 Conclusion

In this paper we analysed and experimentally evaluated the tree edit distance algorithm by Chen [6]. We revised the runtime and improved the space complexity of Chen's algorithm to $O(n \log(n))$ for trees with $O(\sqrt{n})$ leaves. Our experiments showed that Chen beats its competitors for thin and deep zig-zag trees with few leaves. Our analytic results suggest that the recursions of Chen and Zhang can be combined. For the future work, it is interesting to develop new dynamic programming algorithms that can leverage both recursive decompositions. This requires a cost formula for combined Chen and Zhang strategies, and an efficient bottom-up traversal for the dynamic programming implementation of the combined strategy.

Acknowledgments. This work was supported by Austrian Science Fund (FWF): P 29859-N31. We thank Willi Mann and Daniel Kocher for valuable discussions.

References

1. Augsten, N., Böhlen, M., Gamper, J.: The pq-gram distance between ordered labeled trees. ACM Trans. Database Syst. (TODS) **35**(1) (2010)
2. Bringmann, K., Gawrychowski, P., Mozes, S., Weimann, O.: Tree edit distance cannot be computed in strongly subcubic time (unless APSP can). CoRR, abs/1703.08940 (2017)
3. Bringmann, K., Grandoni, F., Saha, B., Williams, V.V.: Truly sub-cubic algorithms for language edit distance and RNA-folding via fast bounded-difference min-plus product. In: IEEE Annual Symposium on Foundations of Computer Science (FOCS), pp. 375–384 (2016)
4. Chan, T.Y.T.: Practical linear space algorithms for computing string-edit distances. In: Huang, D.-S., Li, K., Irwin, G.W. (eds.) ICIC 2006. LNCS, vol. 4115, pp. 504–513. Springer, Heidelberg (2006). doi:10.1007/11816102_54
5. Chen, S., Zhang, K.: An improved algorithm for tree edit distance incorporating structural linearity. In: International Conference on Computing and Combinatorics (COCOON) (2007)
6. Chen, W.: New algorithm for ordered tree-to-tree correction problem. J. Algorithms **40**(2), 135–158 (2001)
7. Demaine, E.D., Mozes, S., Rossman, B., Weimann, O.: An optimal decomposition algorithm for tree edit distance. ACM Trans. Algorithms **6**(1) (2009)
8. Dulucq, S., Touzet, H.: Decomposition algorithms for the tree edit distance problem. J. Discrete Algorithms **3**(2–4), 448–471 (2005)
9. Fong, K.C.K., Li, M., Liang, H., Yang, L., Yuan, H.: Average-case complexity of the min-sum matrix product problem. Theoret. Comput. Sci. **609**, 76–86 (2016)

10. Fredman, M.L.: New bounds on the complexity of the shortest path problem. SIAM J. Comput. **5**(1), 83–89 (1976)

11. Klein, P.N.: Computing the edit-distance between unrooted ordered trees. In: Bilardi, G., Italiano, G.F., Pietracaprina, A., Pucci, G. (eds.) ESA 1998. LNCS, vol. 1461, pp. 91–102. Springer, Heidelberg (1998). doi:10.1007/3-540-68530-8_8

12. Pawlik, M., Augsten, N.: RTED: a robust algorithm for the tree edit distance. Proc. VLDB Endow. (PVLDB) **5**(4), 334–345 (2011)

13. Pawlik, M., Augsten, N.: Efficient computation of the tree edit distance. ACM Trans. Database Syst. (TODS) **40**(1) (2015). Article No. 3

14. Pawlik, M., Augsten, N.: Tree edit distance: robust and memory-efficient. Inf. Syst. **56**, 157–173 (2016)

15. Sleator, D.D., Tarjan, R.E.: A data structure for dynamic trees. J. Comput. Syst. Sci. **26**(3), 362–391 (1983)

16. Tai, H.-C.: The tree-to-tree correction problem. J. ACM (JACM) **26**(3), 422–433 (1979)

17. Wang, L., Zhang, K.: Space efficient algorithms for ordered tree comparison. Algorithmica **51**(3), 283–297 (2008)

18. Zhang, K., Shasha, D.: Simple fast algorithms for the editing distance between trees and related problems. SIAM J. Comput. **18**(6), 1245–1262 (1989)

19. Zwick, U.: All pairs shortest paths using bridging sets and rectangular matrix multiplication. J. ACM (JACM) **49**(3), 289–317 (2002)

Outlier Detection

Good and Bad Neighborhood Approximations for Outlier Detection Ensembles

Evelyn Kirner[1], Erich Schubert[2($\boxtimes$)], and Arthur Zimek[3]

[1] Ludwig-Maximilians-Universität München,
Oettingenstr. 67, 80538 München, Germany
evelyn.kirner@campus.lmu.de
[2] Heidelberg University, INF 205, 69120 Heidelberg, Germany
schubert@informatik.uni-heidelberg.de
[3] University of Southern Denmark, Campusvej 55, 5230 Odense M, Denmark
zimek@imada.sdu.dk

Abstract. Outlier detection methods have used approximate neighborhoods in filter-refinement approaches. Outlier detection ensembles have used artificially obfuscated neighborhoods to achieve diverse ensemble members. Here we argue that outlier detection models could be based on approximate neighborhoods in the first place, thus gaining in both efficiency and effectiveness. It depends, however, on the type of approximation, as only some seem beneficial for the task of outlier detection, while no (large) benefit can be seen for others. In particular, we argue that space-filling curves are beneficial approximations, as they have a stronger tendency to underestimate the density in sparse regions than in dense regions. In comparison, LSH and NN-Descent do not have such a tendency and do not seem to be beneficial for the construction of outlier detection ensembles.

1 Introduction

Any algorithm will have different points of optimization. More often than not, it is not the algorithm that needs to be optimized, but the actual implementation. Implementation details can yield substantial performance differences, in particular when scripting languages such as R and Python or just-in-time optimization such as in Java and Scala are used [29]. An implementation detail often not even mentioned in passing in publications describing a novel outlier detection algorithm is the computation of neighborhoods. Typical outlier detection algorithms compute some property for characterizing outlying behavior based on the nearest neighbors of some object and compare that property for a given object with the corresponding properties of some context of neighboring objects [42]. Because of its complexity, a central bottleneck for all these algorithms is usually the computation of object neighborhoods.

We demonstrate this in a motivating experiment, using the ELKI framework [2] since it offers many algorithms as well as several index structures for acceleration. In Table 1 we give runtime benchmark results running the LOF [10]

© Springer International Publishing AG 2017
C. Beecks et al. (Eds.): SISAP 2017, LNCS 10609, pp. 173–187, 2017.
DOI: 10.1007/978-3-319-68474-1_12

algorithm (local outlier factor) on two larger data sets: the first data set contains all GPS coordinates from DBpedia [31], the second 27 dimensional color histograms for the Amsterdam Library of Object Images (ALOI, [17]). We expect the first to be more amiable to index acceleration using spatial indexes such as the k-d tree [9] and the R*-tree [8,19]. For each data set, we report the runtime broken down into (i) loading the data from text files into memory, (ii) bulk-loading the index, (iii) searching the kNN of each object, and (iv) computing the LOF scores. We repeat the experiments using a linear scan, using a k-d tree, and using a bulk-loaded R*-tree; we also give theoretical results on the complexity of each step.

Table 1. Runtime breakdown of LOF (with $k = 100$) in ELKI 0.6.0

Data set	DBpedia 475.000 instances, 2 dimensions					
Index	linear scan		k-d tree		R*-tree		Theoretical complexity
	(ms)	(%)	(ms)	(%)	(ms)	(%)	
Load Ascii data	990	0.04	1057	4.82	1035	5.99	$\mathcal{O}(n)$
Bulk-load index	0	0.00	829	3.78	768	4.44	$\mathcal{O}(n\log n)$
kNN search	2672128	99.74	15740	71.72	11379	65.85	$\mathcal{O}(n^2)$, maybe $n\log n$
LOF	5879	0.22	4319	19.68	4099	23.72	$\mathcal{O}(nk)$
Data set	ALOI 75.000 instances, 27 dimensions					
Index	linear scan		k-d tree		R*-tree		Theoretical complexity
	(ms)	(%)	(ms)	(%)	(ms)	(%)	
Load Ascii data	2238	0.96	2232	0.50	2231	1.27	$\mathcal{O}(n)$.
Bulk-load index	0	0.00	624	0.14	996	0.56	$\mathcal{O}(n\log n)$
kNN search	230030	98.84	446653	99.28	172791	97.99	$\mathcal{O}(n^2)$, maybe $n\log n$
LOF	468	0.20	372	0.08	321	0.18	$\mathcal{O}(nk)$

Both from a theoretical point of view as well as supported by the empirical results presented here, step (iii), computing the kNN of each object, is the main contributor to total runtime. However, it also becomes evident that the constant factors in the runtime analysis should probably not be as easily dismissed (see also the more extensive discussion by Kriegel et al. [29]). Bulk-loading the index is usually in $\mathcal{O}(\log n)$, while for kNN search with indexes an optimistic empirical estimate is $n\log n$, and the theoretical worst case supposedly is between $\mathcal{O}(n^{4/3})$ and $\mathcal{O}(n^2)$.[1] Effectively these values differ by two to three orders of magnitude, as constant factors with sorting are tiny. With a linear scan, finding the nearest

---

[1] Results from computational geometry indicate that the worst case of nearest neighbor search in more than 3 dimensions cannot be better than $\mathcal{O}(n^{4/3})$ [16]. Empirical results with such indexes are usually much better, and tree-based indexes are often attributed a $n\log n$ cost for searching.

neighbors is in $\Theta(n^2)$. Many implementations will also require $\Theta(n^2)$ memory because of computing a full distance matrix.

In particular on large data sets, finding the kNN is a fairly expensive operation, and traditional indexes such as the k-d tree and the R*-tree only work for low dimensionality (for our 27 dimensional example data set, the k-d tree has become twice as slow as the linear scan, and the R*-tree only yields small performance benefits, as opposed to 2 dimensions, where the speed-up was over 200 fold). Furthermore, neither the k-d tree nor the R*-tree are easy to parallelize in a cluster environment. Therefore, the use of approximate indexes is desirable to reduce runtime complexity.

While there are several attempts to optimize the neighborhood computation for outlier detection algorithms [3,7,14,23,27,36,37,43,46], these aim at computing the *exact* outlier score as fast as possible or at approximating the *exact* outlier score as closely as possible using approximate neighborhoods that are as close to the exact neighborhoods as possible. Note, however, that any outlier score is itself only an approximation of some imprecise statistical property and the "exact" outlier score is therefore an idealization that has probably no counterpart in reality.

Here, we argue that using approximate neighborhoods as such can be beneficial for outlier detection if the approximation has some bias that favors the isolation of outliers, especially in the context of ensemble techniques, that need some diversity among ensemble components anyway [48]. Using approximate neighborhoods as diverse components for outlier ensembles has not been discussed in the literature so far but it seems to be an obvious option. We show, however, that using the approximate neighborhoods can be beneficial or detrimental for the outlier detection ensemble, depending on the type of approximation. There are apparently good and bad kinds of neighborhood approximations for the task of outlier detection (and presumably also for clustering and for other data mining tasks). We take this point here based on preliminary results and suggest to investigate the bias of different neighborhood approximations methods further.

This paper is organized as follows: we review related work in Sect. 2, describe our approach in Sect. 3, and present our experimental results in Sect. 4. We conclude in Sect. 5.

2   Related Work

Existing outlier detection methods differ in the way they model and find the outliers and, thus, in the assumptions they, implicitly or explicitly, rely on. The fundamentals for modern, database-oriented outlier detection methods (i.e., methods that are motivated by the need of being scalable to large data sets, where the exact meaning of "large" has changed over the years) have been laid in the statistics literature. In general, statistical methods for outlier detection (also: outlier identification or rejection) are based on assumptions on the nature of the distributions of objects. The classical textbook of Barnett and Lewis [6] discusses numerous tests for different distributions. The tests are optimized for each distribution dependent on the specific parameters of the corresponding distribution, the number of expected outliers, and the space where to expect an outlier. Different statistical techniques have been discussed by Rousseeuw and Hubert [40].

A broader overview for modern data mining applications has been presented by Chandola et al. [12]. Here, we focus on techniques based on computing distances (and derived secondary characteristics) in Euclidean data spaces.

With the first database-oriented approach, Knorr and Ng [26] triggered the data mining community to develop many different methods, typically with a focus on scalability. A method in the same spirit [39] uses the distances to the k nearest neighbors (kNN) of each object to rank the objects. A partition-based algorithm is then used to efficiently mine top-n outliers. As a variant, the sum of distances to all points within the set of k nearest neighbors (called the "weight") has been used as an outlier degree [4].

Aside from this basic outlier model, they proposed an efficient approximation algorithm, HilOut, based on multiple Hilbert-curves. It is a strongly database oriented technique capable of an efficient on-disk operation. It processes the data set in multiple scans over the data, maintaining an outlier candidate list and thresholds. For every point, its outlier score is approximated with an upper and lower point. Objects whose upper bound becomes less than the global lower bound can be excluded from the candidates. If after a certain number of scans the candidate set has not yet reached the desired size, a final refinement step will compute the pairwise distances from the candidates to the full data set. Hilbert-curves serve a twofold purpose in this method. First, they are used to find good neighbor candidates by comparing each object with its closest neighbors along the Hilbert-curve only. Second, the Hilbert-curves are used to compute lower bounds for the outlierness, as at least for a small radius they can guarantee that there are no missed neighbors. For subsequent scans, the Hilbert-curves are varied by shifting the data set with a multiple of $\frac{1}{d+1}$ on each axis to both create new neighbor candidates and to increase the chance of having a good guarantee on the close neighbor completeness.

The so-called "density-based" approaches consider ratios between the local density around an object and the local density around its neighboring objects, starting with the seminal LOF [10] algorithm. Many variants adapted the original LOF idea in different aspects [42]. Despite those many variants, the original LOF method is still competitive and state of the art [11].

As for other approaches, also for several of the variants of LOF, approximate variants have been proposed. For example the LOCI method [38] came already in the original paper with an approximate version, aLOCI. For aLOCI, the data are preprocessed and organized in (multidimensional) quadtrees. These have the benefit of allowing a simple density estimation based on depth and occupancy numbers alone, i.e., when an object is contained in an area of volume V which contains n objects, the density is estimated to be $\frac{n}{V}$. Since this estimation can be quite inaccurate when an object is close to the fringe of V, aLOCI will generate multiple shifted copies of the data set, and always use the quadtree area where the object is located most closely to the center. Furthermore, as aLOCI considers multiple neighborhood sizes, the algorithm will check multiple such boxes, which may come from different trees. This makes the parallelization of aLOCI hard, while the required random accesses to the quadtree make this primarily an algorithm for data that fits into main memory. Shifting is done by moving the data set along a random vector in each dimension, cyclically wrapping the data within the domain (which may, in turn, cause some unexpected results).

Several approximate approaches use random projection techniques [1,33,45] based on the Johnson/Lindenstrauss lemma [24], especially in the context of high dimensional outlier detection [51]. Wang et al. [46] propose outlier detection based on Locality Sensitive Hashing (LSH) [13,18,22]. The key idea of this method is to use LSH to identify low-density regions, and refine the objects in these regions first, as they are more likely to be in the top-n global outliers. For local outlier detection methods there may be interesting outliers within a globally dense region, though. As a consequence, the pruning rules this method relies upon will not be applicable. Zhang et al. [47] combine LSH with isolation forests [32]. Projection-indexed nearest-neighbours (PINN) [14] shares the idea of using a random projection to reduce dimensionality. On the reduced dimensionality, an exact spatial index is then employed to find neighbor candidates that are refined to k nearest neighbors in the original data space.

Improving efficiency of outlier detection often has been implemented by focussing on the top-n outliers only and pruning objects before refinement that do not have a chance to be among the top-n outliers [3,7,23,27,36]. A broad and general analysis of efficiency techniques for outlier detection algorithms [37] identifies common principles or building blocks for efficient variants of the so-called "distance-based" models [4,26,39]. The most fundamental of these principles is "approximate nearest neighbor search" (ANNS). The use of this technique in the efficient variants studied by Orair et al. [37] is, however, different from the approach we are proposing here in a crucial point. Commonly, ANNS has been used as a filter step to discard objects from computing the *exact* outlier score. The exact kNN distance could only become smaller, not larger, in case some neighbor was missed by the approximation. Hence, if the upper bound of the kNN distance, coming along with the ANNS, is already too small to possibly qualify the considered point as a top-n outlier, the respective point will not be refined. For objects passing this filter step, the *exact neighborhood* is still required in order to compute the *exact outlier score*. All other efficiency techniques, as discussed by Orair et al. [37], are similarly based on this consideration and essentially differ in the exact pruning or ranking strategies. As opposed to using approximate nearest neighborhoods as a filter step, we advocate to *directly* use the resulting set of an approximate nearest neighbor search to compute outlier scores, without any refinement. Schubert et al. [43] based a single outlier model on combinations of several approximate neighborhoods, studying space-filling curves and random projections. Here, we compute the outlier score on each of the k *approximate* nearest neighbors directly, without any refinement, instead of on the exact neighborhood and combine them only afterwards. In addition, we compare different approximate neighborhood search methods: aside from space filling curves we also study LSH (see above) and NN-Descent [15]. The basic idea of NN-Descent is an iterative refinement of neighborhoods, checking the *approximate* neighbors of the *approximate* neighbors (using both forward and reverse neighborhoods). Starting from random neighborhoods, the iteration approximates surprisingly quickly and well the true neighborhoods.

Many more approaches for the computation of approximate neighborhoods could be tested and compared on their suitability for outlier detection (as well as for other data mining tasks). However, most of them focus on a near-perfect recall, and therefore may be unsuitable for our purposes. K-d-trees can be parameterized to give approximation guarantees even when not exploring all branches [5]. For example, randomized k-d-trees [44] build multiple k-d-trees (randomly choosing the split axis amongst the best candidates) and search them in parallel with an approximate search, while the priority search k-means tree [35] uses recursive clustering.

Isolation forests [32] can be seen as an approximate density estimation ensemble, which constructs multiple trees on different samples of the data, where the height of a leaf (which determines "isolation") is implicity used as a kind of density estimate. As it does not find neighbors, but directly estimates density, it cannot be used with methods such as LOF. ALOCI [38] uses a quadtree for a similar purpose. Nevertheless, the idea of building an ensemble of simple outlier detectors is a common idea with our approach, and our observations may yield further insight into this method, too.

Our work here is thus to be seen as a first step towards embracing imprecision of approximate nearest-neighbor search as a source of diversity for ensemble construction.

3   Outlier Detection Ensembles Based on Approximate Neighborhoods

The conclusion we draw from the discussion of related work is to emphasize that, for certain outlier detection models, it does not seem to be of the utmost importance to work on exact neighborhoods. Although the use of approximate neighborhoods for outlier detection was usually an intermediate step, before ultimately neighborhoods are refined to be exact or at least as good as possible, we maintain that approximate neighborhoods can be sufficient or even beneficial (if the approximations exhibit a bias that favors the isolation of outliers), to estimate and compare local densities, in particular if we combine outlier models learned on approximate data to an ensemble. The same reasoning relates to several existing ensemble methods for outlier detection [48], where a better overall judgment is yielded by diversified models. Models are diversified using approximations of different kinds: the results for outlier detection ensembles have been improved by computing neighbors in subsets of features [30], in subsets of the dataset [50], or even by adding *noise* components to the data points in order to yield diverse density-estimates [49]. All these variants can in some sense also be seen as using approximate neighborhoods directly for density estimates (in subspaces, on subsets, or on noisy data), and for some of these approximations it has been argued why the particular approximation technique could even prove beneficial for increasing the gap between outlier and inlier scores [50].

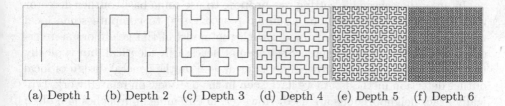

(a) Depth 1   (b) Depth 2   (c) Depth 3   (d) Depth 4   (e) Depth 5   (f) Depth 6

Fig. 1. Hilbert curve approximations at different recursion depth.

Among neighborhood approximation methods, space-filling curves have a particular property w.r.t. outliers that seems to act beneficial. A space-filling curve is recursively cutting the space as visualized in Fig. 1. Neighbors being close in the full space but being separated by such a cut will not be well preserved. In Fig. 2, we showcase why this is of minimal effect on density estimates within a cluster, while the density around outliers is more likely to be underestimated more strongly: losing some neighbor in a low-density area (as around outliers) will incur the identification of approximate neighbors that exhibit larger distances (and thus much smaller local density estimates for the outlier) as compared to losing some neighbor in some high-density area (such as a cluster), where the approximate neighbors will still be rather close. Space-filling curves do exhibit a bias that is actually helpful for outlier detection.

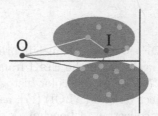

Fig. 2. Approximation error caused by a space filling curve (illustration): black lines indicate neighborhoods not preserved by the space filling curve. Shaded areas are discovered clusters, red lines are approximate 2NN distances, green lines are the real 2NN distances. By the loss of true neighbors, the density estimated based on approximate neighbors will have a stronger tendency to be underestimated for outliers than for cluster points, where the distances do not grow that much by missing some true neighbors.

Neither LSH nor NN-Descent have a similar bias favoring relative underestimation of density around outliers. By using reverse neighborhoods together with forward neighborhoods, NN-Descent is naturally adaptive to different local densities. Kabán [25] pointed out that random projection methods according to the Johnson/Lindenstrauss lemma preserve distances approximately and thus also preserve the distance concentration. Accordingly, LSH, being based on random-projections, tends to preserve distances without bias on higher or lower densities.

In an outlier ensemble setting, we propose to use some basic outlier detector that takes local neighborhoods as input (context set) to compute some local model and that compares this model with the model of the neighbors as reference set. Context set and reference set are not necessarily identical, but typically they are. See the discussion by Schubert et al. [42] on the general design of local outlier detection methods. As we have seen in the overview on related work, typically exact neighborhoods are used. However, in ensemble approaches [48] often special techniques are applied to diversify the models, e.g. by using neighborhood computations in subspaces [30], in subsets of the data [50], or after adding a noise component on the data [49].

Here we propose to not artificially diversify exactly computed neighborhoods but rather to stick to approximate neighborhoods in the first place, which comes obviously with a considerable computational benefit in terms of efficiency. We demonstrate that this approach can also come with a considerable benefit in terms of effectiveness, although this depends on the approximation method chosen. We conjecture that also different outlier detection methods used as ensemble components might react differently to the use of appoximations. In this study, however, we focus on the sketched approximation techniques (space-filling curves, LSH, and NN-Descent) in building outlier ensembles, using LOF [10] as basic outlier detection technique. Outlier scores computed on various approximations are then combined with standard procedures [48], using score normalization [28] and ranking of average scores.

4   Experiments

For experiments, we use LOF as well as the neighborhood approximation methods in the implementation available in the ELKI framework [41].

As data set, we use a 27 dimensional color histogram representation of the Amsterdam Library of Object Images (ALOI) [17], as used before in the outlier detection literature (cf. the collection of benchmark data by Campos et al. [11] and previous usage documented therein). We also take orientation on the results reported by Campos et al. [11] for parameter selection (neighborhood size for LOF), where values larger than 20 do not seem to be beneficial on this data set. We thus test $k = 1, \ldots, 20$.

As space filling curves we use the Z-order [34] and a window size equal to the number of requested neighbors as used by Schubert et al. [43]. We chose the simplest curve because it produces more diversity, and the Hilbert curve [20] is substantially more expensive to compute, although recently some progress has been made on sorting data without transforming it to Hilbert indices [21]. For LSH we use 3 projections based on p-stable distributions [13], 3 hash tables and a projection width of 0.1. For NN-Descent, we restrict the number of iterations to 2 in order to force some diversity in the results. All of these parameters are deliberately chosen to provide a fast, and not overly precise result. Too precise results will obviously be detrimental to building an ensemble afterwards, as ensembles rely on diversity in the ensemble members.

We measure the recall of the delivered neighborhoods (not counting the query point itself) as well as the performance of the outlier detection methods (LOF with exact neighborhoods and ensemble of LOF on approximate neighborhoods) in terms of the area under the ROC curve (ROC AUC).

In Fig. 3, we depict the recall of the three approximation methods as distribution over 25 runs, varying the size k of the requested neighborhood. For NN-Descent we see a strong tendency to achieve better recall for larger neighborhoods (due to the larger neighbors-of-a-neighbor candidate set). Z-order only shows a slight tendency in the same direction, LSH has the opposite tendency, however not very strongly (because of the increasing distance to the k nearest neighbor, these are less likely to be in the same hash bucket). More remarkable is the difference in the variance of achieved recall: Z-order always has a considerable variance, the variance in LSH seems also to depend on the neighborhood size, while NN-Descent has very stable recall over the different runs. If we allowed NN-Descent to perform more iterations, its recall would further improve, but the variance would become even smaller. Note that for the purpose of ensemble method, variance is related to diversity, and therefore desirable.

If we were to compare the approximation methods as such, we would easily notice that LSH achieves very high recall compared to the others, and therefore may be considered to be the best choice. However, as we want to use approximate neighborhoods as input for ensemble members, a low recall might already be sufficient to get good results [43] and the variance is of greater importance.

In Fig. 4 we depict the performance of the resulting ensembles, based on each of the approximation methods and each k. We plot the score distribution of the individual ensemble members using a boxplot, and the performance of the ensemble resulting from the combination. In order to visualize the relationship to recall of the true nearest neighbors, we use the mean recall of the ensemble on the x axis. For comparison, at recall 1, we also plot the results obtained with exact nearest neighbors (multiple points due to multiple choices of k).

For all of the methods, we can observe that the ensemble performs at least as good as 75% of the ensemble members, indicated by the upper quartile of the boxplot. As expected, we see that the combination of LOF based on LSH (high recall) and NN-Descent (low recall), both not exhibiting a beneficial bias for outlier detection, does not improve over the single LOF result based on exact neighborhoods,[2] while the combination of LOF based on space-filling curve approximations (intermediate recall, large variance, beneficial bias) improves also over the exact LOF and shows the best results overall, similar to the observation by Schubert et al. [43].

---

[2] But there may be a performance improvement by nevertheless using these methods.

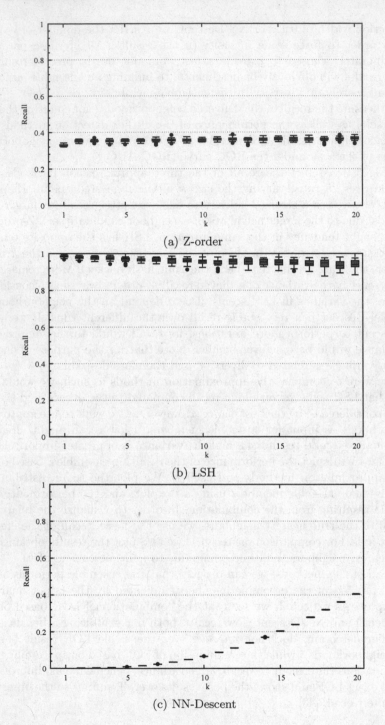

Fig. 3. Recall of the true k nearest neighbors for approximate neighborhood search (distribution over 25 runs), depending on the neighborhood size k. (The query point is not counted as hit in the result.)

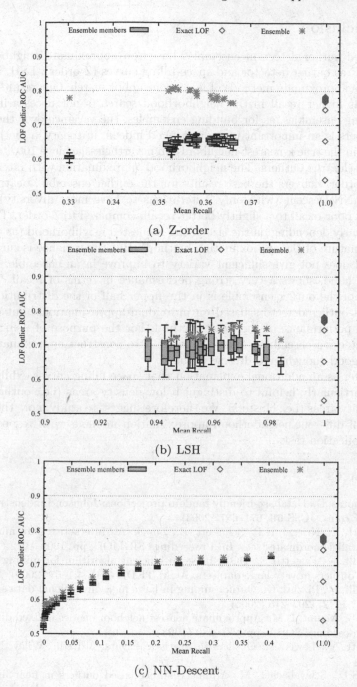

(a) Z-order

(b) LSH

(c) NN-Descent

Fig. 4. ROC AUC of ensemble members (LOF on approximations), ensemble, and exact LOF for different $k = 1 \ldots 20$ (not labeled). Boxplots indicate the ensemble members, the stars indicate the performance of the complete ensemble, diamonds indicate the performance of exact nearest neighbors for comparison.

5   Conclusion

We studied outlier detection ensembles based on approximate neighborhoods, using LOF as outlier detector and space-filling curves (Z-order), LSH, and NN-Descent as approximate methods of nearest neighbor search. Our results demonstrate that higher recall in the neighborhood search is not necessarily better for building ensembles, as for building ensembles, the variance over the ensemble members is an important ingredient. And indeed, in theory, a method with 0% recall in the true k-nearest-neighbors can nevertheless achieve 100% accuracy in finding the true outliers. The neighborhood approximation with intermediate recall, Z-order, delivers the best results for the outlier ensemble, beating exact methods. NN-Descent (with only 2 iterations to have more diversity) reaches from very poor recall to a slightly better recall, compared to Z-order. The recall here is clearly depending on the size of the requested neighborhood (as expected from the nature of the approximation method). But the variance is surprisingly small and does not give sufficient variety to improve in an ensemble. LSH, on the other hand, shows a very strong performance in terms of recall. The performance of the outlier ensemble is in the upper half of the distribution of the individual outlier detectors based on individual approximations, but does not reach the performance of the exact method. For the purpose of using this for outlier detection ensembles, a key challenge is to construct approximation that are both good enough, and diverse enough.

We offer as an additional explanation that space-filling curves exhibit a bias that is particularly helpful to distinguish low-density areas (i.e., outliers) from high-density areas (i.e., clusters). We therefore suggest to study more thoroughly the bias of different neighborhood approximation methods with respect to different application tasks.

References

1. Achlioptas, D.: Database-friendly random projections: Johnson-Lindenstrauss with binary coins. JCSS **66**, 671–687 (2003)
2. Achtert, E., Kriegel, H.P., Schubert, E., Zimek, A.: Interactive data mining with 3D-parallel-coordinate-trees. In: Proceedings SIGMOD, pp. 1009–1012 (2013)
3. Angiulli, F., Fassetti, F.: DOLPHIN: an efficient algorithm for mining distance-based outliers in very large datasets. ACM TKDD **3**(1), 4:1–57 (2009)
4. Angiulli, F., Pizzuti, C.: Outlier mining in large high-dimensional data sets. IEEE TKDE **17**(2), 203–215 (2005)
5. Arya, S., Mount, D.M.: Approximate nearest neighbor queries in fixed dimensions. In: Proceedings SODA, pp. 271–280 (1993)
6. Barnett, V., Lewis, T.: Outliers in Statistical Data, 3rd edn. Wiley, New York (1994)
7. Bay, S.D., Schwabacher, M.: Mining distance-based outliers in near linear time with randomization and a simple pruning rule. In: Proceedings KDD, pp. 29–38 (2003)
8. Beckmann, N., Kriegel, H.P., Schneider, R., Seeger, B.: The R*-Tree: an efficient and robust access method for points and rectangles. In: Proceedings SIGMOD, pp. 322–331 (1990)

9. Bentley, J.L.: Multidimensional binary search trees used for associative searching. Commun. ACM **18**(9), 509–517 (1975)

10. Breunig, M.M., Kriegel, H.P., Ng, R., Sander, J.: LOF: Identifying density-based local outliers. In: Proceedings SIGMOD. pp. 93–104 (2000)

11. Campos, G.O., Zimek, A., Sander, J., Campello, R.J.G.B., Micenková, B., Schubert, E., Assent, I., Houle, M.E.: On the evaluation of unsupervised outlier detection: Measures, datasets, and an empirical study. Data Min. Knowl. Disc. **30**, 891–927 (2016)

12. Chandola, V., Banerjee, A., Kumar, V.: Anomaly detection: a survey. ACM CSUR **41**(3), 1–58 (2009). Article 15

13. Datar, M., Immorlica, N., Indyk, P., Mirrokni, V.S.: Locality-sensitive hashing scheme based on p-stable distributions. In: Proceedings ACM SoCG, pp. 253–262 (2004)

14. de Vries, T., Chawla, S., Houle, M.E.: Density-preserving projections for large-scale local anomaly detection. KAIS **32**(1), 25–52 (2012)

15. Dong, W., Charikar, M., Li, K.: Efficient k-nearest neighbor graph construction for generic similarity measures. In: Proceedings WWW, pp. 577–586 (2011)

16. Erickson, J.: On the relative complexities of some geometric problems. In: Proceedings of the 7th Canadian Conference on Computational Geometry, Quebec City, Quebec, Canada, August 1995, pp. 85–90 (1995)

17. Geusebroek, J.M., Burghouts, G.J., Smeulders, A.W.M.: The amsterdam library of object images. Int. J. Comput. Vis. **61**(1), 103–112 (2005)

18. Gionis, A., Indyk, P., Motwani, R.: Similarity search in high dimensions via hashing. In: Proceedings VLDB, pp. 518–529 (1999)

19. Guttman, A.: R-trees: A dynamic index structure for spatial searching. In: Proceedings SIGMOD, pp. 47–57 (1984)

20. Hilbert, D.: Ueber die stetige Abbildung einer Linie auf ein Flächenstück. Math. Ann. **38**(3), 459–460 (1891)

21. Imamura, Y., Shinohara, T., Hirata, K., Kuboyama, T.: Fast Hilbert Sort Algorithm Without Using Hilbert Indices. In: Amsaleg, L., Houle, M.E., Schubert, E. (eds.) SISAP 2016. LNCS, vol. 9939, pp. 259–267. Springer, Cham (2016). doi:10.1007/978-3-319-46759-7_20

22. Indyk, P., Motwani, R.: Approximate nearest neighbors: towards removing the curse of dimensionality. In: Proceedings STOC, pp. 604–613 (1998)

23. Jin, W., Tung, A.K., Han, J.: Mining top-n local outliers in large databases. In: Proceedings KDD, pp. 293–298 (2001)

24. Johnson, W.B., Lindenstrauss, J.: Extensions of Lipschitz mappings into a Hilbert space. In: Conference in Modern Analysis and Probability, Contemporary Mathematics, vol. 26, pp. 189–206. American Mathematical Society (1984)

25. Kabán, A.: On the distance concentration awareness of certain data reduction techniques. Pattern Recogn. **44**(2), 265–277 (2011)

26. Knorr, E.M., Ng, R.T.: Algorithms for mining distance-based outliers in large datasets. In: Proceedings VLDB, pp. 392–403 (1998)

27. Kollios, G., Gunopulos, D., Koudas, N., Berchthold, S.: Efficient biased sampling for approximate clustering and outlier detection in large datasets. IEEE TKDE **15**(5), 1170–1187 (2003)

28. Kriegel, H.P., Kröger, P., Schubert, E., Zimek, A.: Interpreting and unifying outlier scores. In: Proceedings SDM, pp. 13–24 (2011)

29. Kriegel, H.P., Schubert, E., Zimek, A.: The (black) art of runtime evaluation: are we comparing algorithms or implementations? KAIS **52**(2), 341–378 (2017). doi:10.1007/s10115-016-1004-2

30. Lazarevic, A., Kumar, V.: Feature bagging for outlier detection. In: Proceedings KDD, pp. 157–166 (2005)

31. Lehmann, J., Isele, R., Jakob, M., Jentzsch, A., Kontokostas, D., Mendes, P.N., Hellmann, S., Morsey, M., van Kleef, P., Auer, S., Bizer, C.: DBpedia - a large-scale, multilingual knowledge base extracted from wikipedia. Semant. Web J. **6**(2), 167–195 (2015)

32. Liu, F.T., Ting, K.M., Zhou, Z.H.: Isolation-based anomaly detection. ACM TKDD **6**(1), 3:1–39 (2012)

33. Matoušek, J.: On variants of the Johnson-Lindenstrauss lemma. Random Struct. Algorithms **33**(2), 142–156 (2008)

34. Morton, G.M.: A computer oriented geodetic data base and a new technique in file sequencing. Technical report, International Business Machines Co (1966)

35. Muja, M., Lowe, D.G.: Scalable nearest neighbor algorithms for high dimensional data. IEEE TPAMI **36**(11), 2227–2240 (2014)

36. Nguyen, H.V., Gopalkrishnan, V.: Efficient pruning schemes for distance-based outlier detection. In: Proceedings ECML PKDD, pp. 160–175 (2009)

37. Orair, G.H., Teixeira, C., Wang, Y., Meira, W., Parthasarathy, S.: Distance-based outlier detection: consolidation and renewed bearing. PVLDB **3**(2), 1469–1480 (2010)

38. Papadimitriou, S., Kitagawa, H., Gibbons, P.B., Faloutsos, C.: LOCI: Fast outlier detection using the local correlation integral. In: Proceedings ICDE, pp. 315–326 (2003)

39. Ramaswamy, S., Rastogi, R., Shim, K.: Efficient algorithms for mining outliers from large data sets. In: Proceedings SIGMOD, pp. 427–438 (2000)

40. Rousseeuw, P.J., Hubert, M.: Robust statistics for outlier detection. WIREs DMKD **1**(1), 73–79 (2011)

41. Schubert, E., Koos, A., Emrich, T., Züfle, A., Schmid, K.A., Zimek, A.: A framework for clustering uncertain data. PVLDB **8**(12), 1976–1979 (2015)

42. Schubert, E., Zimek, A., Kriegel, H.P.: Local outlier detection reconsidered: a generalized view on locality with applications to spatial, video, and network outlier detection. Data Min. Knowl. Disc. **28**(1), 190–237 (2014)

43. Schubert, E., Zimek, A., Kriegel, H.-P.: Fast and Scalable Outlier Detection with Approximate Nearest Neighbor Ensembles. In: Renz, M., Shahabi, C., Zhou, X., Cheema, M.A. (eds.) DASFAA 2015. LNCS, vol. 9050, pp. 19–36. Springer, Cham (2015). doi:10.1007/978-3-319-18123-3_2

44. Silpa-Anan, C., Hartley, R.I.: Optimised kd-trees for fast image descriptor matching. In: Proceedings CVPR (2008)

45. Venkatasubramanian, S., Wang, Q.: The Johnson-Lindenstrauss transform: an empirical study. In: Proceedings ALENEX Workshop (SIAM), pp. 164–173 (2011)

46. Wang, Y., Parthasarathy, S., Tatikonda, S.: Locality sensitive outlier detection: a ranking driven approach. In: Proceedings ICDE, pp. 410–421 (2011)

47. Zhang, X., Dou, W., He, Q., Zhou, R., Leckie, C., Kotagiri, R., Salcic, Z.: LSHiForest: A generic framework for fast tree isolation based ensemble anomaly analysis. In: Proceedings ICDE (2017)

48. Zimek, A., Campello, R.J.G.B., Sander, J.: Ensembles for unsupervised outlier detection: challenges and research questions. SIGKDD Explor. **15**(1), 11–22 (2013)

49. Zimek, A., Campello, R., Sander, J.: Data perturbation for outlier detection ensembles. In: Proceedings SSDBM, pp. 13:1–12 (2014)
50. Zimek, A., Gaudet, M., Campello, R., Sander, J.: Subsampling for efficient and effective unsupervised outlier detection ensembles. In: Proceedings KDD, pp. 428–436 (2013)
51. Zimek, A., Schubert, E., Kriegel, H.P.: A survey on unsupervised outlier detection in high-dimensional numerical data. Stat. Anal. Data Min. 5(5), 363–387 (2012)

Intrinsic t-Stochastic Neighbor Embedding for Visualization and Outlier Detection

A Remedy Against the Curse of Dimensionality?

Erich Schubert[(✉)] and Michael Gertz

Heidelberg University, Heidelberg, Germany
{schubert,gertz}@informatik.uni-heidelberg.de

Abstract. Analyzing high-dimensional data poses many challenges due to the "curse of dimensionality". Not all high-dimensional data exhibit these characteristics because many data sets have correlations, which led to the notion of intrinsic dimensionality. Intrinsic dimensionality describes the local behavior of data on a low-dimensional manifold within the higher dimensional space.

We discuss this effect, and describe a surprisingly simple approach modification that allows us to reduce local intrinsic dimensionality of individual points. While this unlikely will be able to "cure" all problems associated with high dimensionality, we show the theoretical impact on idealized distributions and how to practically incorporate it into new, more robust, algorithms. To demonstrate the effect of this adjustment, we introduce the novel Intrinsic Stochastic Outlier Score (ISOS), and we propose modifications of the popular t-Stochastic Neighbor Embedding (t-SNE) visualization technique for intrinsic dimensionality, intrinsic t-Stochastic Neighbor Embedding (it-SNE).

1 Introduction

Analyzing high-dimensional data is a major challenge. Many of our intuitions from low-dimensional space such as distance and density no longer apply in high-dimensional data the same way they do in 2- or 3-dimensional space. For example, the center of a high-dimensional ball contains only very little mass, whereas the majority of the mass of a high-dimensional ball is in its shell. Grid-based approaches do not work well to partition high-dimensional data, because the number of grid cells grows exponentially with the dimensionality, so almost all cells will be empty. We are particularly interested in anomaly detection approaches for high-dimensional data, where many distance-based algorithms are known to suffer from the "curse of dimensionality" [43].

To understand the performance of algorithms, it is advisable to visualize the results, but visualization of high-dimensional data has similar problems because of the sheer number and correlations of attributes to visualize [1]. A promising recent visualization method is t-SNE [35], which embeds data in a way that preserves neighborhoods, but not distances and densities, as seen in Fig. 1, where

C. Beecks et al. (Eds.): SISAP 2017, LNCS 10609, pp. 188–203, 2017.
DOI: 10.1007/978-3-319-68474-1_13

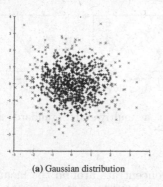
(a) Gaussian distribution

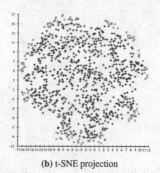
(b) t-SNE projection

Fig. 1. t-SNE projections do not preserve distances or density, but try to preserve neighbors (red x markers indicate points more than 2 standard deviations from the center) (Color figure online)

the density information of the Gaussian distribution is largely lost, but neighborhoods are to a large extend preserved.

In this article, we improve the concept of "stochastic neighbors" which forms the base for SNE [16], t-SNE [35], and the outlier detection method SOS [24]. We study the distance concentration effect and construct a way to avoid the loss of discrimination (although not a universal "cure" for the curse of dimensionality), which we integrate into stochastic neighbors, to construct the improved ISOS outlier detection and it-SNE projection technique for visualizing anomalies in high intrinsic dimensionality.

2 Related Work

2.1 The Curse of Dimensionality

The "curse of dimensionality" was initially coined in combinatorial optimization [4], but now refers to a whole set of phenomena associated with high dimensionality [17,20]. We focus here on the loss of "discrimination" of distances as described by [6]. Intuitively, this curse means that the distances to the closest neighbor and the farthest neighbor become relatively similar, up to the point where they become "indiscernible". This can be formalized as:

$$\lim_{\dim \to \infty} E\left[\frac{\max_{y \neq x} d(x,y) - \min_{y \neq x} d(x,y)}{\min_{y \neq x} d(x,y)}\right] \to 0. \tag{1}$$

This can be proven for idealized distributions, but the effect can be observed in real data, and affects the ability of many distance-based methods, e.g., in outlier detection [43].

Figure 2a visualizes the distribution of distances from the origin of a multivariate standard normal distribution, i.e. $X = (\sum_d Y_i^2)^{1/2}$ with $Y_i \sim \mathcal{N}(0; 1)$. The resulting distance distribution is a Chi distribution with d degrees of freedom. To visualize the concentration of relative distances, we normalize the x-axis

190   E. Schubert and M. Gertz

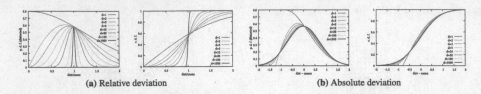

(a) Relative deviation　　　　　　(b) Absolute deviation

Fig. 2. Deviation from the expected value of a multivariate standard normal distribution.

by the mean distance. We can see the p.d.f. concentrate around the mean, and the c.d.f. change abruptly at the mean, as expected by Eq. 1. However, if we look at absolute deviations from the mean in Fig. 2b (by centering the distributions on the mean rather than scaling them), we can no longer see any distance concentration. In terms of deviation from the mean, the distributions appear very similar (for $d > 2$)—so there should be some leeway here against the curse of dimensionality. (But unfortunately, this transformation yields negative values, so it cannot be used as a distance normalization in most applications).

[43] have shown that the distance concentration effect itself is not the main problem, and outliers can still be easy to detect if this effect occurs. [5] have shown that we can discern well-separated clusters in high dimensionality, because we can still distinguish near from far neighbors. [20] show that by ignoring the absolute distance values, but instead counting the overlap of neighborhoods ("shared nearest neighbors"), we can still cluster high-dimensional data, reflecting the observation that the ranking of near points remains meaningful, even when the relative distances do not provide contrast.

There are many other aspects of the curse of dimensionality [17,43], such as hubness [37], which we will not focus on here (and hubness has also been observed in lower dimensional data [33]). Some issues with high dimensionality are very practical in nature: preprocessing, scaling, and weighting of features is often very important for data analysis, but becomes difficult to do with a large number of features of very different nature, such as when combining continuous, discrete, ordinal and categoricial features. Such problems are also beyond the scope of this article.

2.2   Intrinsic Dimensionality

Data on a line in a 10-dimensional space will essentially behave as if it were in a 1-dimensional space. This led to the notion of intrinsic dimensionality, and this intuition has been formally captured for example by the expansion dimension [26].

Text data is often represented in a very high-dimensional data space, where every different word in the corpus corresponds to a dimension. Based on a naive interpretation of the curse of dimensionality, one would assume such a representation to be problematic; yet text search works very well. In the vector space model, text data usually is sparse, i.e., most attributes are zero. Adding additional attributes that are constant, or copies of existing attributes, usually do not increase the difficulty of a data set much.

Therefore, it is good to distinguish between the representation dimensionality—the number of attributes used for encoding the data—and the effective dimensionality for data analysis. [17] establishes the theoretical connection between dimensionality, discriminability, density, and distance distributions; as well as the connection to extreme value theory [18]. Intrinsic dimensionality is often estimated using tail estimators, in particular using the Hill [15] estimator, or a weighted average thereof [21]. More recent approaches involve the expansion dimension [26] and the Generalized Expansion Dimension (GED) [19]. [2] survey and compare several estimation techniques for intrinsic dimensionality. Implementations of several estimators for intrinsic dimensionality can be found in the ELKI data mining toolkit [39]. The Hill maximum-likelihood estimator uses the sorted distances of x to its k-nearest neighbors $y_1 \ldots y_k$ for estimation [2]:

$$\widehat{\text{ID}}_{\text{Hill}}(x) := -\left(\frac{1}{k-1} \sum_{i=1}^{k-1} \log \frac{d(x,y_i)}{d(x,y_k)} \right)^{-1} \tag{2}$$

2.3 Outlier Detection

Distance-based outlier detection is focused around the idea that outliers are in less dense areas of the data space [28], and that distances can be used to quantify density. Since then, many outlier detection methods have been proposed. We focus our comparison on methods that use the full-dimensional k-nearest neighbors, although many other methods exist [43]. [38] use the distance to the k-nearest neighbor, which can be seen as a "curried" version of the original DB-outlier approach by [28]. [3] use the average distance to all k-nearest neighbors instead. LOF [7] introduced the idea of comparing the density of a point to the densities of its neighbors. LoOP [29] attempts to estimate a local outlier probability, while INFLO [25] also takes reverse nearest neighbor relationships into account, while KDEOS [40] uses kernel density estimation instead of the simpler estimate of aforementioned methods. ODIN [14] simply counts how often a point occurs in the nearest-neighbors of others, while SOS [24] (c.f. Sect. 3.3) uses the probability of a point not occurring in stochastic neighborhoods as outlier score. Many more variations of these ideas exist [8,43], and a fair evaluation of such methods is extremely difficult, due to the sensitivity of the methods to data sets, preprocessing, and parameterization [8]. There exist many methods that focus on identifying outliers in feature subspaces [10,27,30,36] or with respect to correlations in the data [32].

2.4 Stochastic Neighbor Embedding

Stochastic neighbor embedding (SNE) [16] and t-distributed stochastic neighbor embedding (t-SNE) [35] are visualization techniques designed for visualizing high-dimensional data in a low-dimensional space (typically 2 or 3 dimensions). These methods originate from computer vision and deep learning research where they are used to visualize large image collections. In contrast to techniques such as principal component analysis (PCA) and multidimensional scaling (MDS), which try to maximize the spread of dissimilar objects, SNE focuses on placing similar objects close to each other, i.e., it preserves locality rather than large distances. But while these methods were developed (and used with great success) on data sets with a high representational dimensionality, [35] noted that the "relatively local nature of t-SNE makes it sensitive to the curse of the intrinsic dimensionality of the data" and that "t-SNE might be less successful if it is applied on datasets with a very high intrinsic dimensionality" [35].

The key idea of these methods is to model the high-dimensional input data with an affinity probability distribution, and use gradient descent to optimize the low-dimensional projection to exhibit similar affinities. By using an affinity which has more weight on nearby points rather than Euclidean distance, one obtains a non-linear projection that preserves local neighborhoods, while away points are mostly independent of each other. In SNE, Gaussian kernels are used in the projected space, whereas t-SNE uses a Student-t distribution. This distribution is well suited for the optimization procedure because it is computationally inexpensive, heavier-tailed, and has a well-formed gradient. The heavier tail of t-SNE is beneficial for visualization, because it increases the tendency of the projection to separate unrelated points in the projected space. But as seen in Fig. 1, t-SNE does not preserve distances or densities well, so we should rather not use the projected coordinates for clustering or outlier detection.

In the input domain, (t-)SNE uses a Gaussian kernel for the input distribution. Given a point i, the conditional probability density $p_{j|i}$ of any neighbor point j is computed as

$$p_{j|i} = \frac{\exp(-\|x_i - x_j\|^2/2\sigma_i^2)}{\sum_{k \neq i} \exp(-\|x_i - x_k\|^2/2\sigma_i^2)} \qquad (3)$$

where $\|x_i - x_j\|$ is the Euclidean distance, and the kernel bandwidth σ_i is optimized for every point to have the desired perplexity h (an input parameter roughly corresponding to the number of neighbors to preserve). The symmetric affinity probability p_{ij} is then obtained as the average of the conditional probabilities $p_{ij} = \frac{1}{2}(p_{i|j} + p_{j|i})$ and is subsequently normalized such that the total sum is $\sum_{i \neq l} p_{ij} = 1$.

SNE uses a Gaussian distribution (similar to Eq. 3, but with constant σ) in the projected space, and t-SNE improved this by using the Student-t distribution instead:

$$q_{ij} = \frac{(1+\|y_i - y_j\|^2)^{-1}}{\sum_{k \neq l}(1+\|y_k - y_l\|^2)^{-1}} \qquad (4)$$

The denominator normalizes the sum to a total of $\sum_{i \neq j} q_{ij} = 1$. The mismatch between the two distributions P and Q (given by p_{ij} and q_{ij}) can now be measured using the Kullback-Leibler divergence [16]:

$$\mathrm{KL}(P \parallel Q) := \sum_i \sum_j p_{ij} \log \frac{p_{ij}}{q_{ij}} \qquad (5)$$

By also using a small constant minimum p_{ij} and q_{ij}, we can prevent unrelated points from being placed too close. To minimize the mismatch of the two distributions, we can use the vector gradient $\frac{\delta C}{\delta y_i}$ (for Student-t/t-SNE, as derived by [35]):

$$\frac{\delta C}{\delta y_i} := 4 \sum_j (p_{ij} - q_{ij}) \, q_{ij} \, Z \, (y_i - y_j) \tag{6}$$

where $Z = \sum_{k \neq l} (1 + \|y_k - y_l\|^2)^{-1}$ (c.f. [34]).

Starting with an initial random solution $Y_0 = \{y_i\}$, the solution is then iteratively optimized using gradient descent with learning rate η and momentum α as used by [35]:

$$Y_{t+1} \leftarrow Y_t - \eta \frac{\delta C}{\delta Y} + \alpha \, (Y_t - Y_{t-1}) \tag{7}$$

The resulting projection y is usually good for visualization, because it preserves neighborhood rather well, but also does not place objects too close to each other. The t-distributed variant t-SNE is often subjectively nicer, because the heavier tail of the student-t distribution leads to a more even tendency to separate points, and thus to more evenly fill the available space. The resulting projections in general tend to be circular.

3   Intrinsic Stochastic Neighbors

3.1   Distance Power Transform for the Curse of Intrinsic Dimensionality

The Stochastic Neighbor Embedding approaches are susceptible to the curse, because they use the distance to the neighbors to compute neighbor weights, which will become too similar to be useful at discriminating neighbors. When we lose distance discrimination, it follows from Eq. 3 that for a data set of size N: $\lim_{d \to \infty} p_{j|i} \to 1/(N-1)$, $\lim_{d \to \infty} p_{ij} \to 1/(N-1)^2$, and that therefore SNE does no longer work well.

Recent advances in understanding intrinsic dimensionality [17] connect intrinsic dimensionality to modeling the near-neighbor tail of the distance distribution with extreme value theory [18]. An interesting property of intrinsic dimensionality is that it changes with certain transformations [18, Table 1], such as the power transform. Let X be a random variable as in [18], and $g(x) := c \cdot x^m$ with c and m constants. Let F_X be the cumulative distribution of X, $Y = g(X)$ and F_Y the resulting cumulative distribution. Then the intrinsic dimensionality changes by $\mathrm{ID}_{F_X}(x) = m \cdot \mathrm{ID}_{F_Y}(c \cdot x^m)$ [18, Table 1]. By choosing $m = \mathrm{ID}_{F_X}(x)/t$ for any $t > 0$, we therefore obtain:

$$\mathrm{ID}_{F_Y}(c \cdot x^m) = \mathrm{ID}_{F_X}(x)/m = t \tag{8}$$

where we can choose $c > 0$ as desired, e.g., for numerical reasons. This variable X serves a theoretical model for the distance distribution on the "short tail" (the nearest neighbors), and ID_{F_X} is the intrinsic dimensionality. This observation means that we can transform our distance distribution of *any* desired dimensionality t.

In Fig. 3, we revisit the theoretical model of a multivariate normal distribution that we used in Sect. 2.1, but this time we transform the x-axis with a power

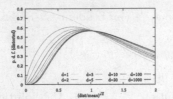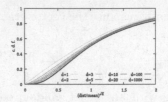

Fig. 3. Deviation from a multivariate standard normal distribution, after power transform.

transform using $m = \sqrt{d}$ and c such that the mean is 1. The power transform yields a transformation that retains the 0 (which the deviation from the mean in Fig. 2b did not), but which allows the numerical discrimination of distances. One may have assumed that $m = d$ would be the best choice in this scenario of a d-dimensional hyperball. This holds true in the limit at the center of the ball, but the decreasing density of the Gaussian yields a smaller expansion rate and therefore a decreasing intrinsic dimensionality as we move outward [19]. Beware that this is a very much idealized model, and that in practical applications, we will simply estimate m from the data.

To improve stochastic neighbor approaches, we propose the following remedy to the distance concentration effect: Based on the k-nearest neighbors of the point of interest x, first estimate the local intrinsic dimensionality $ID(x)$. Then use $d'(x, y_i) := c \cdot d(x, y_i)^m$ (c.f. Eq. 8) with $m = ID(x)/2$ to transform them into squared distances, and $c = 1/\max_y d(x,y)^m$ such that the farthest neighbor always has distance 1. The distances $d(x, y)$ to the neighbors are transformed using

$$d'^2(x, y) = d(x,y)^m / \max_z d(x,z)^m = \left(\frac{d(x,y)}{\max_z d(x,z)}\right)^m \qquad (9)$$

We then use this locally modified distance instead of the squared Euclidean distance to compute $p_{i|j}$ using Eq. 3 (to simplify, we also substitute $\beta_i := -1/2\sigma_i^2$):

$$p'_{j|i} = \frac{\exp(\beta_i d_i'^2(x_i,x_j))}{\sum_{k\neq i} \exp(\beta_i d_i'^2(x_i-x_k))} \qquad (10)$$

We can then continue to optimize β_i by binary search to obtain the desired perplexity h as done for regular SNE and t-SNE.

$$\log_2 \text{Perplexity} = -\sum_{j\neq i} p_{j|i} \log_2 p_{j|i} \qquad (11)$$

On data that was not normalized, regular t-SNE may fail to find a suitable β_i with binary search.[1] This happens when the binary search begins with $\beta_i = -1$ (or $\sigma_i = 1$), but $\exp(\beta_i d_i(x_i, x_j)) = 0$ for all j if the initial distances are too large.

---

[1] The author of t-SNE writes: "Presumably, your data contains some very large numbers, causing the binary search for the correct perplexity to fail. [...] Just divide your data or distances by a big number, and try again." https://lvdmaaten.github.io/tsne/#faq.

With our choice of c this will not happen anymore (as the maximum d' is 1), but the ELKI [39] implementation that we use initializes search with the heuristic estimate $\hat{\beta}_i = -\frac{1}{2}h/\operatorname{mean}_j d(x_i, x_j)^2$ (motivated by $\hat{\sigma}^2 \sim \operatorname{mean}_j d(x_i, x_j)^2$) which usually converges with fewer iterations. Since we only rely on the nearest neighbors, our new approach is compatible with the fast Barnes-Hut approximation [34].

3.2   Consensus Affinity Combination

SNE and t-SNE produce a symmetric affinity by averaging the two asymmetric affinities: $p_{ij} = \frac{1}{2}(p_{i|j} + p_{j|i})$. While this has the desirable property of retaining the total sum, it also tends to pull outliers too close to their neighbors. From a probabilistic point of view, we can interpret this as point x_i and x_j being connected if *either* of them chooses to link. Instead, we may desire them to link only if there is "consensus", by using

$$p'_{ij} := \sqrt{p'_{i|j} \cdot p'_{j|i}}. \tag{12}$$

The resulting affinity matrix will be more sparse, and therefore it is desirable to use a larger perplexity and neighborhood size than for t-SNE. But since the estimation of intrinsic dimensionality suggests to use at least 100 neighbors, whereas t-SNE is often used with a perplexity of about 40, this is not an additional restriction.

Next, the resulting affinities are normalized to have a total sum of 1 (as in regular t-SNE), to balance attractive and repulsive forces during the t-SNE optimization process. We then simply replace p_{ij} in the gradient (Eq. 6) with the new p'_{ij} (Eq. 12).

3.3   Intrinsic Stochastic Outlier Selection

The new outlier detection method Intrinsic Stochastic Outlier Selection (ISOS) is—as the name indicates—a modification of the earlier but rather unknown SOS method published in a technical report [23], a PhD thesis [22], and in a maritime application [24]. The key idea of this approach is that every data point "nominates" its neighbors, and can be seen as a smooth version of ODIN [14].

The original proposal of SOS involved generating random graphs based on an affinity distribution in order to identify frequently unlinked objects as outliers. But the expensive graph sampling process can be avoided, and the probability of a node being disconnected can be computed in closed-form using the simple equation [23]:

$$SOS(x_i) := \prod_{j \neq i} 1 - p_{i|j} \tag{13}$$

The original algorithm, similar to SNE and t-SNE, has quadratic runtime complexity, making it expensive to apply to large data. But because of the exponential function, affinities will quickly drop to a negligible value. Van der Maaten [34]

Algorithm 1. Pseudocode for ISOS

Input: *DB*: Database
Input: *k*: Number of neighbors to use
Data: *logscore*: Outlier scores, initially 1
1 Build a neighbor search index on database *DB* (if not present)
2 **foreach** *point x_i* in *database DB* **do**
3 $\quad$ $kNN(x_i) \leftarrow$ Find k-nearest neighbors (with distances)
4 $\quad$ $ID(x_i) \leftarrow$ Estimate intrinsic dimensionality of $kNN(x_i)$
5 $\quad$ $d'(x_i) \leftarrow$ Adjust squared distances (Eq. 9)
6 $\quad$ Choose β_i such that perplexity $\approx k/3$
7 $\quad$ $p_{j|i} \leftarrow$ Compute normalized affinities (Eq. 10)
8 $\quad$ **foreach** *neighbor x_j* **in** $kNN(x)$ **do**
9 $\quad\quad$ | $\quad$ $logscore(x_j) \leftarrow logscore(x_j) + \log(1 - p_{j|i})$
10 **return** $1 / \left(1 + e^{-x \cdot \log h} \cdot (1 - \varphi)/\varphi\right)$ for each score in *logscore*

---

uses the $k = \lceil 3h \rceil$ nearest neighbors to approximate the $p_{i|j}$. We incorporate this idea into SOS for two reasons: (i) to improve scalability, and (ii) to make it more comparable to k-nearest neighbor based outlier detection algorithms. Instead of the perplexity parameter h, this variant—which we denote as KNNSOS—has the neighborhood size parameter k common to k-nearest neighbor approaches, and uses a derived perplexity of $h = k/3$. Our ISOS method in turn is an extension of this KNNSOS approach, which uses the k-nearest neighbors first to estimate the local intrinsic dimensionality of each point, then uses Eq. 13 with our adjusted affinity $p'_{i|j}$. For $p_{i|j} \ll 1$, Eq. 13 does not give a high numerical precision. We therefore suggest to compute the scores in logspace,

$$\log SOS(x_i) := \sum_{j \neq i} \log(1 - p_{i|j}) \qquad (14)$$

and use the `log1p(-p_i|j)` function if available for increased numerical precision. While SOS yields an outlier probability (which makes the score more interpretable by users [31]), it is not as well-behaved as indicated by its authors [23], because the expected value even for a clear inlier is not 0, since we normalized the $p_{i|j}$ to sum up to 1. Intuitively, every point is supposed to distribute its weight to approximately h (the perplexity) neighbors. Assuming a clique of $h + 1$ objects, each at the same distance such that $p_{i|j} = 1/h$, every point will have the probability

$$E[SOS] := \prod_1^h (1 - 1/h) = \left(\tfrac{h-1}{h}\right)^h \approx_{h \to \infty} 1/e \qquad (15)$$

Alternatively, we can assume the null model that every point is equidistant, and thus every neighbor is chosen with $p_{i|j} = 1/N$, which yields the same limit. Note that $\log 1/e = -1$, if we perform the same computations in logscale. Therefore, we further propose to normalize to the resulting outlier probabilities, by comparing them to the expected value. The likelihood ratio $SOS(x_i)/E[SOS]$ in logspace yields simply the addition of 1 to the log scores. After this transformation, the average score will be about 0.5, but central values will be too frequent.

This is caused by the aggregation over effectively $\approx h$ values, and we can reduce this by multiplication with $\log h$ (in log space). Last but not least, we need to add a prior to reflect that anomalies are rare and not half of the data, but rather the majority of points should have a very low score. We use a desired outlier rate of $\varphi = 1\%$, which yields a prior odds ratio of $(1 - \varphi)/\varphi$ [32, 40].

To convert this back to a probability, we can use the logistic function:

$$l = -(\log SOS'(x_i) + 1) \cdot \log h \qquad (16)$$
$$ISOS(x_i) = 1/\left(1 + \exp(l) \cdot (1 - \varphi)/\varphi\right) \qquad (17)$$

Figure 4 shows (i) the original score before the adjustments on the MNIST test data set, (ii) after adjusting for the expected value (and logistic transformation), (iii) after also taking the perplexity into account, and (iv) with the prior assumption of outliers being 1% rare. The last histogram is the least "informative", but naturally we must expect the majority of outlier scores to be close to zero, so in fact only the exponential-like curve in the final histogram indicates a score that can satisfy the intuition of an "outlier probability". We show the top 50 outliers in Fig. 8b.

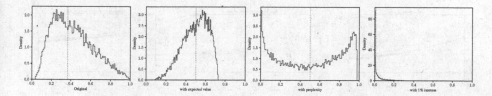

Fig. 4. Histogram of scores on MNIST data. Dotted lines indicate the expected average value.

Algorithm 1 gives the pseudocode for ISOS. Rather than directly computing the score for every point, we initialize all scores with 1 $(= -\log 1/e)$, then iterate over each point x_i and adjust the scores of each neighbor x_j by adding $\log(1 - p_{j|i})$. This reduces the memory requirements from $O(n^2)$ to $O(n)$, and makes the algorithm trivial to distribute except for the nearest neighbor search. For distributed and parallel processing, approximative nearest neighbor search is preferable, and has shown to be surprisingly effective for outlier detection, because errors may be larger for outliers than for inliers [42]. Note that in line 8 we can stop when $p_{j|i}$ is zero, as further away points will no longer change the scores of neighbors. For KNNSOS, do not estimate intrinsic dimensionality in line 4, and use the unmodified distances in line 5 of Algorithm 1.

This is a second-order local outlier detection method (c.f. [41]), where the kNN are used to estimate affinity, and the score depends on the reverse kNN. But because of the efficient message-based algorithm above, we do not need to compute the reverse nearest neighbors (which would require complex indexes for acceleration [9, 11]).

4   Experiments

We implemented our approach in Java as part of the ELKI [39] data mining framework, extending the existing Barnes-Hut approximation [34] t-SNE variant, and using the aggregated Hill estimator [21] for intrinsic dimensionality as default.

4.1   ISOS Outlier Detection

As common when performing a thorough evaluation of outlier detection, the results here remain unconclusive when performed on a large scale of methods and parameters [8,13]. For any method, we can find parameters and data where it performs best, or worst. KNNSOS and ISOS are, not surprisingly, no exception to this rule. Results claiming superior performance on a task as unspecific as anomaly detection are unfortunately often based on too narrow experiments, and unfair parameterization. We will contribute this method to the ELKI data mining toolkit, and submit the entire results for integration into the benchmark repository of [8].

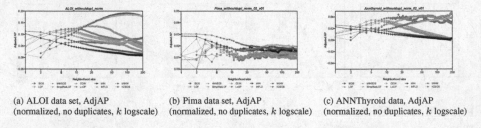

(a) ALOI data set, AdjAP   (b) Pima data set, AdjAP   (c) ANNThyroid data, AdjAP
(normalized, no duplicates, k logscale)   (normalized, no duplicates, k logscale)   (normalized, no duplicates, k logscale)

Fig. 5. Performance of ISOS and related algorithms on selected data sets.

Figure 5a shows anecdotal evidence of the capabilities of ISOS on the popular ALOI data set (color histograms from images of small objects [12], prepared as in [31]) with respect to adjusted average precision. We show the results for the normalized variant with duplicates removed [8], but the results on the other variants and with other evaluation measures are similar. On this data set, ISOS outperforms all other methods by a considerable margin (except for KNNSOS, which it only outperforms a little bit). Furthermore, the proposed method is fairly stable with respect to the choice of k, as long as the values are not chosen too small (for a reliable estimation of intrinsic dimensionality $k \geq 100$ is suggested). This makes it rather easy to choose the parameters. On other data sets such as Pima (Fig. 5b), the simple kNN distance methods work better—although none of the methods really worked well at all. This data set is also likely too small for methods based on intrinsic dimensionality. On ANNThyroid data, KDEOS, LoOP and ODIN compete for the lead, but both KNNSOS and ISOS work reasonably well, too. But again, the results are so low, that the data set must be considered questionable for distance based outlier detection. In Fig. 6, we visualize the data sets with PCA, MDS, t-SNE and it-SNE. In none of these projections, the labeled outlier correspond well to the human intuition of outlierness, and we cannot expect any unsupervised algorithm to perform well. For ANNThyroid, we can see artifacts caused by binary attributes in this data set in each projection. In conclusion of the outlier experiments—and in line with prior research [8,13,31]—there is no clear winner, and ensemble approaches that combine kNN outlier, LOF, but also ISOS, remain the most promising research direction.

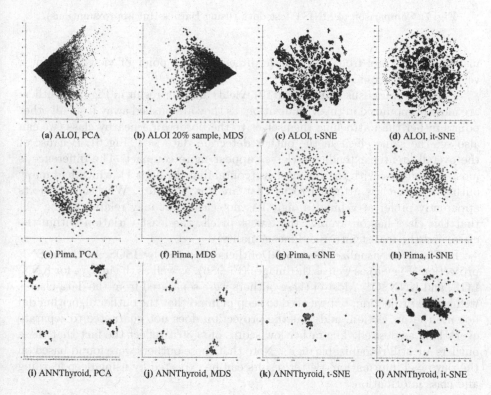

| (a) ALOI, PCA | (b) ALOI 20% sample, MDS | (c) ALOI, t-SNE | (d) ALOI, it-SNE |

| (e) Pima, PCA | (f) Pima, MDS | (g) Pima, t-SNE | (h) Pima, it-SNE |

| (i) ANNThyroid, PCA | (j) ANNThyroid, MDS | (k) ANNThyroid, t-SNE | (l) ANNThyroid, it-SNE |

Fig. 6. Projections of outlier detection data sets. Red x indicate the labeled outliers. (Color figure online)

4.2 it-SNE Visualization

In Fig. 7 we apply t-SNE on the popular MNIST data set, using the smaller "test" data set only. Colors indicate different digits. All runs used the same random seed for comparability. The difference between regular t-SNE (Fig. 7a) and t-SNE with the distances adjusted according to intrinsic dimensionality (Eq. 9, Fig. 7b) is not very big (classes are slightly more compact in the new projection). This can easily be explained with this data set having nominally 784 dimensions (28×28 pixel), but the intrinsic dimensionality is on average

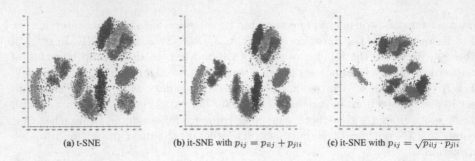

(a) t-SNE   (b) it-SNE with $p_{ij} = p_{i|j} + p_{j|i}$ (c) it-SNE with $p_{ij} = \sqrt{p_{i|j} \cdot p_{j|i}}$

Fig. 7. Comparison of MNIST test data (using Barnes-Hut approximations).

just 6.1. Therefore, from an intrinsic dimensionality point of view, it is not a very high-dimensional data set.

Using the consensus affinity (Eq. 12), yields a better result in Fig. 7c. Outliers are more pronounced in this visualization, as they are pushed away from all other points rather than attaching themselves to the border of a nearby class (we can also see the same effect in the outlier detection data sets, Fig. 6). Because of the overall greater extend, the classes appear more compact. The difference is most pronounced with the yellow class (containing the digit 1), which had many outlier foreign-class attached to it, that are now separate. Why these objects apparently prefer attaching to digit 1 is not clear, but may related to the fact that this class has on average the fewest pixels, the least variation within the class, and the lowest intrinsic dimensionality.

In Fig. 8 we visualize the top 50 outliers detected by ISOS, in the it-SNE projection (Fig. 8a) as well as the images (Fig. 8b), as well as the images for KNN, LOF, and KNNSOS. Most of these outliers were separated from the data classes well by the projection, but we need to keep in mind that the outlier algorithm did not use the projection, and that the projection does not guarantee to separate everything as desired. The rather low scores of $\approx 30\%$ reflect the fact that these outliers are still recognizable digits. Note that these outliers were found based on the raw pixel information. Better results can be expected by using deep learning and class information.

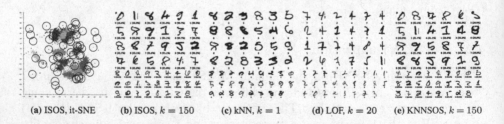

(a) ISOS, it-SNE   (b) ISOS, $k = 150$ (c) kNN, $k = 1$ (d) LOF, $k = 20$ (e) KNNSOS, $k = 150$

Fig. 8. Top 50 outliers in MNIST test data

5   Conclusions

This paper contributes important insights into the distance-based aspects of the curse of dimensionality, contributes a much improved outlier detection method, and modifies the popular t-SNE method for intrinsic dimensionality and use in anomaly detection.

- We have shown that the distance concentration effect of the "curse of dimensionality" sometimes can be avoided with a simple power transform.
- The proposed adjustment for intrinsic dimensionality provides more discriminative affinities when using stochastic neighbor approaches on high-dimensional data.
- The "consensus" affinity separates outliers from nearby clusters better, and thus provides substantially better visualization when used for outlier detection, as regular t-SNE would attach outliers to nearby clusters.
- The SOS outlier detection method was accelerated using the k-nearest neighbors (KNNSOS), a correction for intrinsic dimensionality was added (ISOS), and the resulting outlier scores are normalized such that they can be interpreted as a probability how likely an object belongs to a rare "outlier" class.

The use of the power transform is a promising direction to avoid the distance concentration effect in the later stages of data mining, but it is an open research question how a similar improvement could be achieved to improve for example nearest neighbor search. Thus, it is not a universal "cure" to the curse of dimensionality, yet.

References

1. Achtert, E., Kriegel, H., Schubert, E., Zimek, A.: Interactive data mining with 3D-parallel-coordinate-trees. In: ACM SIGMOD (2013)
2. Amsaleg, L., Chelly, O., Furon, T., Girard, S., Houle, M.E., Kawarabayashi, K., Nett, M.: Estimating local intrinsic dimensionality. In: ACM SIGKDD (2015)
3. Angiulli, F., Pizzuti, C.: Fast outlier detection in high dimensional spaces. In: Elomaa, T., Mannila, H., Toivonen, H. (eds.) PKDD 2002. LNCS, vol. 2431, pp. 15–27. Springer, Heidelberg (2002). doi:10.1007/3-540-45681-3_2
4. Bellman, R.: Adaptive Control Processes. A Guided Tour. Princeton University Press, Princeton (1961)
5. Bennett, K.P., Fayyad, U.M., Geiger, D.: Density-based indexing for approximate nearest-neighbor queries. In: ACM SIGKDD (1999)
6. Beyer, K., Goldstein, J., Ramakrishnan, R., Shaft, U.: When is "nearest neighbor" meaningful? In: Beeri, C., Buneman, P. (eds.) ICDT 1999. LNCS, vol. 1540, pp. 217–235. Springer, Heidelberg (1999). doi:10.1007/3-540-49257-7_15
7. Breunig, M.M., Kriegel, H., Ng, R.T., Sander, J.: LOF: identifying density-based local outliers. In: ACM SIGMOD (2000)
8. Campos, G.O., Zimek, A., Sander, J., Campello, R., Micenková, B., Schubert, E., Assent, I., Houle, M.E.: On the evaluation of unsupervised outlier detection: measures, datasets, and an empirical study. Data Min. Knowl. Discov. 30(4), 891–927 (2016)

9. Casanova, G., Englmeier, E., Houle, M.E., Kröger, P., Nett, M., Schubert, E., Zimek, A.: Dimensional testing for reverse k-nearest neighbor search. VLDB Endowment, vol. 10 (2017, to appear)

10. Dang, X.H., Assent, I., Ng, R.T., Zimek, A., Schubert, E.: Discriminative features for identifying and interpreting outliers. In: IEEE International Conference on Data Engineering, ICDE (2014)

11. Emrich, T., Kriegel, H., Kröger, P., Niedermayer, J., Renz, M., Züfle, A.: On reverse-k-nearest-neighbor joins. GeoInformatica 19(2), 299–330 (2015)

12. Geusebroek, J., Burghouts, G.J., Smeulders, A.W.M.: The amsterdam library of object images. Int. J. Comput. Vis. 61(1), 103–112 (2005)

13. Goldstein, M., Uchida, S.: A comparative evaluation of unsupervised anomaly detection algorithms for multivariate data. PLOS ONE 11(4), e0152173 (2016)

14. Hautamäki, V., Kärkkäinen, I., Fränti, P.: Outlier detection using k-nearest neighbour graph. In: International Conference on Pattern Recognition, ICPR (2004)

15. Hill, B.M.: A simple general approach to inference about the tail of a distribution. Ann. Stat. 3(5), 1163–1174 (1975)

16. Hinton, G.E., Roweis, S.T.: Stochastic neighbor embedding. In: Advances in Neural Information Processing Systems, NIPS, vol. 15 (2002)

17. Houle, M.E.: Dimensionality, discriminability, density and distance distributions. In: ICDM Workshops (2013)

18. Houle, M.E.: Inlierness, outlierness, hubness and discriminability: an extreme-value-theoretic foundation. Technical report NII-2015-002E, National Institute of Informatics, Tokyo, Japan (2015)

19. Houle, M.E., Kashima, H., Nett, M.: Generalized expansion dimension. In: ICDM Workshops (2012)

20. Houle, M.E., Kriegel, H.-P., Kröger, P., Schubert, E., Zimek, A.: Can shared-neighbor distances defeat the curse of dimensionality? In: Gertz, M., Ludäscher, B. (eds.) SSDBM 2010. LNCS, vol. 6187, pp. 482–500. Springer, Heidelberg (2010). doi:10.1007/978-3-642-13818-8_34

21. Huisman, R., Koedijk, K.G., Kool, C.J.M., Palm, F.: Tail-index estimates in small samples. Bus. Econ. Stat. 19(2), 208–216 (2001)

22. Janssens, J.H.M.: Outlier selection and one-class classification. Ph.D. thesis, Tilburg University (2013)

23. Janssens, J.H.M., Huszár, F., Postma, E.O., van den Herik, H.J.: Stochastic outlier selection. Technical report TiCC TR 2012–001, Tilburg Center for Cognition and Communication (2012)

24. Janssens, J., Postma, E., van den Herik, J.: Density-based anomaly detection in the maritime domain. In: van de Laar, P., Tretmans, J., Borth, M. (eds.) Situation Awareness with Systems of Systems, pp. 119–131. Springer, New York (2013). doi:10.1007/978-1-4614-6230-9_8

25. Jin, W., Tung, A.K.H., Han, J., Wang, W.: Ranking outliers using symmetric neighborhood relationship. In: Ng, W.-K., Kitsuregawa, M., Li, J., Chang, K. (eds.) PAKDD 2006. LNCS, vol. 3918, pp. 577–593. Springer, Heidelberg (2006). doi:10.1007/11731139_68

26. Karger, D.R., Ruhl, M.: Finding nearest neighbors in growth-restricted metrics. In: ACM Symposium on Theory of Computing, STOC (2002)

27. Keller, F., Müller, E., Böhm, K.: HiCS: high contrast subspaces for density-based outlier ranking. In: IEEE International Conference on Data Engineering (2012)

28. Knorr, E.M., Ng, R.T.: Algorithms for mining distance-based outliers in large datasets. In: Very Large Data Bases, VLDB (1998)

29. Kriegel, H., Kröger, P., Schubert, E., Zimek, A.: Loop: local outlier probabilities. In: ACM Conference on Information and Knowledge Management (2009)
30. Kriegel, H.-P., Kröger, P., Schubert, E., Zimek, A.: Outlier detection in axis-parallel subspaces of high dimensional data. In: Theeramunkong, T., Kijsirikul, B., Cercone, N., Ho, T.-B. (eds.) PAKDD 2009. LNCS, vol. 5476, pp. 831–838. Springer, Heidelberg (2009). doi:10.1007/978-3-642-01307-2_86
31. Kriegel, H., Kröger, P., Schubert, E., Zimek, A.: Interpreting and unifying outlier scores. In: SIAM Data Mining, SDM (2011)
32. Kriegel, H., Kröger, P., Schubert, E., Zimek, A.: Outlier detection in arbitrarily oriented subspaces. In: IEEE International Conference on Data Mining (2012)
33. Low, T., Borgelt, C., Stober, S., Nürnberger, A.: The hubness phenomenon: fact or artifact? In: Borgelt, C., Gil, M., Sousa, J., Verleysen, M. (eds.) Towards Advanced Data Analysis by Combining Soft Computing and Statistics. STUDFUZZ, vol. 285, pp. 267–278. Springer, Heidelberg (2013). doi:10.1007/978-3-642-30278-7_21
34. van der Maaten, L.: Accelerating t-SNE using tree-based algorithms. J. Mach. Learn. Res. **15**(1), 3221–3245 (2014)
35. van der Maaten, L., Hinton, G.: Visualizing data using t-SNE. J. Mach. Learn. Res. **9**(11), 2579–2605 (2008)
36. Nguyen, H.V., Gopalkrishnan, V., Assent, I.: An unbiased distance-based outlier detection approach for high-dimensional data. In: Yu, J.X., Kim, M.H., Unland, R. (eds.) DASFAA 2011. LNCS, vol. 6587, pp. 138–152. Springer, Heidelberg (2011). doi:10.1007/978-3-642-20149-3_12
37. Radovanovic, M., Nanopoulos, A., Ivanovic, M.: Hubs in space: popular nearest neighbors in high-dimensional data. J. Mach. Learn. Res. **11**, 2487–2531 (2010)
38. Ramaswamy, S., Rastogi, R., Shim, K.: Efficient algorithms for mining outliers from large data sets. In: ACM SIGMOD (2000)
39. Schubert, E., Koos, A., Emrich, T., Züfle, A., Schmid, K.A., Zimek, A.: A framework for clustering uncertain data. VLDB Endowment **8**(12), 1976–1979 (2015). https://elki-project.github.io/
40. Schubert, E., Zimek, A., Kriegel, H.: Generalized outlier detection with flexible kernel density estimates. In: SIAM Data Mining (2014)
41. Schubert, E., Zimek, A., Kriegel, H.: Local outlier detection reconsidered: a generalized view on locality with applications to spatial, video, and network outlier detection. Data Min. Knowl. Discov. **28**(1), 190–237 (2014)
42. Schubert, E., Zimek, A., Kriegel, H.-P.: Fast and scalable outlier detection with approximate nearest neighbor ensembles. In: Renz, M., Shahabi, C., Zhou, X., Cheema, M.A. (eds.) DASFAA 2015. LNCS, vol. 9050, pp. 19–36. Springer, Cham (2015). doi:10.1007/978-3-319-18123-3_2
43. Zimek, A., Schubert, E., Kriegel, H.: A survey on unsupervised outlier detection in high-dimensional numerical data. Stat. Anal. Data Min. **5**(5), 363–387 (2012)

Indexing and Applications

Scalable Similarity Search for Molecular Descriptors

Yasuo Tabei[1]([⊠]) and Simon J. Puglisi[2]

[1] RIKEN Center for Advanced Intelligence Project, Tokyo, Japan
`yasuo.tabei@riken.jp`
[2] Department of Computer Science, Helsinki Institute for Information Technology,
University of Helsinki, Helsinki, Finland
`puglisi@cs.helsinki.fi`

Abstract. Similarity search over chemical compound databases is a fundamental task in the discovery and design of novel drug-like molecules. Such databases often encode molecules as non-negative integer vectors, called *molecular descriptors*, which represent rich information on various molecular properties. While there exist efficient indexing structures for searching databases of *binary* vectors, solutions for more general integer vectors are in their infancy. In this paper we present a time- and space-efficient index for the problem that we call the *succinct intervals-splitting tree algorithm for molecular descriptors (SITAd)*. Our approach extends efficient methods for binary-vector databases, and uses ideas from succinct data structures. Our experiments, on a large database of over 40 million compounds, show SITAd significantly outperforms alternative approaches in practice.

1 Introduction

Molecules that are chemically similar tend to have a similar molecular function. The first step in predicting the function of a new molecule is, therefore, to conduct a similarity search for the molecule in huge databases of molecules with known properties and functions. Current molecular databases store vast numbers of chemical compounds. For example, the PubChem database in the National Center for Biotechnology Information (NCBI) files more than 40 million molecules. Because the size of the whole chemical space [6] is said to be approximately 10^{60}, molecular databases are growing and are expected to grow substantially in the future. There is therefore a strong need to develop scalable methods for rapidly searching for molecules that are similar to a previously unseen target molecule.

A *molecular fingerprint*, defined as a binary vector, is a standard representation of molecules in chemoinformatics [18]. In practice the fingerprint representation of molecules is in widespread use [2,3] because it conveniently encodes

---

This work was supported the JST PRESTO program and the Academy of Finland through grants 294143 and 2845984.

C. Beecks et al. (Eds.): SISAP 2017, LNCS 10609, pp. 207–219, 2017.
DOI: 10.1007/978-3-319-68474-1_14

the presence or absence of molecular substructures and functions. Jaccard similarity, also called Tanimoto similarity, is the *de facto* standard measure [11] to evaluate similarities between compounds represented as fingerprints in chemoinformatics and pharmacology. To date, a considerable number of similarity search methods for molecular fingerprints using Jaccard similarity have been proposed [9,12,16,17]. Among them, the *succinct intervals-splitting tree* (SITA) [16] is the fastest method that is also capable of dealing with large databases. Despite the current popularity of the molecular fingerprint representation in cheminformatics, because it is only a binary feature vector, it has a severely limited ability to distinguish between molecules, and so similarity search is often ineffective [10].

A *molecular descriptor*, defined as a non-negative integer vector, is a powerful representation of molecules and enables storing richer information on various properties of molecules than a fingerprint does. Representative descriptors are LINGO [19] and KCF-S [7]. Recent studies have shown descriptor representations of molecules to be significantly better than fingerprint representations for predicting and interpreting molecular functions [8] and interactions [15]. Although similarity search using descriptor representations of molecules is expected to become common in the near future, no efficient method for a similarity search with descriptors has been proposed so far. Kristensen et al. [10] presented a fast similarity search method for molecular descriptors using an inverted index. The inverted index however consumes a large amount of memory when applied to large molecular databases. Of course one can compress the inverted index to reduce memory usage, but then the overhead of decompression at query time results in slower performance. An important open challenge is thus to develop similarity search methods for molecular descriptors that are simultaneously fast and have a small memory footprint.

We present a novel method called SITAd by modifying the idea behind SITA. SITAd efficiently performs similarity search of molecular descriptors using generalized Jaccard similarity. By splitting a database into clusters of descriptors using upperbound information of generalized Jaccard similarity and then building binary trees that recursively split descriptors on each cluster, SITAd can effectively prune out useless portions of the search space. While providing search times as fast as inverted index-based approaches, SITAd requires substantially less memory by exploiting tools from succinct data structures, in particular rank dictionaries [5] and wavelet trees [4]. SITAd efficiently solves range maximum queries (RMQ) many times in similarity searches by using fast RMQ data structures [1] that are necessary for fast and space-efficient similarity searches. By synthesizing these techniques, SITAd's time complexity is *output-sensitive*. That is, the greater the desired similarity with the query molecule is, the faster SITAd returns answers.

To evaluate SITAd, we performed retrieval experiments over a huge database of more than 40 million chemical compounds from the PubChem database. Our results demonstrate SITAd to be significantly faster and more space efficient than state-of-the-art methods.

2   Similarity Search Problem for Molecular Descriptors

We now formulate a similarity search problem for molecular descriptors. A molecular descriptor is a fixed-dimension vector, each dimension of which is a nonnegative integer. It is conceptually equivalent to the set that consists of pairs $(d : f)$ of index d and weight f such that the d-th dimension of the descriptor is a non-zero value f. Let D be a dimension with respect to the vector representation of descriptors. For clarity, notations x_i and q denote D dimension vector representation of molecular descriptors, while W_i and Q correspond to their set representation. $|W_i|$ denotes the cardinality of W_i, i.e., the number of elements in W_i. The Jaccard similarity for two vectors x and x' is defined as $J(x, x') = \frac{x \cdot x'}{||x||_2^2 + ||x'||_2^2 - x \cdot x'}$ where $||x||_2$ is the L_2 norm. For notational convenience, we let $J(W, W')$ represent $J(x, x')$ of x and x' that correspond respectively to sets W and W'. Given a query compound Q, the similarity search task is to retrieve from the database of N compounds all the identifiers i of descriptors W_i whose Jaccard similarity between W_i and Q is no less than ϵ, i.e., the set $I_N = \{i \in \{1, 2, ..., N\}; J(W_i, Q) \geq \epsilon\}$.

3   Method

Our method splits a database into blocks of descriptors with the same squared norm and searches descriptors similar to a query in a limited number of blocks satisfying a similarity constraint. Our similarity constraint depends on Jaccard similarity threshold ϵ. The larger ϵ is, the smaller the number of selected blocks is. A standard method is to compute the Jaccard similarity between the query and each descriptor in the selected blocks, and then check whether or not the similarity is larger than ϵ. However, such pairwise computation of Jaccard similarity is prohibitively time consuming. Our method builds an intervals-splitting tree for each block of descriptors and searches descriptors similar to a query by pruning useless portions of the search space in the tree.

3.1   Database Partitioning

We relax the solution set I_N for fast search using the following theorem.

Theorem 1. *If $J(x, q) \geq \epsilon$, then $\epsilon ||q||_2^2 \leq ||x||_2^2 \leq ||q||_2^2 / \epsilon$.*

Proof. $J(x, q) \geq \epsilon$ is equivalent to $|x \cdot q| \geq \frac{\epsilon}{1+\epsilon}(||x||_2^2 + ||q||_2^2)$. By the Cauchy-Schwarz inequality $||x||_2 ||q||_2 \geq |x \cdot q|$, we obtain $||x||_2 ||q||_2 \geq \frac{\epsilon}{1+\epsilon}(||x||_2^2 + ||q||_2^2)$. When $||x||_2 \geq ||q||_2$, we get $||x||_2^2 \geq \epsilon ||q||_2^2$. Otherwise, we get $||q||_2^2 / \epsilon \geq ||x||_2^2$. Putting these results together, the theorem is obtained.

The theorem indicates that $I_1 = \{i \in \{1, 2, ..., N\}; \epsilon ||q||_2^2 \leq ||x_i||_2^2 \leq ||q||_2^2 / \epsilon\}$ must contain all elements in I_N, i.e., $I_N \subseteq I_1$. This means a descriptor identifier (ID) that is not in I_1 is never a member of I_N. Such useless descriptors can be efficiently excluded by partitioning the database into blocks, each of which contains descriptor IDs with the same squared norm. More specifically, let block $B^c = \{i \in \{1, 2, ..., N\}; ||x_i||_2^2 = c\}$ be the block containing all the descriptors in the database with squared norm c. Searching descriptors for a query needs to examine no element in B^c if either $c < \epsilon||q||_2^2$ or $c > ||q||_2^2/\epsilon$ holds.

3.2   Intervals-Splitting Tree for Efficient Similarity Search

Once blocks B^c satisfying $\epsilon||q||_2^2 \leq c \leq ||q||_2^2/\epsilon$ are selected, SITAd is able to bypass one-on-one computations of Jaccard similarity between each descriptor in B^c and a query q.

A binary tree T^c called an intervals-splitting tree is built on each B^c beforehand. When a query q is given, T^c is traversed with a pruning scheme to efficiently select all the descriptor IDs with squared norm c whose Jaccard similarity to query q is no less than ϵ. Each node in T^c represents a set of descriptor IDs by using an interval of B^c. Let $B^c[i]$ be the i-th descriptor ID in B^c and I_v^c be the interval of node v. Node v with interval $I_v^c = [s, e]$ contains descriptor IDs $B^c[s], B^c[s + 1], \cdots, B^c[e]$. The interval of a leaf is of the form $[s, s]$, indicating that the leaf has only one ID. The interval of the root is $[1, |B^c|]$.

Let $left(v)$ and $right(v)$ be the left and right children of node v with interval $I_v^c = [s, e]$, respectively. When these children are generated, $I_v^c = [s, e]$ is partitioned into disjoint segments $I_{left(v)}^c = [s, \lfloor(s + e)/2\rfloor]$ and $I_{right(v)}^c = [\lfloor(s + e)/2\rfloor + 1, e]$. The procedure of splitting the interval is recursively applied from the root to the leaves (see the middle and right of Fig. 1 illustrating intervals and sets of descriptors at the root and its children).

Each node v is identified by a bit string (e.g., $v = 010$) indicating the path from the root to v; "0" and "1" denote the selection of left and right children, respectively. At each leaf v, the index of B^c is calculated by $int(v) + 1$, where $int(\cdot)$ converts a bit string to its corresponding integer (see the middle of Fig. 1).

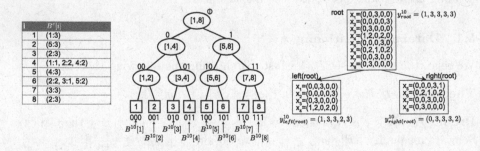

Fig. 1. Descriptors in block B^{10} (left), intervals-splitting tree T^c (middle) and T^c's first two levels (right). The root interval $[1, 8]$ is split into $[1, 4]$ and $[5, 8]$ for the left and right children. Each node v has a summary descriptor y_v for the descriptors in its interval.

3.3   Pruning the Search Space Using Summary Descriptors

Given query q, SITAd recursively traverses T^c from the root in a depth-first manner. If SITAd reaches a leaf and its descriptor is similar to q, the ID of that descriptor is included as one solution. To avoid traversing the whole T^c, we present a scheme to prune subtrees of nodes if all the descriptors for the nodes are deemed not to be sufficiently similar to query q.

The pruning is performed on node v by using D dimension descriptor y_v, which summarizes the information on descriptors in I_v, and is used for computing the upperbound of the Jaccard similarity between query q and $X_{B^c[i]}$ for any $i \in I_v$. The d-th dimension $y_v[d]$ of y_v is defined as the maximum value among $x_{B^c[i]}[d]$ for any $i \in I_v$, i.e., $y_v[d] = \max_{i \in I_v} x_{B^c[i]}[d]$. Thus $y_v = (\max_{i \in I_v} x_{B^c[i]}[1], \max_{i \in I_v} x_{B^c[i]}[2], ..., \max_{i \in I_v} x_{B^c[i]}[D])$. When T^c is built, y_v is computed. (see the right of Fig. 1, which represents y_v in the first two-level nodes of T^{10}).

Assume that SITAd checks descriptors in B^c and traverses T^c in the depth-first manner. $||x_{B^c[i]}||^2 = c$ holds in any descriptor in B^c. The following equivalent constraint is derived from Jaccard similarity:

$$J(x_{B^c[i]}, q) = \frac{x_{B^c[i]} \cdot q}{||x_{B^c[i]}||_2^2 + ||q||_2^2 - x_{B^c[i]} \cdot q} \geq \epsilon$$

$$\iff \quad x_{B^c[i]} \cdot q \geq \frac{\epsilon}{1+\epsilon}(||x_{B^c[i]}||_2^2 + ||q||_2^2) = \frac{\epsilon}{1+\epsilon}(c + ||q||_2^2)$$

Since $y_v^c \cdot q \geq x_{B^c[i]} \cdot q$ holds for any $i \in I_v^c$, SITAd examines the constraint at each node v in T^c and checks whether or not the following condition,

$$\sum_{(d:f) \in Q} y_v^c[d] f \geq \frac{\epsilon}{1+\epsilon}(c + ||q||^2), \tag{1}$$

holds at each node v. If the inequality does not hold at node v, SITAd safely prunes the subtrees rooted at v, because there are no descriptors similar to q in leaves under v. As we shall see, this greatly improves search efficiency in practice. Algorithm 1 shows the pseudo-code of SITAd.

3.4   Search Time and Memory

SITAd efficiently traverses T^c by pruning its useless subtrees. Let τ be the numbers of traversed nodes. The search time for query Q is $O(\tau|Q|)$. In particular, SITAd is efficient for larger ϵ, because more nodes in T^c are pruned.

A crucial drawback of SITAd is that T^c requires $O(D \log M |B^c| \log (|B^c|))$ space for each c, the dimension D of descriptors and the maximum value M among all weight values in descriptors. Since D is large in practice, SITAd consumes a large amount of memory. The next two subsections describe approaches to reduce the memory usage while retaining query-time efficiency.

Algorithm 1. Algorithm for finding similar descriptors to query q.

1: **function** SEARCH(q)
2:     **for** c satisfying $\epsilon||q||_2^2 \leq c \leq ||q||_2^2/\epsilon$ **do**
3:         $k \leftarrow \frac{\epsilon}{1+\epsilon}(c + ||q||_2^2)$, $I_{root}^c \leftarrow [1, |B^c|]$, $v \leftarrow \phi$
4:         Recursion(v, I_{root}^c, q, c)
5:     **end for**
6: **end function**
7: **function** RECURSION(v, I_v^c, q, c)
8:     **if** $\sum_{(d:f)\in Q} y_v^c[d]f < k$ **then**   ▷ Q : set representation of q
9:         **return**
10:    **end if**
11:    **if** $|v| = \lceil \log |B_c| \rceil$ **then**   ▷ Leaf Node
12:        Output index $B^c[int(v) + 1]$
13:    **end if**
14:    Recursive($v +' 0', [s, \lfloor (s+e)/2 \rfloor], q, c$)   ▷ To left child
15:    Recursive($v +' 1', [\lfloor (s+e)/2 \rfloor + 1, e], q, c$)   ▷ To right child
16: **end function**

3.5   Space Reduction Using Inverted Index

To reduce the large amount of space needed to store summary descriptors, we use an inverted index that enables computing an upperbound on descriptor similarity. The inverted index itself does not always reduce the memory requirement. However, SITAd compactly maintains the information in a *rank dictionary*, significantly decreasing memory usage.

We use two kinds of inverted indexes for separately storing index and weight pairs in descriptors. One is an associative array that maps each index d to the set of all descriptor IDs that contain pairs $(d : f)$ of index d and any weight $f(\neq 0)$ at each node v. Let $Z_{vd}^c = \{i \in I_v^c; (d : f) \in W_{B^c[i]}$ for any $f(\neq 0)\}$ for index d, (i.e., all IDs of a descriptor containing d with any weight f in any pair $(d : f)$ within I_v^c. The inverted index for storing indexes at node v in T^c is a one-dimensional array that concatenates all Z_{vd}^c in ascending order of d and is defined as $A_v^c = Z_{v1}^c \cup Z_{v2}^c \cup \cdots \cup Z_{vD}^c$. Figure 2 shows Z_{rootd}^{10} and the first two levels of the inverted indexes A_{root}^{10}, $A_{left(root)}^{10}$ and $A_{right(root)}^{10}$ in T^{10} in Fig. 1.

The other kind of inverted index is also an associative array that maps each index d to the set of all weights that are paired with d. Let $F_{vd}^c = \{f; (d, f) \in W_{B^c[i]}, i \in I_v^c\}$ for index d (i.e., all weights that are paired with d within I_v^c). The inverted index for storing weights at node v in T^c is a one-dimensional array that concatenates all F_{vd}^c in ascending order of d and is defined as $E_v^c = F_{v1}^c \cup F_{v2}^c \cup \cdots \cup F_{vD}^c$. We build E_v^c at only the root, i.e., $E_{root}^c = F_{root1}^c \cup F_{root2}^c \cup \cdots \cup F_{rootD}^c$. Figure 2 shows an example of E_{root}^{10} in T^{10} in Fig. 1.

Let P_{vd}^c indicate the ending position of Z_{vd}^c and F_{vd}^c on A_v^c and E_v^c for each $d \in [1, D]$, i.e., $P_{v0}^c = 0$ and $P_{vd}^c = P_{v(d-1)}^c + |Z_{vd}^c|$ for $d = 1, 2, ..., D$. If all descriptors at node v do not have any pair $(d : f)$ of index d and any weight $f(\neq 0)$, then $P_{vd}^c = P_{v(d+1)}^c$ holds.

When searching for descriptors similar to query $Q = (d_1 : f_1, d_2 : f_2, ..., d_m : f_m)$ in T^c, we traverse T^c from the root. At each node, we set $s_{vj} = P^c_{v(d_{(j-1)})} + 1$ and $t_{vj} = P^c_{vd_j}$ for $j = 1, 2, ..., m$. If $s_{vj} \leq t_{vj}$ holds, there is at least one descriptor that contains d_j because of A^c_v's property. Otherwise, no descriptor at v contains d_j. We check the following constraint, which is equivalent to condition (1) as $\sum_{(d:f) \in Q} y^c_v[d]f \geq \frac{\epsilon}{(1+\epsilon)}(c + ||q||^2)$ at each node v,

$$\sum_{j=1}^{m} I[s_{vj} \leq t_{vj}] \cdot \max E^c_{root}[s_{vj}, t_{vj}] \cdot f_j \geq \frac{\epsilon}{1+\epsilon}(c + ||q||^2_2), \qquad (2)$$

where $I[cond]$ is the indicator function that returns one if $cond$ is true and zero otherwise and $\max E^c_{root}[s_{vj}, t_{vj}]$ returns the maximum value in subarray $E^c_{root}[s_{vj}, t_{vj}]$. For example in Fig. 2, for $Q = (1 : 3, 3 : 1, 4 : 2)$ and A^{10}_{root}, $I[1 \leq 2] \cdot \max\{3, 1\} \cdot 3 + I[7 \leq 9] \cdot \max\{1, 1, 3\} \cdot 1 + I[10 \leq 12] \cdot \max\{3, 2, 3\} \cdot 2 = 18$.

A crucial observation is that computing constraint (2) needs s_{vj}, t_{vj} and $\max D^c_{root}[s_{vj}, t_{vj}]$ at each node v. If we compute s_{vj} and t_{vj} at each node v, we can omit A^c_v, resulting in a huge memory reduction. We compute s_{vj} and t_{vj} using rank dictionaries. The problem of computing $\max D^c_{root}[s_{vj}, t_{vj}]$ is called a *range maximum query (RMQ)*. Rank dictionaries and RMQ data structures are reviewed in the next section.

d	Z^{10}_{rootd}
1	1 4
2	3 4 6 8
3	1 6 7
4	2 4 5
5	3 5 6 7

d	F^{10}_{rootd}
1	3 1
2	3 2 2 3
3	1 1 3
4	3 2 3
5	1 1 2 1

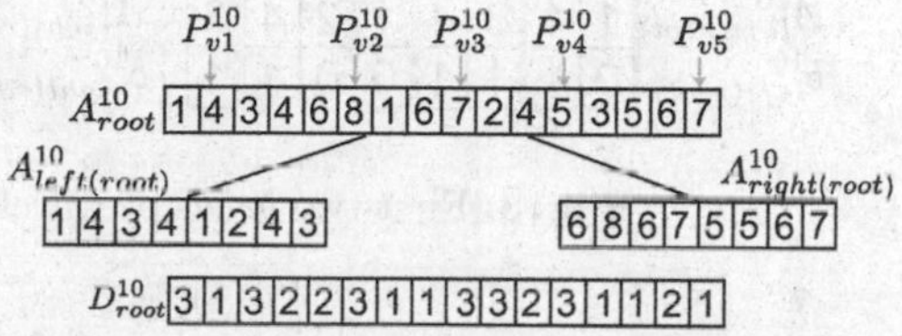

$P^{10}_{v1} \quad\quad P^{10}_{v2} \quad P^{10}_{v3} \quad P^{10}_{v4} \quad\quad P^{10}_{v5}$

A^{10}_{root} | 1 | 4 | 3 | 4 | 6 | 8 | 1 | 6 | 7 | 2 | 4 | 5 | 3 | 5 | 6 | 7 |

$A^{10}_{left(root)}$ | 1 | 4 | 3 | 4 | 1 | 2 | 4 | 3 | $A^{10}_{right(root)}$ | 6 | 8 | 6 | 7 | 5 | 5 | 6 | 7 |

D^{10}_{root} | 3 | 1 | 3 | 2 | 2 | 3 | 1 | 1 | 3 | 3 | 2 | 3 | 1 | 1 | 2 | 1 |

Fig. 2. Example of Z^c_{vd}, F^c_{vd}, A^c_v and D^c_v for T^c.

3.6   Rank Dictionaries and RMQ Data Structures

Rank Dictionary. A rank dictionary [14] is a data structure built over a bit array B of length n. It supports the rank query $rank_c(B, i)$, which returns the number of occurrences of $c \in \{0, 1\}$ in $B[1, i]$. Although naive approaches require the $O(n)$ time to compute a rank, several data structures with only $n + o(n)$ bits storage have been presented to achieve $O(1)$ query time [13]. We employ *hybrid bit vectors* [5] (which are compressed rank dictionaries) to calculate $I[s_{vj} \leq t_{vj}]$ in Eq. (2) with $O(1)$ time and only $n + o(n)$ bits (and sometimes much less).

RMQ Data Structures. The RMQ problem for an array D of length n is defined as follows: for indices i and j between 1 and n, query $RMQ_E[i, j]$ returns the index of the largest element in subarray $E[i, j]$. An RMQ data structure is built by preprocessing E and is used for efficiently solving the RMQ problem.

A naive data structure is simply a table storing $RMQ_E(i, j)$ for all possible pairs (i, j) such that $1 \leq i < j \leq n$. This takes $O(n^2)$ preprocessing time and space, and it solves the RMQ problem in $O(1)$ query time. An $O(n)$ preprocessing time and $O(1)$ query time data structure has been proposed [1] that uses $\frac{n \log n}{2} + n \log M + 2n$ bits of space. RMQ data structure U^c for each $c \in [1, D]$ is built for E^c_{root} in $O(N^c)$ preprocessing time where N^c is the total number of pairs $(d : f)$ in B^c, i.e., $N^c = \sum_{i \in B^c} |W_i|$. Then, $\max E^c_{root}[s_{vj}, t_{vj}]$ in Eq. (2) can be computed using U^c in $O(1)$ time.

3.7 Similarity Search Using Rank Dictionaries and RMQs

At the heart of SITAd is the wavelet tree, a succinct data structure usually applied to string data [4]. SITAd stores only rank dictionaries and an RMQ data structure without maintaining A^c_v and E^c_v in memory. Thus, SITAd can compute constraint (2) in a space-efficient manner.

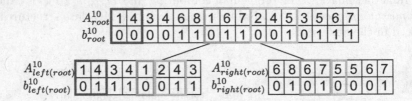

Fig. 3. First two levels of wavelet tree in Fig. 2.

A wavelet tree is a collection of rank dictionaries to update interval at each node. Let b^c_v be a bit array of length $|A^c_v|$ at node v. Let $I^c_v = [a, b]$ be an interval of node v, and let $left(v)$ (resp. $right(v)$) be the left child (resp. the right child) of node v. We build rank dictionaries for the bit arrays b^c_v.

$A^c_{left(v)}$ and $A^c_{right(v)}$ are constructed by moving each element of A^c_v to either $left(v)$ or $right(v)$ while keeping the order of elements in A^c_v. This is performed by taking into account the fact that each element of A^c_v is a descriptor ID in I^c_v satisfying two conditions: (i) $I^c_{left(v)} \cup I^c_{right(v)} = I^c_v$ and (ii) $I^c_{left(v)} \cap I^c_{right(v)} = \emptyset$. Bit $b^c_v[k]$ indicates $A^c_v[k]$ moves to whether $left(v)$ or $right(v)$. $b^c_v[k] = 0$ indicates $A^c_v[k]$ moves to $A^c_{left(v)}[k]$ and $b^c_v[k] = 1$ indicates $A^c_{right(v)}[k]$ inherits $A^c_v[k]$. Bit $b^c_v[k]$ is computed by $A^c_v[k]$ as follows:

$$b^c_v[k] = \begin{cases} 1 \text{ if } A^c_v[k] > \lfloor (a+b)/2 \rfloor \\ 0 \text{ if } A^c_v[k] \leq \lfloor (a+b)/2 \rfloor. \end{cases}$$

Fig. 3 shows bit b^{10}_{root}, $b^{10}_{left(root)}$ and $b^{10}_{right(root)}$ computed from A^{10}_{root}, $A^{10}_{left(root)}$ and $A^{10}_{right(root)}$, respectively. For example, $b^{10}_{root}[7] = 0$ indicates $A^{10}_{root}[7] = A^{10}_{left(root)}[5] = 1$. $A^{10}_{root}[8] = A^{10}_{right(root)}[3] = 6$ is indicated by $b^{10}_{root}[8] = 1$.

To perform a similarity search for query of m non-zero weights $Q = (d_1 : f_1, d_2 : f_2, \cdots, d_m : f_m)$, SITAd computes s_{vj} and t_{vj} at each node v and checks constraint (2) by computing $I[s_{vj} \leq t_{tj}]$ and $\max E_{root}(s_{vj}, t_{vj})$ on RMQ data structure U^c. SITAd sets $s_{vj} = P_v^c[d_j - 1] + 1$ and $t_{vj} = P_v^c[d_j]$ only at the root v. Using s_{vj} and t_{vj}, SITAd computes $s_{left(v)j}, t_{left(v)j}, s_{right(v)j}$ and $s_{right(v)j}$ by using rank operations in $O(1)$ time as follows:

$$s_{left(v)j} = rank_0(b_v^c, s_{vj} - 1), \quad t_{left(v)j} = rank_0(b_v^c, t_{vj})$$
$$s_{right(v)j} = rank_1(b_v^c, s_{vj} - 1) + 1, \quad t_{right(v)j} = rank_1(b_v^c, t_{vj}).$$

Note that P_v^c is required at the root for maintaining s_{vj} and t_{vj}. Thus, SITAd keeps P_v^c only at the root.

The memory for storing b_v^c for all nodes in T^c is $N^c \log |B^c| + o(N^c \log |B^c|)$ bits. Thus, SITAd needs $\sum_{c=1}^{D} N^c (\log |B^c| + \frac{N^c \log N^c}{2} + N^c \log M + 2N^c + o(N^c \log |B^c|))$ bits of space for storing b_v^c at all nodes v and an RMQ data structure U^c for all $c \in [1, D]$. The memory requirement of SITAd is much less than that for storing summary descriptors y_v^c using $D \log M \sum_{c=1}^{D} |B^c| \log (|B^c|)$ bits. In our experiments $D = 642,297$. Although storing P_{root}^c needs $D \sum_{c=1}^{D} \log N^c$ bits, this is not an obstacle in practice, even for large D.

4   Experiments

4.1   Setup

We implemented SITAd and compared its performance to the following alternative similarity search methods: one-vs-all search (OVA); an uncompressed inverted index (INV); an inverted index compressed with variable-byte codes (INV-VBYTE); an inverted index compressed with PForDelta codes (INV-PD). All experiments were carried out on a single core of a quad-core Intel Xeon CPU E5-2680 (2.8GHz). OVA is a strawman baseline that computes generalized Jaccard similarity between the query and every descriptor in a database. INV was first proposed as a tool for cheminformatics similarity search of molecular descriptors by Kristensen et al. [10] and is the current state-of-the-art approach. INV-VBYTE and INV-PD are the same as INV except that the inverted lists are compressed using variable-byte codes and PForDelta, respectively, reducing space requirements. We implemented these three inverted indexes in C++. For computing rank operations in SITAd we used an efficient implementation of hybrid bitvector [5] downloadable from https://www.cs.helsinki.fi/group/pads/hybrid_bitvector.html.

Our database consisted of the 42,971,672 chemical compounds in the Pub-Chem database [2]. We represented each compound by a descriptor with the dimension of 642,297 constructed by the KCF-S algorithm [7]. We randomly sampled 1,000 compounds as queries.

4.2   Results

Figure 4 shows the preprocessing time taken to construct the SITAd index. The construction time clearly increases linearly as the number of descriptors increases, and takes only eight minutes for the whole database of around 42 million compounds. We should emphasize that index construction is performed only once for a given database and that phase does not need to be repeated for each query. Indeed, this fast construction time is an attractive and practical aspect of SITAd.

Table 1 shows the results of each algorithm for $\epsilon \in \{0.9, 0.95, 0.98\}$. The reported search times are averages taken over 1,000 queries (standard deviations are also provided, and as well as small deviations), where $\#B^c$ is the number of selected blocks per query, $\#TN$ is the number of traversed nodes in SITAd,

Table 1. Performance summary showing average search time, memory in megabytes (MB), number of selected blocks per query ($\#B^c$), average number of traversed nodes ($\#TN$), and average number of rank computations (#Ranks), when processing the database of 42,971,672 descriptors.

	Time (sec)			Memory (MB)
	$\epsilon = 0.98$	$\epsilon = 0.95$	$\epsilon = 0.9$	
INV	1.38 ± 0.46			$33,012$
INV-VBYTE	5.59 ± 2.66			$1,815$
INV-PD	5.24 ± 2.45			$1,694$
OVA	9.58 ± 2.08			$8,171$
SITAd	0.23 ± 0.23	0.61 ± 0.57	1.54 ± 1.47	$2,470$
$\#B^c$	2	6	12	
$\#TN$	$43,206$	$118,368$	$279,335$	
#Ranks	$1,063,113$	$2,914,208$	$6,786,619$	
$|I_N|$	31	132	721	

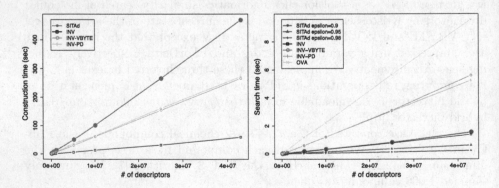

Fig. 4. Index construction time.   **Fig. 5.** Average search time.

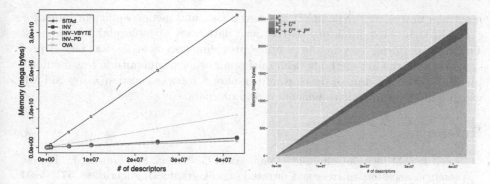

Fig. 6. Index size.   **Fig. 7.** Space usage of SITAd components.

#Ranks is the number of rank operations performed, and $|I_N|$ is the size of the answer set.

Unsurprisingly OVA had the slowest search time among the tested methods, requiring 9.58 s per query on average and using 8 GB of main memory. In line with previously reported results [10], INV provided faster querying than OVA but used more memory. The average search time of INV was faster than that of SITAd when the latter system had $\epsilon = 0.9$, but became significantly slower than SITAd with large ϵ of 0.95 and 0.98. INV required 33 GB of main memory, the most of any system. The compressed inverted indexes INV-VBYTE and INV-PT used much smaller amounts of memory — 1.8 GB and 1.7 GB, respectively. This space saving comes at a price, however; the average search time of INV-VBYTE and INV-PT is 4-5 times slower than that of INV.

Overall, SITAd performed well; its similarity search was fastest for $\epsilon = 0.95$ and 0.98 and its memory usage was low. In fact, SITAd with $\epsilon = 0.98$ was 20 times faster than INV-VBYTE and INV-PD with almost the same memory consumption. It took only 0.23 and 0.61 s for $\epsilon = 0.98$ and 0.95, respectively, and it used small memory of only 2 GB, which fits into the memory of an ordinary laptop computer. Its performance of SITAd was validated by the values of $\#B^c$, $\#TN$ and $\#Ranks$. The larger the threshold ϵ was, the smaller those values were, which demonstrates efficiency in the methods for pruning the search space in SITAd.

Figure 5 shows that for each method, the average search time per query increases linearly as the number of descriptors in the database increases. Figure 6 shows a similar linear trend for index size. Figure 7 illustrates that for SITAd, rank dictionaries of bit strings b_v^c and RMQ data structure U^c are the most space consuming components of the index.

5   Conclusion

We have presented a time- and space-efficient index for for solving similarity search problems that we call the *succinct intervals-splitting tree algorithm for molecular descriptors (SITAd)*. It is a novel, fast, and memory-efficient index for generalized-Jaccard similarity search over databases of molecular compounds. The index performs very well in practice providing speeds at least as fast as previous state-of-the-art methods, while using an order of magnitude less memory. In future work we aim to develop and deploy a software system using SITAd, which will be of immediate benefit to practitioners.

References

1. Bender, A.M., Farah-Colton, M., Pemmasani, G., Skiena, S., Sumazin, P.: Lowest common ancestors in trees and directed acyclic graphs. J. Algorithms **57**, 75–94 (2005)
2. Chen, B., Wild, D., Guha, R.: PubChem as a source of polypharmacology. J. Chem. Inf. Model. **49**, 2044–2055 (2009)
3. Chen, J., Swamidass, S., Dou, Y., Bruand, J., Baldi, P.: ChemDB: a public database of small molecules and related chemoinformatics resources. Bioinformatics **21**, 4133–4139 (2005)
4. Grossi, R., Gupta, A., Vitter, J.: High-order entropy-compressed text indexes. In: Proceedings of the 14th Annual ACM-SIAM Symposium on Discrete Algorithms, pp. 636–645 (2003)
5. Kärkkäinen, J., Kempa, D., Puglisi, S.J.: Hybrid compression of bitvectors for the FM-index. In: Proceedings of Data Compression Conference, pp. 302–311 (2014)
6. Keiser, M., Roth, B., Armbruster, B., Ernsberger, P., Irwin, J., Shoichet, B.: Relating protein pharmacology by ligand chemistry. Nat. Biotechnol. **25**(2), 197–206 (2007)
7. Kotera, M., Tabei, Y., Yamanishi, Y., Moriya, Y., Tokimatsu, T., Kanehisa, M., Goto, S.: KCF-S: KEGG chemical function and substructure for improved interpretability and prediction in chemical bioinformatics. BMC Syst. Biol. **7**, S2 (2013)
8. Kotera, M., Tabei, Y., Yamanishi, Y., Tokimatsu, T., Goto, S.: Supervised de novo reconstruction of metabolic pathways from metabolome-scale compound sets. Bioinformatics **29**, i135–i144 (2013)
9. Kristensen, T.G., Nielsen, J., Pedersen, C.N.S.: A tree based method for the rapid screening of chemical fingerprints. In: Proceedings of the 9th International Workshop of Algorithms in Bioinformatics, pp. 194–205 (2009)
10. Kristensen, T.G., Nielsen, J., Pedersen, C.N.S.: Using inverted indices for accelerating LINGO calculations. J. Chem. Inf. Model. **51**, 597–600 (2011)
11. Leach, A., Gillet, V.: An Introduction to Chemoinformatics, Revised edn. Kluwer Academic Publishers, The Netherlands (2007)
12. Nasr, R., Vernica, R., Li, C., Baldi, P.: Speeding up chemical searches using the inverted index: the convergence of chemoinformatics and text search methods. J. Chem. Inf. Model. **52**, 891–900 (2012)
13. Okanohara, D., Sadakane, K.: Practical entropy-compressed rank/select dictionary. In: Proceedings of the 9th Workshop on Algorithm Engineering and Experiments, pp. 60–70 (2007)
14. Raman, R., Raman, V., Rao, S.S.: Succinct indexable dictionaries with applications to encoding k-ary trees and multisets. In: Proceedings of the 13th Annual ACM-SIAM Symposium on Discrete Algorithms, pp. 232–242 (2002)
15. Sawada, R., Kotera, M., Yamanishi, Y.: Benchmarking a wide range of chemical descriptors for drug-target interaction prediction using a chemogenomic approach. J. Chem. Inf. Model. **33**, 719–731 (2014)

16. Tabei, Y.: Succinct multibit tree: compact representation of multibit trees by using succinct data structures in chemical fingerprint searches. In: Proceedings of the 12th Workshop on Algorithms in Bioinformatics, pp. 201–213 (2012)
17. Tabei, Y., Kishimoto, A., Kotera, M., Yamanishi, Y.: Succinct interval-splitting tree for scalable similarity search of compound-protein pairs with property constraints. In: Proceedings of the 19th ACM SIGKDD Conference on Knowledge Discovery and Data Mining, pp. 176–184 (2013)
18. Todeschini, R., Consonni, V.: Handbook of Molecular Descriptors. Wiley-VCH Verlag GmbH, Weinheim (2002)
19. Vida, D., Thormann, M., Pons, M.: LINGO: an efficient holographic text-based method to calculate biophysical properties and intermolecular similarities. J. Chem. Inf. Model. **45**, 386–393 (2005)

Self-indexed Motion Planning

Angello Hoyos[1], Ubaldo Ruiz[1,3], Eric Tellez[2,3], and Edgar Chavez[1(✉)]

[1] Department of Computer Science, CICESE, Ensenada, Mexico
{ajhoyos,uruiz,elchavez}@cicese.mx
[2] INFOTEC, Mexico, Mexico
eric.tellez@infotec.com.mx
[3] Cátedra CONACyT, Mexico, Mexico

Abstract. Motion planning is a central problem for robotics. The PRM algorithm is, together with the asymptotically optimal variant PRM*, the standard method to maintain a (collision-free) roadmap in the configuration space. The PRM algorithm is randomized, and requires a large number of high-dimensional point samples generated online, hence a subproblem to discovering and maintaining a collision-free path is inserting new sample points connecting them with the k-nearest neighbors in the previous set. A standard way to speedup the PRM is by using an *external* index for making the search. On the other hand, a recent trend in object indexing for proximity search consists in maintaining a so-called *Approximate Proximity Graph* (APG) connecting each object with its approximate k-nearest neighbors. This hints the idea of using the PRM as a self-index for motion planning. Although similar in principle, the graphs have two incompatible characteristics: (1) The APG needs long-length links for speeding up the searches, while the PRM avoids long links because they increase the probability of collision in the configuration space. (2) The APG requires to connect a large number of neighbors at each node to achieve high precision results which turns out in an expensive construction while the PRM's goal is to produce a roadmap as fast as possible. In this paper, we solve the above problems with a counter-intuitive, simple and effective procedure. We reinsert the sample points in the configuration space, and compute a collision-free graph after that. This simple step eliminates long links, improves the search time, and reduce the total space needed for the algorithm. We present simulations, showing an improvement in performance for high-dimensional configuration spaces, compared to standard techniques used by the robotics community.

1 Introduction

Motion planning is a fundamental research topic in robotics and it has received considerable attention over the last decade [1–3]. The problem of navigating through a complex environment appears in almost all robotics applications. This problem also appears and is relevant in other domains such as computational biology, search and rescue, autonomous exploration, etc.

© Springer International Publishing AG 2017
C. Beecks et al. (Eds.): SISAP 2017, LNCS 10609, pp. 220–233, 2017.
DOI: 10.1007/978-3-319-68474-1_15

Sampling-based methods are widely used for motion and path planning. Instead of using an explicit representation of the obstacles in the configuration space, which may result in an excessive computational cost, sampling-based methods rely on a collision-checking module providing information about the feasibility of the computed trajectories. Two popular sampling-based techniques are Probabilistic Road Maps (PRMs) [4] and Rapidly-exploring Random Trees (RRTs) [5] which have been shown to work well in practice and possess theoretical guarantees such as probabilistic completeness. One of the major drawbacks of those algorithms is that they made no claims about the optimality of the solution. Recently, Karaman et. al. [6] proved that both algorithms are not asymptotically optimal. To address that limitation, they proposed a new class of asymptotically optimal planners named PRM* and RRT*, and proved that both are probabilistically complete and asymptotically optimal.

The PRM and RRT algorithms, and its asymptotically optimal variants PRM* and RRT*, respectively, maintain a collision-free roadmap in the configuration space. Both algorithms require a large number of samples for high-dimensional problems (many degrees of freedom), and for problems with large configuration spaces including many obstacles. To add a new configuration in the roadmap, it is important to compute the k-nearest neighbors of that configuration in the roadmap. For low-dimensional spaces and a small number of vertices in the roadmap, the k-nearest neighbor search is commonly performed using a naive brute force approach, where the distances to all vertices are computed to find the neighbors.

For a large number of vertices, typically a tree based subdivision of the configuration space is used to compute the k-nearest neighbors. In this category, the kd-tree is the most widely used, which works by creating a recursive subdivision of the space into two half spaces [7]. The kd-tree can be constructed in $O(n \log(n))$ operations and the query time complexity is $O(dn^{1-\frac{1}{d}})$, where n is the number of vertices and d is the dimension of the space.

In the literature, it is well-know that the algorithms for k-nearest neighbors search degrade their performance to a sequential search as the intrinsic dimensionality of the data increases [8]. In addition, most of the algorithms for k-nearest neighbor search are designed to perform fast queries but they do not consider the time it takes to process the database and create the search data structure. This is a problem for motion-planning algorithms, which need to process the data as fast as possible.

The key idea of our contribution is avoiding the construction of auxiliary data structures for k-nearest neighbors searching, as it is traditionally done. We propose to use the roadmap itself. This will lower the overall time to obtain the roadmap. We borrow the idea of Approximate Proximity Graphs [9,10] that have been proposed for searching in general metric spaces [8].

For motion-planning problems it is important to maximize the number of added vertices during a fixed time period. Usually, a large number of vertices in the roadmap implies a better quality of the trajectories in the configuration space. Also, since these algorithms are probabilistically complete, a large number of vertices implies a higher probability of finding a solution [11,12]. If the performance of the k-nearest neighbors search is improved then, more vertices can be added to the roadmap in a fixed time period, increasing the overall performance of the PRM-based algorithms.

2   Related Work

A key component in sampling-based motion planners is nearest neighbor search. Usually, the planners use nearest neighbor search data structures to find and connect configurations in order to compute a motion plan. Several approaches have been proposed in the literature. One of the most popular is spatial partitioning of the data. Example of this type of algorithms are kd-trees [7], quadtrees [7] and vp-trees [13]. These data structures can efficiently handle exact nearest neighbor searching in lower dimensions.

Yershova et. al. [14] proposed an extension of the kd-tree to handle $\mathbb{R}$, S and $SO(3)$, and the cartesian product of any number of these spaces. Similar to kd-trees for $\mathbb{R}^m$, their approach splits $SO(3)$ using rectilinear axis-aligned planes created by a quaternion representation of rotations. In [15], the authors report that the previous approach performs well in many cases but rectilinear splits produce inefficient partitions of $SO(3)$ near the corners of the partitions. They propose a method that eschews rectilinear splits in favor of splits along the rotational axes, resulting in a more uniform partition of $SO(3)$.

An approach to improve the efficiency of kd-trees is presented in [16]. The authors describe a box-based subdivision of the space that allows to focus the searches only in specific regions of the subdivision. They show that the computational complexity is lowered from a theoretical point of view.

Non-Euclidean spaces, including $SO(3)$, can be searched by general nearest neighbor search data structures such a GNAT [17], cover-trees [18], and M-trees [19]. These data structures generally perform better than a sequential search. However, these methods are usually outperformed by kd-trees in practice [20].

Plaku et. al. [21] present a quantitively analysis of the performance of exact and approximate nearest-neighbors algorithms on increasingly high-dimensional problems in the context of sampling-based motion planning. Their analysis shows that after a critical dimension, exact nearest neighbors algorithms examine almost all samples thus they become impractical for sampling-based algorithms when a large number of samples is required. This behavior motivates the use of approximate algorithms [20] which trade off accuracy for efficiency.

In [22], Kleinbort et. al. adapt the Randomly Transformed Grids (RTG) algorithm [23], for finding all-pairs r-nearest-neighbors in Euclidean spaces, to sampling-based motion planning algorithms.

3   Background

In this section, we review the standard algorithm for constructing the PRM. The following definitions and functions are going to be used in the description of the algorithms.

Let $\mathcal{X}$ be the *configuration space* where $d \in \mathbb{N}$ is the dimension of the configuration space. We denote as $\mathcal{X}_{obs}$ to the *obstacle region*, and $\mathcal{X}_{free}$ as the *obstacle-free space*. The initial configuration x_{init} is an element of $\mathcal{X}_{free}$, and the goal region $\mathcal{X}_{goal}$ is a subset of $\mathcal{X}_{free}$. A *collision-free path* $\sigma : [0, 1] \rightarrow \mathcal{X}_{free}$ is a continuous mapping to the free space. It is *feasible* if $\sigma(0) = x_{init}$ and $\sigma(1) \in \mathcal{X}_{goal}$. The goal of motion-planning algorithms is to compute a feasible collision-free path.

Given a graph $G = (V, E)$ where $V \subset \mathcal{X}$, a vertex $v \in G$, and $k \in \mathbb{N}$, the function Nearest_Neighbors (G, v, k) returns the k-nearest vertices in G to v.

Given two points $x_1, x_2 \in \mathcal{X}$, a boolean function Collision_Check(x_1, x_2) returns *true* if the line segment between x_1 and x_2 lies in $\mathcal{X}_{free}$ and *false* otherwise.

The PRM constructs a roadmap represented as a graph $G = (V, E)$ whose vertices are samples from $\mathcal{X}_{free}$ and the edges are collision-free trajectories between vertices. The PRM initializes the vertex set with n samples from $\mathcal{X}_{free}$ and attempts to connect the k-nearest points. The PRM is described in Algorithm 1. From [6], we have that for the asymptotically optimal variant PRM*, $k = e(1 + 1/d) \log(|V|)$ where d is the dimension of the configuration space, thus $k = 2e \log(|V|)$ is a good choice for all applications. Here, $|V|$ denotes the number of vertex in G.

Algorithm 1. k-nearest PRM

Input: n samples from X_{free}.
Output: A roadmap $G = (V, E)$ in X_{free}.
  $V \leftarrow \{sample\_free_i\}_{i=1,\ldots,n}$
  $E \leftarrow \emptyset$
  for each $v \in V$ **do**
    $U \leftarrow$ Nearest_Neighbors(G, v, k)
    for each $u \in U$ **do**
      if Collision_Check(v, u) **then**
        $E \leftarrow E \cup \{(v, u), (u, v)\}$
      end if
    end for
  end for
  return G

A popular variant of the PRM is the lazy-collision PRM [24, 25] or its asymptotically optimal version lazy-collision PRM*. In those variants, the collision check is omitted during the construction stage. Thus, the lazy-collision PRM or lazy-collision PRM* are built just connecting each node to its k-nearest neighbors. During the query stage, every-time a path between two vertices is searched and found in the graph, it is validated to detect if at least one of its edges is in collision with an obstacle. If one of the edges in the path is in collision then that edge is erased from the graph and a new path is searched again. We will use this approach of the PRM* latter in our paper.

4   Approximate Proximity Graph

In this section, we describe the Approximate Proximity Graph (APG), a search data structure introduced by Malkov et. al. in [9,10]. This data structure has attracted a lot of interest in the similarity search community because of its simplicity. It is constructed with succesive insertions, rendering excellent searching times in high-dimensional spaces. The main idea behind the APG is to build a search graph where each node is connected to its approximate k-nearest neighbors.

A take out from studying the APG and other generalizations of the same idea is that multiple local searches give a good global approximation to the true nearest neighbors of the query. Local searches use two types of links, long and short. Long links are responsible for speedy searches, while local links are responsible for accuracy. The former are naturally obtained with earlier inserts (when the number of sample points is small) and the later are also natural when the density of the graph is larger (when inserting the most recent points). This becomes more clear in the construction depicted in Algorithm 2. Note that this algorithm requires a definition of the function Nearest_Neighbors which computes the nearest neighbors in the graph. From [9], the nearest neighbors can be searched following a greedy approach using the graph itself, as it is described in Algorithm 4.

Algorithm 2. APG

Input: A set $\mathcal{U}$ of n elements.
Output: A graph $G = (V, E)$ containing the k-nearest neighbors of each element in $\mathcal{U}$.

    $V \leftarrow \emptyset$
    $E \leftarrow \emptyset$
    for each $u \in \mathcal{U}$ **do**
        $X_{near} \leftarrow$ Nearest_Neighbors(G, u, k)
        $V \leftarrow V \cup \{u\}$
        for each $v \in X_{near}$ **do**
            $E \leftarrow E \cup \{(u, v), (v, u)\}$
        end for
    end for
    return G

Greedy search does not guarantee the true k-nearest neighbor. The result depends on the initial vertex where the search started. To amplify the probability of finding the true nearest neighbor m local searches, initiated from random vertices of the graph, can be used. This method is described in Algorithm 4. This algorithm requires the function Random_Vertex(G) which randomly samples a vertex from G.

Since the initial vertex is chosen at random, there is a probability p of finding the true nearest neighbor for a particular element q. This probability is non-zero because it is always possible to choose the exact nearest neighbor as the

Algorithm 3. Greedy_Search

Input: A graph $G = (V, E)$, an initial vertex v_{init} and a query q.
Output: A vertex $v_{min} \in V$ whose distance to q is a local minimum.

$v_{min} \leftarrow v_{init}$
$d_{min} \leftarrow \text{Distance}(v_{min}, q)$
$v_{next} \leftarrow NIL$
$X_{friends} \leftarrow \{u \in V | (u, v_{min}) \in E\}$
for each $u_{friend} \in X_{friends}$ **do**
    $d_{friend} \leftarrow \text{Distance}(u_{friend}, q)$
    if $d_{friend} < d_{min}$ **then**
        $d_{min} \leftarrow d_{friend}$
        $v_{next} \leftarrow u_{friend}$
    end if
end for
if v_{next} is not NIL **then**
    return Greedy_Search(G, v_{next}, q)
else
    return v_{min}
end if

initial vertex. Let us simplify the model by assuming independent identically distributed random variables; thus if for a fixed query element q the probability of finding the true nearest neighbor in a single search attempt is p, then the probability of finding the true nearest neighbor after m attempts is $1 - (1 - p)^m$. Therefore, the precision of the search increases exponentially with the number of search attempts. If m is comparable to $|V|$, the algorithm becomes an exhaustive search, assuming the entry points are never reused. If G has the small-world properties [28] then it is possible to choose a vertex in a random number of steps proportional to $\log(n)$, maintaining an overall logarithmic search complexity.

Algorithm 4. Multi_Search

Input: A graph $G = (V, E)$, a query q, and a number m of restarts.
Output: A candidate set U of nearest neighbors of q in G.

$U \leftarrow \emptyset$
for $i = 1, \ldots, m$ **do**
    $v_{init} \leftarrow \text{Random\_Vertex}(G)$
    $v_{min} \leftarrow \text{Greedy\_Search}(G, v_{init}, q)$
    if $v_{min} \notin U$ **then**
        $U \leftarrow U \cup \{v_{min}\}$
    end if
end for
return U

An important parameter of the APG is the number of pseudo nearest neighbors connected to each newly added vertex. A large number of pseudo neighbors increases the accuracy, while at the same time decreases the search speed. Note that this parameter is also closely related to the time it takes to build the data structure. Malkov et. al. suggest that $k = 3d$ where d is the dimension of the search space is a good choice for database applications where the cost to build the APG is amortized by the number of queries that are going to be solved after the construction.

In [10], Malkov et. al. propose a more sophisticated algorithm to perform the search in the APG. In this algorithm, a different condition is used. The algorithm iterates on not previously visited elements close to the query, i.e., those for which the edge list has not been verified. The algorithm stops when at the next iteration the k-nearest results to the query do not change. The list of previously visited elements during the search is shared preventing repeated distance evaluations. The search algorithm is described in Algorithm 5.

Algorithm 5. Tabu_Search

Input: A graph $G = (V, E)$, a query q, and a number m of restarts.
Output: A candidate set U of nearest neighbors of q in G.
1: Let U be an empty min-queue of fixed size k.
2: Let C be an empty min-queue.
3: Let r be the updated distance of the furthest element to q in U. An empty U defines $r = \infty$.
4: $S \leftarrow \emptyset$
5: **for** $i = 1, \ldots, m$ **do**
6:     $c \leftarrow$ Random_Vertex$(V - S)$
7:     $S \leftarrow S \cup \{c\}$
8:     Append (Distance$(c, q), c$) into U and C
9:     **loop**
10:        Let $(r_b, best)$ be the nearest pair in C
11:        Remove $best$ from C
12:        **if** $r_b > r$ **then**
13:           **break** loop
14:        **end if**
15:        $X_{friends} \leftarrow \{u \in V | (u, best) \in E\}$
16:        **for each** $u_{friend} \in X_{friends}$ **do**
17:           **if** $u_{friend} \notin S$ **then**
18:              $S \leftarrow S \cup \{u_{friend}\}$
19:              Append (Distance$(q, u_{friend}), u_{friend}$) to U and C
20:           **end if**
21:        **end for**
22:     **end loop**
23: **end for**

---

5 Our Work

A natural connection can be made between the APG and the PRM. Both the APG and the PRM aim at connecting the k-nearest neighbors of the sample points. In the literature the PRM is built using an auxiliary data structure, which needs to be constructed beforehand. In this work, we will not use an auxiliary data structure, we will instead use the same PRM for searching, adapting the APG algorithm for our purposes. Special care is needed for the PRM algorithm, because two neighbors are connected if and only if a free-collision path exists between them. Note that in the APG algorithm there is no notion of obstacles in the space.

Please notice that in the configuration space, long-length edges have a higher probability of collision with obstacles, thus most of them are not considered in the graph created by the PRM algorithm. On the other hand, long-length edges are important to maintain the navigation small world properties of the graph in the APG, which are related to the logarithmic scaling of the search. To maintain the small world properties, we propose to use the *lazy-collision* [24, 25] version of the PRM*. In this case, the Collision-Check procedure is not applied during the execution of Algorithm 1, thus for a given set of vertices the constructed graph using the lazy-collision PRM* is the same than the one computed using the APG, if the same number of neighbors are connected to each node. The edges in the graph, constructed using the lazy-collision PRM*, are only removed once a path between two vertices in the graph is founded and validated.

Another aspect to take into account is the number of nearest neighbors that are connected to each vertex. In the case of the PRM*, the value of $k = 2e \log(n)$ has been suggested to achieve good results in all applications [6]. For the APG, in order to have a good precision in the search, $k = 3d$ where d is the dimension of the search space is recommended by [10]. As k increases the quality of the search improves but the time to construct the graph also increases. In our case, we are interested in maximizing the number of vertices added to the graph in a fixed amount of time, but also keep a good precision for finding the k-nearest neighbors of each added vertex, since we are using an approximate search algorithm. Thus, we need to find a suitable value of k to achieve our goals.

To tackle the problems described above we propose a two-step procedure. Firstly, we build a standard APG, which naturally contains long links, and then *reinsert* all the nodes. This counter-intuitive step removes long links, just because when reinserting an old node in the final graph, the density is higher and the precision of the local search increased. After the reinsertion, it is very likely that the long-length edges will be removed. The lazy-collision PRM* obtained with this procedure will be closer to the true k-nearest neighbors graph. This strategy also allows to use a value of k smaller than the one suggested by Malkov et. al. in [10] improving the time of construction and maintaining a good precision. In this paper two approaches for reinserting the nodes are tested. The details about the selection of the parameters and the experiments that were performed will be discussed in the next section.

6 Experiments

In this section, we present the results of using the proposed heuristics for the k-nearest neighbors search in the construction of the lazy-collision PRM*. Our results are compared with the ones obtained using a sequential search or kd-tree.

To study the performance of our proposal, we define three metrics. The first one is called the *speedup*, which is the ratio between the time t_s to construct a lazy-collision PRM* using a sequential search (ground truth) for finding the k-nearest neighbors and the time t_a to construct the same data structure using an alternative algorithm for finding the k-nearest neighbors, we have that

$$speedup = \frac{t_s}{t_a}$$

Note that a speedup bigger than one implies that using the alternative algorithm is faster than performing a sequential search while a speedup smaller than one implies that it is slower.

The second metric is called the *precision* and it gives a measure of the quality of the solution obtained with the approximate search algorithms presented in this paper. Let A_i be the set of nearest neighbors computed by one of the alternative algorithms for a vertex i, and B_i be the set of nearest neighbors computed by the sequential search (ground truth). We denote as knn_i^c to the number of elements in $A_i \cap B_i$ and as knn_i^s to the number of elements in B_i. We have that

$$precision = \frac{1}{n} \sum_{i=1,\dots,n} \frac{knn_i^c}{knn_i^s}$$

As the value of the precision gets closer to one implies that a better approximation of the solution has been computed.

The third metric is called the *proximity ratio*, and it is a comparison between the distances from the approximate nearest neighbors to a vertex i and the distances from the true nearest neighbors to that vertex. The average distance from a vertex i and its k-nearest neighbors in a set N_i is given by

$$avg\_dist(i, N_i) = \frac{1}{k} \sum_{j \in N_i} dist(i,j)$$

Let A_i be the set of nearest neighbors computed by one of the alternative algorithms for a vertex i, and B_i be the set of nearest neighbors computed by the sequential search. The proximity ratio is given by

$$proximity\_ratio = \frac{1}{n} \sum_{i=1,\dots,n} \frac{avg\_dist(i, A_i)}{avg\_dist(i, B_i)}$$

Note that as the proximity ratio approaches 1, the approximate nearest neighbors converge the true nearest neighbors.

In the next experiments, we use uniform randomly generated samples from a space $\mathcal{X} = (0, 100)^d$ where d is the dimension of the space.

We start our analysis with a characterization of the kd-tree performance in high-dimensional spaces. Figure 1 shows the behavior of using the kd-tree in the construction of a lazy-collision PRM* in different space dimensions as the number of samples grows. In this figure, we can observe that as the space dimension increases the speedup of the lazy-collision PRM* based on a kd-tree decreases, becoming worse than the lazy-collision PRM* based on a sequential search. Note that in most cases, after reaching a space dimension of 12, the kd-tree starts to present an overhead compared to a sequential search due to the additional logical operations involved in the search.

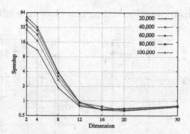

Fig. 1. The speedup of constructing a lazy-collision PRM* using a kd-tree for the k-nearest neighbor search. Each curve corresponds to a different sample set size.

In the next experiment, we analyze the performance of Algorithms 4 (Greedy multisearch) and 5 (Tabu search) for computing the k-nearest neighbors in the construction of a lazy-collision PRM*. Following the recommendation of Malkov et. al., we set $k = 3d$, where d is the space dimension. We also tested the value $k = 2e \log(n)$ by Karaman et. al. in [6], for the construction of a PRM*. We use a set of 10000 samples in $4, 8, 12, 16, 20, 25, 30, 40$ and 50 dimensions. The number m of restarts is set to 1 for Tabu search and to $10, 20$ and 30 for Greedy multisearch. Figures 2, 3 and 4 show the results of this experiment. From these figures, we can observe that for both algorithms, as the dimension increases, the best precision and the best proximity ratio are obtained for $k = 3d$. However, that value of k also produces the lowest speedup, in many cases worst than a sequential search, which makes it impractical for our purpose. The results also confirm that Algorithm 5 produces a better precision and a better proximity ratio in comparison to Algorithm 4. On the other hand, Algorithm 5 has the worst speedup.

Figures 5a, b, and c show the results of constructing an initial lazy-collision PRM* and refine the graph by reinserting all nodes updating the information about their nearest neighbors. We present the results of using two approaches for reinserting the nodes. In the first one, Tabu search is used to update the neighbors of each node. In the second approach, each node verifies if the neighbors of its neighbors can be considered as a better approximation. In this experiment, 100000 random samples were used for constructing lazy-collision PRMs* in $4, 8, 12, 16, 20, 25, 30, 40$ and 50. In the construction of the graph we have

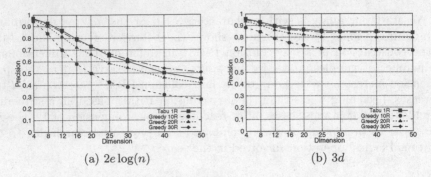

Fig. 2. Precision of the approximation using the Greedy multisearch and Tabu search.

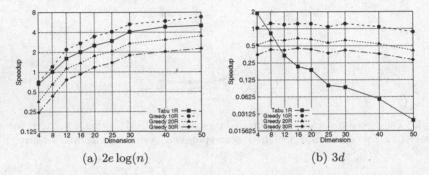

Fig. 3. Speedup of the approximation using the Greedy multisearch and Tabu search.

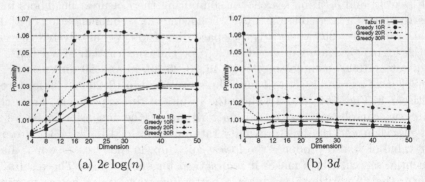

Fig. 4. Proximity ratio of the approximation using the Greedy multisearch and Tabu search.

selected $k = 2e \log(n)$. From Figs. 5a, b and c, it is possible to conclude that the algorithm has a better performance than using a sequential search or a kd-tree for finding the nearest neighbors. Another interesting property is that even if the precision decreases as the dimension increases, the value of the proximity ratio in our proposal remains closer to one compared to the results of Algorithms 4 and 5. This means that the approximate nearest neighbors computed by our algorithm are closer to the true nearest neighbors. We can expect that the trajectories

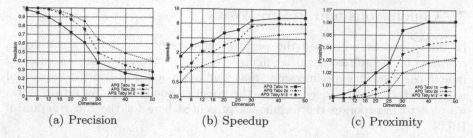

(a) Precision   (b) Speedup   (c) Proximity

Fig. 5. Precision, speedup, and proximity ratio of the two approaches for reinserting the nodes using a neighborhood $k = 2e \log(n)$.

in the lazy-collision PRM* produced with our proposal will be more similar to the ones obtained using an exact method for the nearest neighbors search like a kd-tree and sequential search.

In Figs. 6 and 7, we can observe that our proposal achieves similar results for the precision and proximity ratio at each dimension as the number of sample increases. Figures 6b and 7b shows an improvement of the speedup as the number of samples increases since it takes more time to compute the k-nearest neighbors using a sequential search.

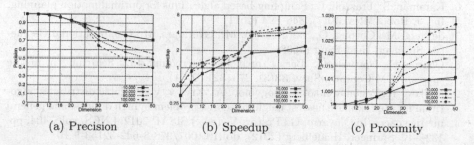

(a) Precision   (b) Speedup   (c) Proximity

Fig. 6. Precision, speedup, and proximity ratio of the approximation for different sample sizes using a neighborhood $k = 2e \log(n)$ and Tabu search reinserting the nodes (APG 2x).

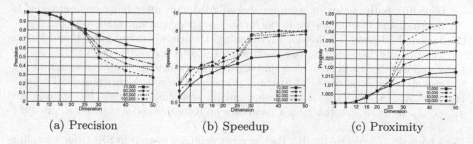

(a) Precision   (b) Speedup   (c) Proximity

Fig. 7. Precision, speedup, and proximity ratio of the approximation for different sample sizes using a neighborhood $k = 2e \log(n)$ and verifying for each node if the neighbors of its neighbors can be considered as a better approximation.

7 Conclusions and Future Work

In this paper we addressed the problem of constructing and maintaining a roadmap in high-dimensional configuration spaces, without using an external index. We showed experimentally that our approach outperforms sequential search and kd-trees, widely used in the robotics community. As future work, we plan to incorporate the techniques presented here for the construction of RRTs*.

References

1. Latombe, J.C.: Robot Motion Planning. Academic Publishers, Boston (1991)
2. Choset, H., Lynch, K., Hutchinson, S., Kantor, G., Burgard, W., Kavraki, L., Thrun, S.: Principles of Robot Motion: Theory, Algorithms, and Implementations. MIT Press, Boston (2005)
3. LaValle, S.M.: Planning Algorithms. Cambridge University Press, Cambridge (2006)
4. Kavraki, L.E., Svestka, P., Latombe, J.C., Overmars, M.H.: Probilistic roadmaps for path planning in high dimensional configuration spaces. IEEE Trans. Robot. **12**(4), 566–580 (1996)
5. LaValle, S.M., Kuffner, J.J.: Randomized kinodynamic planning. Int. J. Robot. Res. **20**(5), 378–400 (2001)
6. Karaman, S., Frazzoli, E.: Sampling-based algorithms for optimal motion planning. Int. J. Robot. Res. **30**(7), 846–894 (2011)
7. de Berg, M., Cheong, O., van Kreveld, M., Overmars, M.: Computational Geometry. Springer, Heidelberg (2008)
8. Chavez, E., Navarro, G., Baeza-Yates, R., Marroquin, J.L.: Searching in metric spaces. ACM Comput. Surv. **33**(3), 273–321 (2001)
9. Malkov, Y., Ponomarenko, A., Logvinov, A., Krylov, V.: Scalable distributed algorithm for approximate nearest neighbor search problem in high dimensional general metric spaces. In: Navarro, G., Pestov, V. (eds.) SISAP 2012. LNCS, vol. 7404, pp. 132–147. Springer, Heidelberg (2012). doi:10.1007/978-3-642-32153-5_10
10. Malkov, Y., Ponomarenko, A., Logvinov, A., Krylov, V.: Approximate nearest neighbor algorithm based on navigable small world graphs. Inf. Syst. **45**(2014), 61–68 (2014). Elsevier
11. Dobson, A., Moustakides, G.V., Bekris, K.E.: Geometric probability results for bounding path quality in sampling-based roadmaps after finite computation. In: International Conference on Robotics and Automation, pp. 4180–4186 (2015)
12. Janson, L., Ichter, B., Pavone, M., Planning, Deterministic Sampling-Based Motion : Optimality, Complexity, and Performance. CoRR abs/1505.00023 (2015)
13. Yianilos, P.N.: Data structures and algorithms for nearest neighbor search in general metric spaces. In: ACM-SIAM Symposium Discrete Algorithms (1993)
14. Yershova, A., LaValle, S.M.: Deterministic sampling methods for spheres and SO(3). In: IEEE International Conference on Robotics and Automation, pp. 3974–3980 (2004)
15. Ichnowski, J., Alterovitz, R.: Fast nearest neighbor search in SE(3) for sampling-based motion planning. In: Akin, H.L., Amato, N.M., Isler, V., van der Stappen, A.F. (eds.) Algorithmic Foundations of Robotics XI. STAR, vol. 107, pp. 197–214. Springer, Cham (2015). doi:10.1007/978-3-319-16595-0_12

16. Svenstrup, M., Bak, T., Andersen, H.J.: Minimising computational complexity of the RRT algorithm a practical approach. In: International Conference on Robotics and Automation, pp. 5602–5607 (2011)
17. Brin, S.: Near neighbor search in large metric spaces. In: International Conference on Very Large Databases (VLDB), pp. 574–584 (1995)
18. Beygelzimer, A., Kakade, S., Langford, J.: Cover trees for nearest neighbor. In: International Conference on Machine Learning, pp. 97–104 (2006)
19. Ciaccia, P., Patella, M., Zezula, P.: M-tree: an efficient access method for similarity search in metric spaces. In: International Conference on Very Large Databases, pp. 426–435 (1997)
20. Yershova, A., LaValle, S.M.: Improving motion-planning algorithms by efficient nearest-neighbor searching. IEEE Trans. Robot. **23**(1), 151–157 (2007)
21. Plaku, E., Kavraki, L.E.: Quantitative analysis of nearest-neighbors search in high-dimensional sampling-based motion planning. In: Akella, S., Amato, N.M., Huang, W.H., Mishra, B. (eds.) Algorithmic Foundation of Robotics VII. Springer Tracts in Advanced Robotics, vol. 47, pp. 3–18. Springer, Heidelberg (2008). doi:10.1007/978-3-540-68405-3_1
22. Kleinbort, M., Salzman, O., Halperin, D.: Efficient high-quality motion planning by fast all-pair r-nearest-neighbors. In: International Conference on Robotics and Automation, pp. 2985–2990 (2015)
23. Aiger, D., Kaplan, H., Sharir, M.: Reporting neighbors in high-dimensional Euclidean space. SIAM J. Comput. **43**(4), 1363–1395 (2014)
24. Bohlin,R., Kavraki, L.: Path planning using lazy PRM. In: IEEE Conference on Robotics and Automation, pp. 521–528 (2000)
25. Sanchez, G., Latombe, J.C.: On delaying collision checking in PRM planning: application to multi-robot coordination. Int. J. Robot. Res. **21**(1), 5–26 (2002)
26. Aurenhammer, F.: Voronoi diagrams - a survey of a fundamental geometric data structure. ACM Comput. Surv. (CSUR) **23**(3), 345–405 (1991)
27. Navarro, G.: Searching in metric spaces by spatial approximation. VLDB J. **11**(1), 28–46 (2002)
28. Kleinberg, J.: The small-world phenomenon: an algorithmic perspective. Ann. ACM Symp. Theory Comput. **32**, 163–170 (2000)

Practical Space-Efficient Data Structures for High-Dimensional Orthogonal Range Searching

Kazuki Ishiyama$^{(\boxtimes)}$ and Kunihiko Sadakane

Graduate School of Information Science and Technology,
The University of Tokyo, Tokyo, Japan
{kazuki_ishiyama,sada}@mist.i.u-tokyo.ac.jp

Abstract. We consider the orthogonal range search problem: given a point set P in d-dimensional space and an orthogonal query region Q, return some information on $P \cap Q$. We focus on the counting query to count the number of points of P contained in Q, and the reporting query to enumerate all points of P in Q.

For 2-dimensional case, Bose et al. proposed a space-efficient data structure supporting the counting query in $O(\lg n / \lg \lg n)$ time and the reporting query in $O(k \lg n / \lg \lg n)$ time, where $n = |P|$ and $k = |P \cap Q|$. For high-dimensional cases, the KDW-tree [Okajima, Maruyama, ALENEX 2015] and the data structure of [Ishiyama, Sadakane, DCC 2017] have been proposed. These are however not efficient for very large d.

This paper proposes practical space-efficient data structures for the problem. They run fast when the number of dimensions d' used in queries is smaller than the data dimension d. This kind of queries are typical in database queries.

1 Introduction

1.1 Orthogonal Range Searching

Consider a set P of n points in the d-dimensional space $\mathbb{R}^d$. Given a d-dimensional orthogonal region $Q = \left[l_1^{(Q)}, u_1^{(Q)}\right] \times \left[l_2^{(Q)}, u_2^{(Q)}\right] \times \cdots \times \left[l_d^{(Q)}, u_d^{(Q)}\right]$, the problem of answering some information on the points of P contained in Q ($P \cap Q$) is called *orthogonal range searching*, which is one of the important problems in computational geometry.

The information on $P \cap Q$ to be reported depends on the query. In this paper, we consider the *counting* query, which is to answer the number of points $|P \cap Q|$, and the *reporting* query, which is to enumerate all points of $P \cap Q$. Other queries are the *emptiness* query to determine if $P \cap Q$ is empty or not, the *aggregate* query to compute the summation or the variant of weights in the query region if each point $p \in P$ has weight $w(p)$, etc.

This work was supported by JST CREST Grant Number JPMJCR1402, Japan.

C. Beecks et al. (Eds.): SISAP 2017, LNCS 10609, pp. 234–246, 2017.
DOI: 10.1007/978-3-319-68474-1_16

A typical application of the orthogonal range searching in high dimensions is database search [16]. For example, in an employee database of a company, the query to count the number of people whose service years are between x_1 and x_2, whose ages are between y_1 and y_2, and whose annual incomes are between z_1 and z_2 is expressed as an orthogonal range searching. There are also other applications in GIS (Geographic Information Systems), CAD, and computer graphics.

1.2   Existing Work

It is known that if the space complexity of the data structure can be superlinear, counting and reporting queries are done in polylog(n) time. For example, the range tree [3] can perform counting in $O\left(\lg^{d-1} n\right)$ time and reporting in $O\left(\lg^{d-1} n + k\right)$ time, where k is the number of points output by the query, that is, $k = |P \cap Q|$, using $O\left(n \lg^{d-1} n\right)$-word space. It is however desirable that the data structure uses less space if the number of points n is huge.

There are several linear space data structures for the problem. The kd-tree [2] supports counting in $O\left(n^{(d-1)/d}\right)$ time and reporting in $O\left(n^{(d-1)/d} + k\right)$ time. For 2-dim., Chazelle [5] proposed a linear space data structure supporting counting in $O(\lg n)$ time and reporting in $O(\lg n + k \lg^\varepsilon n)$ time $(0 < \varepsilon < 1)$.

Here *linear space* means $O(n)$ words, which are actually $O(n \log n)$ bits and may not be optimal. *Succinct data structures* are data structures which use the minimum number of bits to store data (information-theoretic lower bound) and support efficient queries. *Wavelet trees* [10] are succinct data structures for compressing suffix arrays, which are used for full-text search. Later it became clear that wavelet trees can be used for other problems [14]. For the orthogonal range searching, it was proved that counting is done in $O(\lg n)$ time and reporting is done in $O\left((1 + k) \lg \frac{n}{1+k}\right)$ time [8]. Bose et al. [4] proposed, for 2-dim. cases, succinct data structure supporting counting and reporting queries in $O(\lg n / \lg \lg n)$ time and $O((k \lg n / \lg \lg n)$ time, respectively.

For general dimensions, Okajima and Maruyama [15] proposed KDW-tree. Query time complexities of KDW-tree are better than those of the kd-tree; counting in $O\left(n^{(d-2)/d} \lg n\right)$ time and reporting in $O\left((n^{(d-2)/d} + k) \lg n\right)$ time. Ishiyama and Sadakane [11] improved the time complexity so that counting in $O\left(n^{(d-2)/d} \lg n / \lg \lg n\right)$ time and reporting in $O\left((n^{(d-2)/d} + k) \lg n / \lg \lg n\right)$ time using the same space complexity. They also extended the data structure so that the coordinate values can take integers from 0 to $U - 1$ for some fixed U, while in existing succinct data structures the coordinate values are restricted to integers from 0 to $n - 1$. Table 1 shows existing data structures for orthogonal range searching in general dimensions.

Table 1. Comparison of existing data structures. KDW-tree and [11] are for $d \geq 3$. Here k is the number of points to be reported. The counting time complexities coincide with those of reporting with $k = 1$

Data structures	Point space	Space complexity	Reporting time
kd-tree [2]	$\mathbb{R}^d$	$O(n)$ words	$O\left(n^{(d-1)/d} + k\right)$
KDW-tree [15]	$[n]^d$	$dn \lg n + o(n \lg n)$ bits	$O\left((n^{(d-2)/d} + k) \lg n\right)$
[11]	$[U]^d$	$dn \lg U + o(n \lg n)$ bits	$O\left((n^{(d-2)/d} + k) \lg n / \lg \lg n\right)$

1.3   Our Contribution

It is well-known that because of the curse of dimensionality any range search data structure including the kd-tree and the KDW-tree will search most of the points if the dimension d is large. Thus in this paper we propose space-efficient data structures for orthogonal range searching whose worst-case query time complexity is large but which are expected to run fast in practice. Especially our data structures are expected to search points fast if the number of dimensions d' used in queries is smaller than the data dimension d. That is, in the query region $Q = \left[l_1^{(Q)}, u_1^{(Q)}\right] \times \left[l_2^{(Q)}, u_2^{(Q)}\right] \times \cdots \times \left[l_d^{(Q)}, u_d^{(Q)}\right]$, ranges are bounded in only d' dimensions and in other dimensions the ranges are $[-\infty, \infty]$. This kind of queries are common in high-dimensional databases.

Our data structures are succinct; that is, they use the minimum number of bits to store the input point set P. If n points are in $[U]^d$, there are $\binom{U^d}{n}$ distinct sets. Therefore the minimum number of bits to represent a set is $\lg \binom{U^d}{n} = dn \lg U - n \lg n + O(n)$ bits. The size of our data structure asymptotically matches this lower bound.

Theorem 1. *There exists a data structure for orthogonal range searching for n points in $[U]^d$ using $dn \lg U - n \lg n + o(n \lg n)$ bits. Let c_i be the number of points whose i-th coordinates are in the query region $\left[l_i^{(Q)}, u_i^{(Q)}\right]$, $c_{\min} = \min_{1 \leq i \leq d} c_i$, and d' be the number of dimensions used in the query. Then the data structure performs a reporting query in $O(d' c_{\min} \lg n + d' \lg d')$ time.*

The proof of this theorem is described in Sect. 3.4.

2   Wavelet Trees

Wavelet tree is a succinct data structure which supports many queries on a string, an integer sequence, and so on. It was originally proposed to represent compressed suffix arrays [10]. However, it turned out gradually that a wavelet tree can answer many queries in some contexts [14]. Two-dimensional orthogonal range searching is one of the such queries [13].

In this chapter, we describe how wavelet trees are constructed in Sect. 2.1. Then, we explain the algorithm for orthogonal range searching by using wavelet tree in Sect. 2.2.

2.1 Construction

Two-dimensional point set P which are directly described by a wavelet tree must satisfy the following two conditions. First, the coordinates of the points must be integers from 1 to n. Second, the x-coordinates of the points must be distinct. Although they seem to be strong restrictions, it is known that a point set on $\mathbb{R}^d$ can be reduced to a point set on $[n]^d$ [7] and it does not lose generality. Thus, we consider two-dimensional point set P which satisfies these conditions. For a point set P, we place the y-coordinates of the points of P in increasing order of the x-coordinates and make an integer sequence C. For example, we make an integer sequence $3, 5, 0, 2, 7, 4, 6, 1$ for the point set of Fig. 1. Then, we construct a wavelet tree for C in the following way.

First, we associate the root of the wavelet tree with the integer sequence C^1. Then, we represent each integer in binary and make a bit sequence by arranging the most significant bit of each integer. We store this bit sequence in the root. Then, we distribute the integers to the two children while keeping the order. Specifically, the integers whose most significant bit is 0 are assigned to the left child, and the integers whose most significant bit is 1 are assigned to the right child. For example, in the Fig. 1, the integers between 0 and 3 are distributed to the left child. Therefore, the left child is associated with the integer sequence $3, 0, 2, 1$.

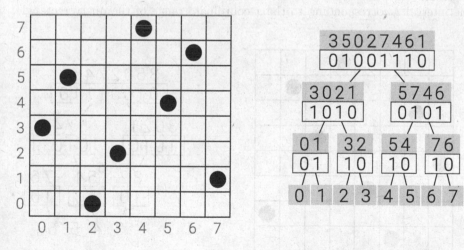

Fig. 1. Two-dimensional point set P (left) and the corresponding wavelet tree (right).

---

[1] Although we don't store C in the root, we consider that the root corresponds to C.

Subsequently, for each child node, we make a bit sequence by arranging the second significant bit of each integer, and store it. Recursively, we distribute the integers to the two children according to the second significant bits. We repeat this process until each leaf is associated with a sequence of a particular integer. However, as we described before, we do not store the integer sequences. We only store the bit sequences, and they represent which child each integer is distributed to. Therefore, as you can see in the Fig. 1, we do not store the bit sequence in the leaves.

Additionally, to answer the query fast, we have to support efficient operation on each bit sequence. There are two representative queries on the bit sequence: the rank query and the select query. In the rank query, we answer the number of 0/1 before the i-th bit, and in the select query, we answer the position of the i-th 0/1 bit. The following result [6,12] is known regarding these queries.

Lemma 1. *For a bit sequence of length n, there is a succinct data structure. It supports rank queries and select queries in constant time and uses $n + o(n)$ bits of space.*

By using this result for each bit sequence, wavelet tree uses $n \lg n + o(n \lg n)$ bits of space and can answer queries efficiently. In the following section, we describe the algorithm to answer the orthogonal range searching.

2.2   Range Search Algorithm

We explain the searching algorithm using the wavelet tree with the example of Fig. 2.

In the searching algorithm, we descend the tree from the root while sustaining the interval I corresponding to the x-coordinate range of the query rectangle.

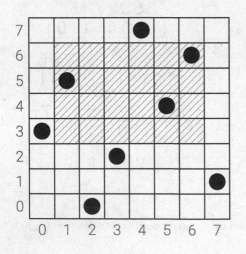
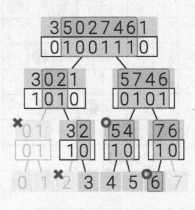

Fig. 2. Illustration of searching algorithm.

In the example of Fig. 2, we focus on the interval $I = [1,6]$ in the root[2]. Then, we descend to the left child, and at this time, we have to calculate a new interval corresponding to the left child. To get the new interval, we need to count the number of 0 s before the left end and the right end of the interval. In other words, if the interval at the root is $[l,r]$, we count the number of 0 s in $[0, l-1]$ and in $[0,r]$. It can be done efficiently by using rank queries. In the example of Fig. 2, by counting the number of 0 s in $[0,0]$ $(= 1)$ and in $[0,6]$ $(= 3)$, we find the new interval corresponding to the left child is $[1,2]$. When we descend to the right child, we have to count the number of 1 s.

We repeat this process and descend the tree. If we descend until reaching a leaf, we can judge whether the point is included in the query rectangle or not. However, we can stop descending the tree at an internal node in some cases. For example, in the Fig. 2, after we descend to the left child from the root, we don't have to search the left child of that node. This is because the range of y-coordinate corresponding to that node (= the left child of the left child of the root) is $[0,1]$, and it does not intersect the y-coordinate range of the query rectangle. Also, we can stop when we descend to the right from the root and then to the left. This is because the range of y-coordinate corresponding to the visiting node is $[4,5]$, and it is included in the range of y-coordinate of the query rectangle. Therefore, the points contained in the I at the visiting node are included in the query rectangle.

After we find the points which are contained in the query rectangle, we sum up the number of such points in the counting query. However, in the reporting query, we have to calculate the coordinates of points. About the y-coordinates, we can get them by descending the tree by using rank queries as we described before. On the other hand, to get the x-coordinates, we have to ascend the tree by using select queries on the bit sequences. For example, in the Fig. 2, let suppose we visit the node whose corresponding y-coordinate range is $[4,5]$ (i.e. the left child of the right child of the root), and find the x-coordinate of the point corresponding to the second position in the visiting node. First we need to find the corresponding position of that point in the parent node. Since the visiting node is the left child of the parent, the position is the second 0-bit in the parent node and we can find it efficiently by using a select query. In this example, we find the position is third. Then, we find the corresponding position of the point in the root. In this case, the visiting node is the right child of the parent (= the root), the position is the third 1-bit in the parent and we find it is sixth. Therefore, we can conclude that the x-coordinate of the point is 5[3].

As described above, by using rank queries and select queries, we can move in the wavelet tree efficiently. In the range searching, we descend the tree from the root while sustaining the interval corresponding to the x-coordinate range of the query rectangle, and when we find the node whose corresponding y-coordinate range is contained in the query rectangle, we stop descending and count the number of points or calculate the coordinates of the points.

---

[2] Here, we use zero-based indexing.
[3] As we mentioned before, we use zero-base indexing here.

3 Proposed Scheme

In our scheme, we use $d-1$ wavelet trees to represent a point set in $[n]^d$ and perform searches. In Sect. 3.1, we explain how to construct the data structure (index). In Sect. 3.2, we explain the algorithm for orthogonal range searching. In Sect. 3.3, we show how to extend the data space from $[n]^d$ to $[U]^d$. In Sect. 3.4, we analyze the space complexity of the data structure and the query time complexity.

3.1 Index Construction

We first consider a point set P in $[n]^d$ space in which the 0-th coordinates[4] of points are distinct. The case there are two points with the same 0-th coordinate is explained in Sect. 3.3.

First we make $d-1$ integer arrays $A_1, \ldots, A_{d-1}$ of length n each, corresponding the first to the $(d-1)$-st dimension, respectively. The array A_i stores the i-th coordinates of the points in the increasing order of the 0-th coordinate values. For the arrays $A_1, \ldots, A_{d-1}$, we construct wavelet trees $W_1, \ldots, W_{d-1}$. This can be also regarded as constructing a wavelet tree W_i for a two-dimensional point set P_i generated from the d-dimensional point set P by orthographically projecting it to the plane spanned by the 0-th and i-th axes.

3.2 Range Search Algorithm

We explain the algorithm (Algorithm 1) for orthogonal range searching using the above data structure. We are given a query region $Q = \left[l_0^{(Q)}, u_0^{(Q)}\right] \times \cdots \times \left[l_{d-1}^{(Q)}, u_{d-1}^{(Q)}\right]$. First, for $i = 1, \ldots, d-1$ such that $\left[l_i^{(Q)}, u_i^{(Q)}\right] \neq [0, n-1]$, that is, for each dimension i used in the search, we count the number of points of the two-dimensional point set P_i in a region $\left[l_0^{(Q)}, u_0^{(Q)}\right] \times \left[l_i^{(Q)}, u_i^{(Q)}\right]$ using the wavelet tree W_i. Let $m(=|D|)$ be the number of dimensions i $(1 \leq i \leq d-1)$ such that $\left[l_i^{(Q)}, u_i^{(Q)}\right] \neq [0, n-1]$, and $i_1, \ldots, i_m$ be the dimensions sorted in the increasing order of the answers of the counting queries. Next, using the wavelet tree W_{i_1}, enumerate the x-coordinates of points of P_{i_1} contained in the region $\left[l_0^{(Q)}, u_0^{(Q)}\right] \times \left[l_{i_1}^{(Q)}, u_{i_1}^{(Q)}\right]$, and store them in set A. Then, for each $i = i_2, \ldots, i_m$ in order, for each $a \in A$, check if the i-th coordinate of the point whose 0-th coordinate is a is in the query region. The remaining set A contains 0-th coordinates of points contained in the original query region. Therefore for the reporting query, we compute all coordinates and output them. For the counting query, the cardinality of A is the answer. Note that $m = d'$ if $\left[l_0^{(Q)}, u_0^{(Q)}\right] = [0, n-1]$ and $m = d'-1$ otherwise.

---

[4] From now on we call the d dimensions as 0-th dimension to $(d-1)$-st dimension.

Algorithm 1. REPORT(Q)

Input: A query region $Q = \left[l_0^{(Q)}, u_0^{(Q)}\right] \times \cdots \times \left[l_{d-1}^{(Q)}, u_{d-1}^{(Q)}\right]$.

Output: Coordinates of points of P contained in Q.

1: $D := \varnothing$
2: **for** $i = 1$ to $d - 1$ **do**
3:    **if** $\left[l_i^{(Q)}, u_i^{(Q)}\right] \subsetneq [0, n-1]$ **then**
4:       $D = D \cup \{i\}$
5:       $c_i := $ COUNT $\left(P_i, \left[l_0^{(Q)}, u_0^{(Q)}\right] \times \left[l_i^{(Q)}, u_i^{(Q)}\right]\right)$
6:    **end if**
7: **end for**
8: Sort $i_1, \ldots, i_{|D|} \in D$ in the increasing order of c_i.
9: $A := $ REPORTX $\left(P_{i_1}, \left[l_0^{(Q)}, u_0^{(Q)}\right] \times \left[l_{i_1}^{(Q)}, u_{i_1}^{(Q)}\right]\right)$
10: **for** $i = i_2$ to $i_{|D|}$ **do**
11:    **for all** $a \in A$ **do**
12:       **if** The i-th coordinate of the point whose 0-th coordinate is a is not contained in $\left[l_i^{(Q)}, u_i^{(Q)}\right]$ **then**
13:          $A = A \setminus \{a\}$
14:       **end if**
15:    **end for**
16: **end for**
17: **for all** $a \in A$ **do**
18:    Compute coordinates of the point whose 0-th coordinate is a and output them.
19: **end for**

This algorithm is based on the SvS algorithm [1]. The reasons why we first count the number of points contained in each dimension are twofold. Firstly, to output x-coordinates (0-th dimension) of points in a query region in Line 9, it is faster if we enumerate smaller number of points. Secondly, in the double loop in Lines 10 to 16, it is faster if we can reduce the cardinality of A as quickly as possible. With high chance, we can make A small if we examine dimensions with fewer points in the query region first.

3.3 Extension from $[n]^d$ Space to $[U]^d$ Space

Here we explain how to extend the data structure so that coordinates of points are in $[U]^d$ based on [11].

First we create a point set P' in $[n]^d$ space. The set P' consists of n points, which have one-to-one correspondence with those in P. Let p' be the point in P' corresponding to $p \in P$. Then the i-th coordinate p'_i of p' is defined as

$$p'_i = \#\{q \in P \mid q_i < p_i\} \tag{1}$$

where p_i is the i-th coordinate of p. The value is called the rank of p with respect to the i-th coordinate. Two points may have the same rank w.r.t. an i-th coordinate. It is however easy to have distinct values; if k points have rank r, we change them as $r, r + 1, \ldots, r + k - 1$. Then we can apply the data structure of Sect. 3.1 to store P'.

We also store d integer arrays $C_0, \ldots, C_{d-1}$ of length n each. The entry $C_i[r]$ stores the i-th coordinate of $p \in P$ whose rank is r w.r.t. the i-th coordinate. Using this array, we can transform a query region Q in $[U]^d$ space into a query region Q' in $[n]^d$ space in $O(\lg n)$ time if k is considered as a constant. For a reporting query, we first find points of P', then convert them into coordinates values in $[U]^d$ space. This is done in constant time for each point. If we store each C_i naively, it occupies $n \lg U$ bits and the data structure is not succinct. We can however compress it into $n \lg(U/n) + O(n)$ bits without any sacrifice in access time.

3.4 Complexity Analyses

We discuss space and time complexities of the proposed scheme.

As for the space complexity, for points in $[n]^d$ space, we use $d - 1$ wavelet trees and therefore the size is $(d - 1) \lg n + o(\lg n)$ bits. For points in $[U]^d$ space, we additionally use d arrays C_i in a compressed form. In total, the space complexity is $dn \lg U - n \lg n + o(n \lg n)$ bits.

Next we consider the query time complexity. Let $m \le d'$ be the number of wavelet trees used in the query. Then the time to perform m counting queries in the wavelet trees is $O(m \lg n)$. Then we sort m integers obtained by the counting queries in $O(m \lg m)$ time. We enumerate the x-coordinate (the 0-th coordinate) values of points which are contained in the query region with smallest number of points in it in $O((1 + c_{\min}) \lg(n/(1 + c_{\min})))$ time. Then we check if each point is also contained in query regions for other dimensions in $O((m - 1)c_{\min} \lg n)$ time. Then, in total, the query time complexity is $O(d'c_{\min} \lg n + d' \lg d')$. This proves Theorem 1.

4 Experimental Evaluation

We experimentally evaluated our scheme on data structure size and query time with existing ones. We implemented our data structure by using the succinct data structure library [9]. We compared our scheme with:

- naïve: We store points in an array. Queries are done by sequential searches.
- kd-tree [2]: We use our own implementation.
- DCC [11]: A succinct variant of the kd-tree.
- LB: The information-theoretic lower bound to store the point set

For our scheme, we used two variants. One uses uncompressed bit vectors to construct wavelet trees, and the other uses rrr [17] to compress them. The latter is slower but uses less space.

4.1   Machines and Compilers

We used a Linux machine with 128GB of main memory and an Intel Xeon CPU E5-2650 v2 @ 2.60GHz running CentOS 7. We used only single core. The compiler we used was g++ version 4.8.5.

4.2   Space Usage

We show memory usage of the algorithms. We used uniformly random point sets with $n = 2^{26}$, $d = 3, 4, 6, 8$, and $U = 2^{16}, 2^{24}, 2^{32}$.

Table 2 shows space usage if U is fixed to 2^{32} and d is varied from 3 to 8. The data structure of the naïve scheme is just an array of coordinates. The kd-tree uses $O(n)$ words in addition to the array of coordinates. The proposed schemes use less space than others. In particular, the one using rrr for compressing bit vectors of the wavelet tree uses space close to the information-theoretic lower bound.

Table 3 shows space usage if d is fixed to 8 and d is varied from 2^{16} to 2^{32}. Because the kd-tree and the naïve scheme store point coordinates in 32-bit integers, the space usage does not change if U varies. On the other hand, the memory usage of the proposed schemes increases according to U, but it is smaller than that of kd-tree.

Table 2. Comparison of memory usage for various d. $n = 2^{26}$, $U = 2^{32}$.

Data structures	Memory usage (MiB)			
	$d = 3$	$d = 4$	$d = 6$	$d = 8$
lower bound	427.542	635.542	1051.55	1467.55
naïve	768	1024	1536	2048
kd-tree	3328	3584	4096	4608
Proposed	635.485	944.712	1563.08	2181.42
Proposed (rrr)	462.258	684.811	1129.92	1575.03

Table 3. Comparison of memory usage for various U. $n = 2^{24}$, $d = 8$.

Data structures	Memory usage (MiB)		
	$U = 2^{16}$	$U = 2^{24}$	$U = 2^{32}$
lower bound	210.885	338.885	466.885
naïve	512	512	512
kd-tree	1152	1152	1152
Proposed	333.113	503.722	686.744
Proposed (rrr)	239.163	363.125	546.803

4.3   Query Time

We created uniformly random point sets with $n = U = 2^{26}$ and $d = 3, 4, 6, 8$. All query regions are hypercubes. We call the ratio between the volume of a query region and the volume of the entire space as the *selectivity*. We used query regions with different selectivities from 10^{-7} to 0.5.

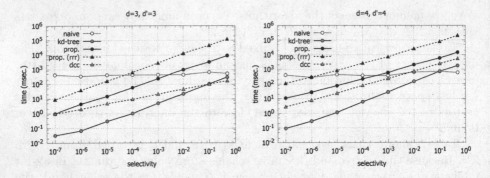

Fig. 3. Query time and selectivity. Left: $d = d' = 3$, Right: $d = d' = 4$.

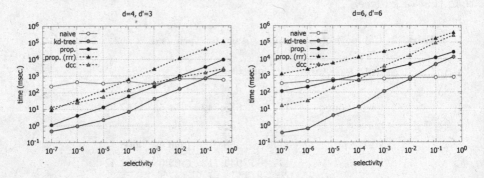

Fig. 4. Query time and selectivity. Left: $d = 4, d' = 3$, Right: $d = d' = 6$.

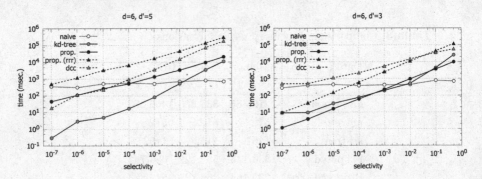

Fig. 5. Query time and selectivity. Left: $d = 6, d' = 5$, Right: $d = 6, d' = 3$.

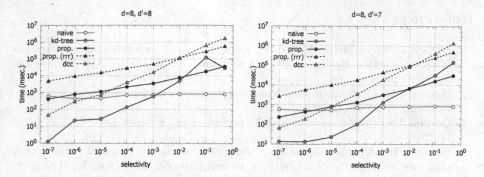

Fig. 6. Query time and selectivity. Left: $d = 8, d' = 8$, Right: $d = 8, d' = 7$.

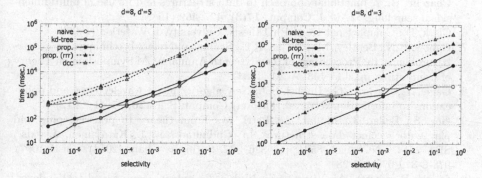

Fig. 7. Query time and selectivity. Left: $d = 8, d' = 5$, Right: $d = 8, d' = 3$.

Figures 3, 4, 5, 6 and 7 show the results. In all schemes except naïve, queries are done quickly if the selectivity is small. If d' is close to d, that is, the query region is bounded in most of the dimensions, the kd-tree is the fastest. However, if d' is much smaller than d, the proposed schemes are faster than the kd-tree, as we expected.

5   Conclusion

We have proposed practical space-efficient data structures for high-dimensional orthogonal range searching. The space complexity asymptotically matches the information-theoretic lower bound for representing the input point set, which is also confirmed by computational experiments. On the other hand, query time is not always faster than existing methods such as the kd-tree. Our schemes are faster if the dimension of point sets is large, whereas the number of dimensions used in search is small.

References

1. Barbay, J., López-Ortiz, A., Lu, T., Salinger, A.: An experimental investigation of set intersection algorithms for text searching. ACM J. Exp. Algorithmics **14**, 3.7–3.24 (2009)
2. Bentley, J.S.: Multidimensional binary search trees used for associative searching. Commun. ACM **18**(9), 509–517 (1975)
3. Bentley, J.S.: Decomposable searching problems. Inf. Process. Lett. **8**(5), 244–251 (1979)
4. Bose, P., He, M., Maheshwari, A., Morin, P.: Succinct orthogonal range search structures on a grid with applications to text indexing. In: Dehne, F., Gavrilova, M., Sack, J.-R., Tóth, C.D. (eds.) WADS 2009. LNCS, vol. 5664, pp. 98–109. Springer, Heidelberg (2009). doi:10.1007/978-3-642-03367-4_9
5. Chazelle, B.: A functional approach to data structures and its use in multidimensional searching. SIAM J. Comput. **17**(3), 427–462 (1988)
6. Clark, D.: Compact pat trees. PhD thesis, University of Waterloo (1997)
7. Gabow, H.N., Bentley, J.L., Tarjan, R.E.: Scaling and related techniques for geometry problems. In: Proceedings of the Sixteenth Annual ACM Symposium on Theory of Computing, pp. 135–143. ACM (1984)
8. Gagie, T., Navarro, G., Puglisi, S.J.: New algorithms on wavelet trees and applications to information retrieval. Theoret. Comput. Sci. **426**, 25–41 (2012)
9. Gog, S., Beller, T., Moffat, A., Petri, M.: From theory to practice: plug and play with succinct data structures. In: Gudmundsson, J., Katajainen, J. (eds.) SEA 2014. LNCS, vol. 8504, pp. 326–337. Springer, Cham (2014). doi:10.1007/978-3-319-07959-2_28
10. Grossi, R., Gupta, A., Vitter, J.S.: High-order entropy-compressed text indexes. In: Proceedings of the Fourteenth Annual ACM-SIAM Symposium on Discrete Algorithms, pp. 841–850. Society for Industrial and Applied Mathematics (2003)
11. Ishiyama, K., Sadakane, K.: A succinct data structure for multidimensional orthogonal range searching. In: Proceedings of Data Compression Conference 2017, pp. 270–279. IEEE (to be published, 2017)
12. Jacobson, G.J.: Succinct static data structures. PhD thesis, Carnegie Mellon University (1988)
13. Mäkinen, V., Navarro, G.: Position-restricted substring searching. In: Correa, J.R., Hevia, A., Kiwi, M. (eds.) LATIN 2006. LNCS, vol. 3887, pp. 703–714. Springer, Heidelberg (2006). doi:10.1007/11682462_64
14. Navarro, G.: Wavelet trees for all. J. Discrete Algorithms **25**, 2–20 (2014)
15. Okajima, Y., Maruyama, K.: Faster linear-space orthogonal range searching in arbitrary dimensions. In: ALENEX, pp. 82–93. SIAM (2015)
16. O'Rourke, J., Goodman, J.E.: Handbook of Discrete and Computational Geometry. CRC Press (2004)
17. Raman, R., Raman, V., Satti, S.R.: Succinct indexable dictionaries with applications to encoding k-ary trees, prefix sums and multisets. ACM Trans. Algorithms (TALG) **3**(4), 43 (2007)

Semantic Similarity Group By Operators for Metric Data

Natan A. Laverde[(✉)], Mirela T. Cazzolato, Agma J.M. Traina, and Caetano Traina Jr.

Institute of Mathematics and Computer Sciences, University of Sao Paulo, Av. Trabalhador Sancarlense, 400, Sao Carlos, SP, Brazil
{laverde,mirelac}@usp.br, {agma,caetano}@icmc.usp.br

Abstract. Grouping operators summarize data in DBMS arranging elements in groups using identity comparisons. However, for metric data, grouping by identity is seldom useful, since adopting the concept of similarity is often a better fit. There are operators that can group data elements using similarity. However, the existing operators do not achieve good results for certain data domains or distributions. The major contributions of this work are a novel operator called the *SGB-Vote* that assign groups using an election involving already assigned groups and an extension for current operators bounds each group to a maximum amount of the nearest neighbors. The operators were implemented in a framework and evaluated using real and synthetic datasets from diverse domains considering both quality of and execution time. The results obtained show that the proposed operators produce higher quality groups in all tested datasets and highlight that the operators can efficiently run inside a DBMS.

Keywords: Similarity Group By · Grouping · Similarity comparison · Metric data

1 Introduction

Complex data such as images, videos, and spatial coordinates are increasingly frequent in several applications such as in social networks, geographic information systems, medicine, and others. To efficiently store, index, and search this kind of data, database systems can leverage on a concept of similarity between objects, which is usually more suitable than simply comparing them by identity [19]. Identity comparisons are employed by the *Group By* operator when dealing with traditional data. The *Group By* operator is often used to aggregate and summarize information in Relational Database Management Systems (RDBMS). It selects the subsets of tuples from a relation that have the same value for specified relation attributes, creating groups by comparing the values by identity. By enabling the creation of similarity-based grouping, it is possible to define analytical queries that better capture the rich semantics provided by complex data. One application of *Group By* is data clustering, what is usually done

© Springer International Publishing AG 2017
C. Beecks et al. (Eds.): SISAP 2017, LNCS 10609, pp. 247–261, 2017.
DOI: 10.1007/978-3-319-68474-1_17

through ad-hoc processes that are not meant to be executed in integrated and query intensive environments, such as RDBMS. By doing this operation inside RDBMS, all the operator inherent the relational operators would be available to the user when defining the queries. Therefore, a similarity-based grouping operator in the relational environment is useful and desirable. Such operator must join similar elements into a single group and place dissimilar ones into distinct groups, which are then subject to aggregation operations. The whole process can benefit from the RDBMS native indexing and the query optimizer.

Similarity Group By (SGB) operators were proposed to group data by similarity inside RDBMS, focused either on one-dimensional data [16] or, as in the case of *SGB-Any* and *SGB-All*, on multidimensional data [18]. Each of these operators evaluates the distances between objects and determines a range of interest to group them, given a set of constraints. However, they can be applied only to a few applications, and it is necessary to develop new operators. For example, a group could be chosen using a more flexible constraint, such as the election between the objects that have already an assigned group and are within a distance. Each vote should be weighted inversely by the distance, so closer objects have more impact than farther ones. Likewise, instead of using a distance threshold, a group could be selected using the number of neighbors.

Furthermore, the existing operators are focused on multidimensional data, and existing algorithms as well as data structures do not handle non-dimensional metric data. To solve these problems, here we present:

- a new SGB operator called the *SGB-Vote*, which group the objects using an election criterion with weighted votes;
- a new way of using the SGB operators, taking advantage of the number of neighbors in place of a distance range;
- algorithms to execute the new operators, which are sufficiently generic to support the existent operators and corresponding semantics.

We thoroughly evaluated the proposed operators, comparing them to existing ones over different data domains, showing *SGB-Vote* superior performance and flexibility.

The rest of the paper proceeds as follows: Sect. 2 provides background concepts to understand this work. Section 3 describes significant works related to this one. Section 4 presents the proposed operators and algorithms. Section 5 presents the experiments performed to evaluate the proposed operators. Finally, Sect. 6 summarizes the paper and presents future works.

2   Background

Aggregation Functions summarize the values of an attribute in a relation. When applied to the whole relation, the result is a single tuple containing a column for each employed aggregation. The value of each column summarizes all the values in the attribute. Aggregation functions are usually statistical functions, such as sum, average, maximum, minimum or count [4].

It is often required to apply aggregation functions over subsets of the relation tuples. In these cases, the *Grouping Operator* creates the subsets, grouping the tuples that share the same value in some attributes, so the aggregation functions produce one tuple for each partition. In the Relational algebra, the notation $\gamma_{\{L_g, L_a\}}(\mathsf{T})$, represents the grouping operator, where T is the relation to be grouped, L_g is the list of *grouping attributes*, and L_a is the list of *aggregated attributes*. The operator generates one group for each distinct value (or value combination) of the attributes in L_g – that is, it groups the tuples that have the same values in the grouping attributes. Each attribute in list L_a is associated with an aggregation function [4]. After the grouping operation, the aggregation functions are evaluated over the tuples of each group, generating one tuple for each partition, which has the resulting aggregation results.

In an RDBMS, the `GROUP BY` clause represents the grouping operator, where the grouping attributes are listed. The `SELECT` clause lists the aggregation attributes as arguments of the respective aggregation functions. SQL also provides a `HAVING` clause that can execute a filter (select operations) after the grouping operator.

Traditional applications deal with simple data types, such as numbers and small character strings. Those data are called *scalar data*, as each element corresponds to a single value that cannot be decomposed. Scalar data meet the properties of *Identity Relations* (IR) and *Order Relations* (OR); thus they can be compared using the comparison operators $=$ and $\neq$ (due to the IR properties), and $<, \leq, \geq$ and $>$ (due to the OR properties). Complex data types, such as multidimensional and metric data called *complex data*, in general, do not meet the OR property, so the $<, \leq, \geq$ and $>$ comparisons cannot be used. Moreover, although they can be compared using IR properties, this is usually meaningless [19]. In fact, comparing such data by *similarity* is the usual approach.

Distance functions are frequently employed to evaluate the (dis)similarity between a pair of objects. To this intent, whenever the data is in a metric space, such comparison is possible. A metric space is defined as a pair $<\mathbb{S}, d>$, where $\mathbb{S}$ is the data domain, and d is a distance function. It is assumed that the closer a pair of objects is, the more similar the objects are. To define a metric space, the function $d : \mathbb{S} \times \mathbb{S} \to \mathbb{R}^+$ must meet the following properties [19]:

- $d(s_1, s_2) = 0 \Leftrightarrow s_1 = s_2$ (identity);
- $0 < d(s_1, s_2) < \infty, \forall s_1 \neq s_2$ (non-negativity);
- $d(s_1, s_2) = d(s_2, s_1)$ (symmetry);
- $d(s_1, s_2) \leq d(s_1, s_3) + d(s_3, s_2)$ (triangle inequality).

Examples of metric distance functions are *Manhattan*, *Euclidean*, and *Jaccard* distance.

The most common types of similarity queries are the *similarity range* and the *k-nearest neighbors query*, both corresponding to similarity selections. Both of them select objects that are similar to a given *query center* s_q. A range query retrieves the objects found within a distance ξ from s_q. A *k*-nearest neighbor query retrieves the k objects nearest to s_q [19]. There are also similarity joins, which are defined over two sets X and Y sampled from the same domain $\mathbb{S}$ and retrieves every pair $(x \in X, y \in Y)$ that satisfies a similarity constraint [19]. The most common kinds of similarity joins are the *Range Join*, the *k-Nearest Neighbors Join*, and the *k-Closest Pairs join* [1].

The quality of a grouping method is assessed by evaluating its result. Quality evaluation methods are classified following two approaches, according to the availability of a ground truth. When a ground truth is available, an *extrinsic* method is used, such as the Mutual Information (MI); otherwise, an *intrinsic* method is used [6]. MI is a similarity measure which quantifies the amount of information between two labels existing in the same data. The Adjusted Mutual Information (AMI) is the corrected-for-chance version for MI. The AMI score is 1 when the groups are perfectly matched, and it is around 0 when the groups separated.

In this paper, we propose grouping algorithms that can be implemented over existing data structures, such as *union-find*, *multiset*, and *map*. The *union-find* data structure, also named *disjoint-set* or *merge-find set*, can maintain non-overlapping subsets and support the operations, *makeSet*, *union* and *find* that respectively creates a set containing only singleton elements, merge two subsets into one and determine which subset contains an element [14]. A *multiset* is a generalized notion of a set that allows duplicated values [14]. Some useful operations are applicable over multisets, including *count*, which returns the number of occurrences of an element in the multiset. A *map* is also named *associative array*, *symbol table* or *dictionary*, and have the purpose of associating a value with a key. A map provides the *insert* and *search* operations [14].

3   Related Works

The literature presents several extensions to the GROUP BY operator defined in the SQL standard. One that made its way into several DBMS is the CUBE operator [5], which can generate the same result of several grouping queries in a single, more efficient operation. Some works provide grouping extensions focused on the duplicate elimination for data integration. Schallehn et al. developed interfaces for user-defined grouping and aggregation as an SQL extension and provided concepts for duplicate detection and elimination using similarity [13]. *Cluster-Rank* was proposed by Li et al. to generalize the GROUP BY as a fuzzy grouping operator, integrating the operator with ranking and Boolean filters [9]. *Cluster By* was proposed by Zhang and Huang as an extension to SQL to support cluster algorithms in spatial databases, grouping spatial attributes and applying aggregation functions over spatial and others attributes [20].

Silva et al. proposed the SGB operators, focused on one-dimensional data. They provide three techniques to create groups: Unsupervised Similarity Group-by (SGB-U), Supervised Similarity Group Around (SGB-A) and Supervised SGB using Delimiters (SGB-D) [16]. Tang et al. extended the SGB operators to handle multidimensional data, providing the operators SGB-Any and SGB-All [18]. For SGB-Any, an object belongs to a group if it is closer or at the same distance to any other objects in this group. For the SGB-All operator, an object belongs to a group if it is closer or at the same distance to every other object in this group. Following SGB-All, an object may belong to more than one group, whenever it satisfies overlapping constraints. The SGB-All operator requires an overlap clause, which expresses different semantics for the operation and could assume three distinct values. These values arbitrate the action to be taken over the overlapped object. Using `JOIN-ANY` the object is randomly inserted in one of the overlapping groups, with value `ELIMINATE`, discard the overlapping object and `FORM-NEW-GROUP` creates groups for overlapping objects. The SGB operators were implemented in PostgreSQL using an SQL extension.

There are also some works focused on other similarity operators. Pola et al. provided the concept of similarity sets [12]. Marri et al. extended by similarity the set intersection operator [10]. There are also works focused on Similarity Joins [3,7,17].

Several works are aimed at including similarity in the DBMS execution environment. The *SImilarity Retrieval ENgine* (SIREN) provides an extended SQL and executes similarity using a layer between an application and the DBMS [2]. The similarity-aware database management system (SimDB) adds new keywords and operations to execute similarity queries in PostgreSQL [15]. The *Features, Metrics, and Indexes for Similarity Retrieval* (FMI-SiR) is a similarity module to enable Oracle DBMS to execute similarity queries using user-defined functions for feature extraction and indexing [8]. The *Kiara* platform is an extension for PostgreSQL that works with user-defined functions and supports an extended SQL through a parser that rewrites queries to the standard operators [11].

4   Proposed Methods

In this section, we present the new *Similarity Group-By* operators and the proposed algorithms to execute them. In Sect. 4.1, we describe a new SGB operator to group objects using a weighted election. In Sect. 4.2, we propose new operators to generalize the existing SGB operators so the grouping threshold can also be based on quantities of nearest neighbors, besides distance limits. In Sect. 4.3, we provide algorithms that can perform the proposed and existing operators.

4.1   The SGB-Vote Operator

In some cases, none of the existing operators, SGB-Any and SGB-All are adequate to spot the groups properly. Consider the example in Fig. 1(a), which shows nine elements as points in a bi-dimensional Euclidean space. Visually it can be noticed that there are two groups of points. One group composed of points p_1 to p_5 and another consisting of points p_6 to p_9. The pairs of objects that satisfy the distance threshold ξ are linked (only two covering rings are shown to not clutter the figure). The SGB-Any operator identifies the groups correctly.

If a new point p_{10} arrives as shown in Fig. 1(b), the SGB-Any operator will join the two groups, and the SGB-All operator will return at least three groups

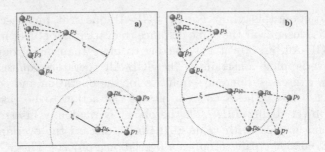

Fig. 1. Points in a bi-dimensional Euclidean space. (a) two well-separable groups; (b) a new point p_{10} placed between the groups.

because there is no way to create two groups where each of which is a clique over threshold ξ. Other threshold values can be used, but, SGB-All will only discover the right groups when the threshold is sufficiently large so that the entire group forms a clique. That is, the SGB-Any only properly discovers the groups when the threshold is smaller than the inter-group distance.

However, an approach based on the election among the groups that has objects satisfying the constraint can properly detect the groups in broader scenarios. Aiming at this objective, we developed a new SGB operator, called the *SGB-Vote* that achieves a better quality in the results, whereas executing at a speed equivalent to SGB-All and SGB-Any. *SGB-Vote* chooses the most appropriate group for an object counting the votes of the objects that satisfy each constraint, like a *KNN-classifier*. The votes are weighted, so votes of closer objects contribute more than the vote from further objects. The weight assigned to each vote is $1/d$, where d is the distance between the pair of objects p_i and p_j. When $d(p_i, p_j) = 0$, the weight assigned is ∞, then both objects are in the same group. When no objects satisfy the threshold constraint, or when the objects that satisfy it are both not assigned to a group, then a new group is created for them. Using threshold ξ, only the *SGB-Vote* operator can properly assign the point p_{10} in Fig. 1(b) to group of points p_6 to p_9. The *SGB-Vote* will choose the group p_6 to p_9 instead of p_1 to p_5, because the points p_6 and p_8 are closer than p_4, then the weight assigned to group p_6 to p_9 is greater than the assigned to p_1 to p_5.

4.2   Nearest Neighbors-Based SGB: The KNN-SGB Operators

The literature about similarity queries usually considers two main similarity predicates: the distance-based (or range-based) predicate and the quantity-based (or k-nearest neighbors) predicates. The similarity predicate of both SGB-Any and SGB-All operators are distance-based. When the data space presents a variable density of object distribution, they cannot generate consistent grouping over all the space. In this situation, an SGB operator based on quantities will be better suited.

Therefore, we generalize the SGB operators (the SGB-Any, SGB-All, and SGB-Vote) to use either a range-based or a quantity-based predicate.

Therefore for each data object, the operators will consider the set of k-nearest neighbors to partition the data, instead of considering the set of objects within a given distance. The quantity-based variations are defined as follows.

- The SGB-Any operator groups the objects that are one of the k-nearest neighbors of object p_i, instead of the objects which are closer than a radius ξ from p_i.
- The SGB-All generates the cliques using k-nearest neighbors and uses the same overlap clauses as its distance-based variation.
- The SGB-Vote operator groups the objects using weighted votes of the k-nearest neighbors; when none of the k-nearest neighbors have a group assigned yet, the object is placed into a new group.

To identify the six operators, we prefix the distance-based methods with *Range* and the quantity-based ones with *KNN*, as in range and KNN queries. For example, the *SGB-Any* operator exists in the *Range-SGB-Any* and *KNN-SGB-Any* variations. When we do not specify a *Range* or *KNN* prefix, we are referring to the generic methods, independently of is threshold predicate. When we do not specify an *All*, *Any* or *Vote* suffix, we are talking about the predicate criterion, regardless of the grouping technique.

4.3   Algorithms

The existing algorithms for *Range-SGB* are not applicable to *KNN-SGB*, so in this section, we present algorithms that can perform both *KNN-SGB* and *Range-SGB*. The algorithms follow a nested loop approach, analogous to a join. The inner loop uses a function $Query(\delta, d, \xi)$, which represents either a KNN query or a Range query, and can be executed either following a sequential scan or taking advantage of an indexing method (e.g. R-tree or Slim-Tree). This approach can employ optimization techniques existing for similarity selection queries as an additional benefit. The function *getGroup* can use a *map* structure to associate an object with a group.

Algorithm 1 shows the pseudo-code for *SGB-Any*, which performs either the KNN or Range query for each object in the data. Every object that satisfies the constraint is merged in the same group of the query center. The function result is a union-find structure where the connected elements are in a single group.

Algorithm 2 presents the pseudo-code for *SGB-All*. It performs a similarity query for each object d in the input data. For each object in the *result* of the similarity query with a group already assigned, the group identifier is inserted in an *overlap* multiset. Thereafter, the algorithm compares the number of occurrences of each partition in the *overlap* multiset with the number of elements contained in the groups, where equal values imply that all the similarity constraints are satisfied for every object in the group, so the group identifier is inserted in the *candidates* set. Thereafter, if the *candidates* set is empty, a new group is created for the object. Whether the *candidates* set has only one satisfying group, the element is placed in it. When the *candidates* set has more than one satisfying

Input: D: Set of objects, ξ: Similarity threshold, δ: Distance function
Result: Set of Groups G

1    $G \leftarrow makeSet(D.size())$;
2   **foreach** $d \in D$ **do**
3       $result \leftarrow Query(\delta, d, \xi)$;
4   **foreach** $r \in result$ **do** $G.union(G.find(d), r.id)$;
5   **end**
6   **return** G

Algorithm 1. SGB-Any

group, an action is taken according to the value of the CLS clause. The possible actions are JOIN-ANY that insert the object in one of the overlapping groups, ELIMINATE that discard the object, and FORM-NEW-GROUP that creates a new group.

Input: D: set of data objects, ξ: similarity threshold, δ: distance function, CLS: overlap clause
Result: Set of Groups G

1    $G \leftarrow NULL$;
2   **foreach** $d \in D$ **do**
3       $candidates \leftarrow NULL$; $overlap \leftarrow NULL$; $result \leftarrow Query(\delta, d, \xi)$;
4   **foreach** $r \in result$ **do**
5         $g \leftarrow getGroup(r, G)$;
6   **if** g *is not empty* **then** $overlap.insert(g)$;
7   **end**
8   **foreach** $distinct(o) \in overlap$ **do**
9   **if** $overlap.count(o) = G[o].size()$ **then** $candidates.insert(o)$;
10   **end**
11   **if** *candidates is empty* **then**
12   Create a new group and insert d
13   **else if** $candidates.size() = 1$ **then**
14   Insert d in the group contained in the candidates set;
15   **else**
16   **switch** CLS **do**
17   **case** $JOIN\text{-}ANY$ Insert d in any group of candidates set;
18   **case** $FORM\text{-}NEW\text{-}GROUP$ Create a new group and insert d;
19   **case** $ELIMINATE$ Discard the data point;
20   **endsw**
21   **end**
22   **end**
23   **return** G

Algorithm 2. SGB-All

Algorithm 3 shows the pseudo-code for *SGB-Vote*. It performs a similarity query for each object d in the input data. For each object r in the result of the query that already has an assigned group, an entry in the map W is updated by the inverse of the distance between d and r (or is assigned ∞ if $distance = 0$), so that a closer object contributes more than a farther object. When the map of weights is $NULL$, it means that none of the objects that satisfy the similarity predicates has a group already assigned, so a new group is created to place object d. Otherwise, d is assigned to a group that has the highest weight value.

Input: D: set of data objects, ξ: similarity threshold, δ: distance function
Result: Set of Groups G

1 $G \leftarrow NULL$;
2 **foreach** $d \in D$ **do**
3      $weights \leftarrow NULL$; $result \leftarrow Query(\delta, d, \xi)$;
4   **foreach** $r \in result$ **do**
5          $g \leftarrow getGroup(r, G)$;
6   **if** g *is not empty* **then**
7   **if** $r.distance = 0$ **then** $weights[g] \leftarrow \infty$;
8   **else** $weights[g] \leftarrow weights[g] + (1/r.distance)$;
9   **end**
10   **end**
11   **if** *weights is empty* **then** Create a new group and insert d ;
12   **else** Insert d into group with highest weight value ;
13 **end**
14 **return** G

Algorithm 3. SGB-Vote

---

5   Evaluation

We implemented all the grouping operators in a framework written in the C++ language using the *Arboretum* library[1]. The operators were implemented using an indexed nested-loop approach as presented in the previous section, using the *Slim-Tree* to index the data, as it is well-suitable for all the operators. The experiments are performed on an Intel ® Core ™ i5-5200U running at 2.2 GHz with 8GB of RAM under Fedora 23.

First, we used datasets whose grouping ground truth is available to evaluate the quality of the groups obtained using the AMI score. Some operators are processing-order dependent, that is, they can produce different results when the data is processed in distinct orders. Thereby, for fair comparisons, we executed the operators ten times shuffling the input data and calculated the average value and the standard deviation for the AMI score. After that, we compared the impact of the parameters in the execution time, executing the operators ten times and calculating the average time.

---

[1] https://bitbucket.org/gbdi/arboretum.

5.1   The Datasets

We used real and synthetic datasets to validate the proposed methods and to compare them with the existing methods. Table 1 presents a comparison between all datasets containing the cardinality, number of dimensions, number of existing groups, distance function used and a short description of the dataset.

Table 1. Description of datasets.

Name	Size	Dim	Groups	Function	Description
Aggregation	788	2	7	Euclidean	Groups of different shapes and sizes.
Compound	399	2	6	Euclidean	Groups of different shapes and sizes.
D31	3100	2	31	Euclidean	Gaussian groups randomly placed.
R15	600	2	15	Euclidean	Gaussian groups positioned in rings.
Corel	10000	128	100	Manhattan	Images represented using the MPEG-7 Color Structure Descriptor.
GHIM	10000	128	500	Manhattan	Images represented using the MPEG-7 Color Structure Descriptor.
Sport	737	–	5	Jaccard	Documents from BBC Sport website represented using a binary bag-of-words model.
News	2225	–	5	Jaccard	Documents from BBC News website represented using a binary bag-of-words model

The first experiment compared the results obtained from four synthetic datasets containing bi-dimensional Euclidean points distributed in an Euclidean space[2]: The Aggregation dataset contains seven different groups of points, with different shapes and sizes. Dataset Compound contains six different groups of points, with different shapes and sizes. The D31 dataset contains 31 Gaussian groups randomly placed. The R15 dataset is composed of 15 Gaussian groups, 14 of them distributed in 2 circles around a central group. Figure 2 shows a graphical representation of the datasets.

To evaluate the grouping operators applied to image datasets, we used the Corel and the GHIM datasets[3]. The Corel dataset has 100 groups of 100 images

[2] http://cs.joensuu.fi/sipu/datasets/.
[3] http://www.ci.gxnu.edu.cn/cbir/dataset.aspx.

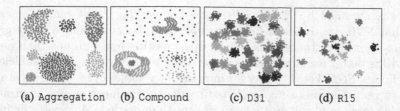

(a) Aggregation   (b) Compound   (c) D31   (d) R15

Fig. 2. Shape datasets.

each. The GHIM has 10,000 images distributed in 20 groups with diverse image contents, where each group contains 500 images. Figure 3 shows some images from the datasets. We represent these datasets using the MPEG-7 Color Structure Descriptor, which results in a feature vector with 128 features compared by the Manhattan distance.

(a) Corel   (b) GHIM

Fig. 3. Example of the images included in datasets Corel and GHIM.

To evaluate the operators over textual data, we used the News and Sport[4] datasets, which contains, respectively, 2225 documents extracted from the BBC news website and 737 from the BBC Sport website. Each document is represented using a binary bag-of-words model with the words it contains. The documents were compared using the *Jaccard* distance measure. The News documents are distributed in 5 groups (business, entertainment, politics, sport, and tech), and the vocabulary for all documents consists of 9635 distinct words. The documents in the Sport dataset are distributed in 5 groups (athletics, cricket, football, rugby, and tennis), and the vocabulary consists of 4613 distinct words.

5.2   Results

We use the AMI quality evaluation technique to compare the obtained results with the known ground truth, so a score closer to 1 means better groups. Figure 4 shows the results for the synthetic datasets using *Range-SGB*, where the horizontal axis is the ξ threshold while the vertical axis is the average of the AMI scores. In *Range-SGB-All* and *Range-SGB-Vote*, the vertical lines represent the standard deviation for AMI scores. For all tested shape datasets, the highest

---

[4] http://mlg.ucd.ie/datasets/bbc.html.

score was obtained by *Range-SGB-Vote*. For varying range radius, it can be seen that *SGB-Any* is better for very small radii, because using a range greater than inter-group distance joins the groups, creating fewer groups than needed. *SGB-All* is better for large radii because using a small radius is not sufficient to form cliques. *SGB-Vote* is better for intermediary values of radii, the most useful range of radii because using small radii values form more groups than needed and using large radii will create fewer groups than needed. Moreover, the impact of more objects from distant groups in weight can be greater than few objects of closer groups.

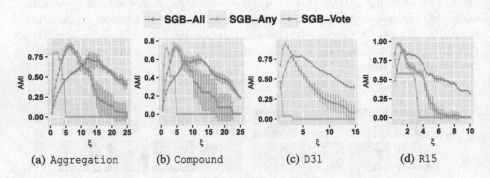

Fig. 4. AMI score results for shape datasets using *Range-SGB*.

Figure 5 presents the results for the textual and images datasets using the *Range-SGB*. For textual datasets, the graphs show ξ varying between 0.6 and 0.975, because this interval achieved better results for all operators. *Range-SGB-Vote* achieved the bests results with ξ around 0.9. For the image datasets with ξ varying from 0.25 to 5.0, the best results were also obtained by *Range-SGB-Vote* with ξ around 4.0. For those datasets, we can see that *SGB-Any* is almost useless, and *SGB-Vote* is the best for small and medium radii, whereas *SGB-All* is better for large radii.

Figure 6 shows the results for textual and images datasets using *KNN-SGB*, with k varying between 1 and 120 in textual datasets and between 1 and 50 in images datasets, the horizontal axis is the k value, and the vertical axis is the average of the AMI scores. For both textual and images datasets, the bests results were achieved using the *KNN-SGB-Vote* despite the greater varying in results. In general, for textual and images datasets tested, the results obtained by *KNN-SGB-Vote* are better than the results obtained by *Range-SGB*.

Figure 7 presents a comparison of the execution time for all three *Range-SGB* operators for varying radii over the textual and image datasets. The horizontal axis represents the threshold ξ, and the vertical axis is the average of 10 executions for total processing time in seconds, with the standard deviation represented by vertical lines. Figure 8 presents a comparison between the k and the

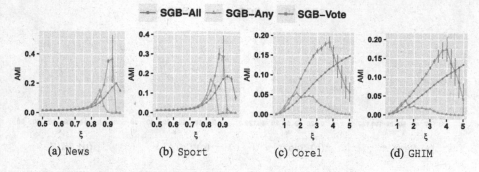

Fig. 5. AMI score results (News and Sport) and image datasets (Corel and GHIM) using *Range-SGB*, considering varying ξ value.

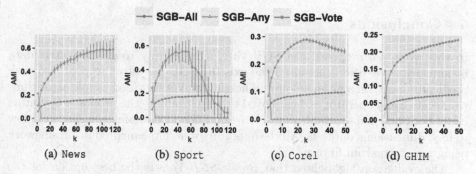

Fig. 6. AMI score results for textual (News and Sport) and image datasets (Corel and GHIM) using *KNN-SGB*, considering varying k value.

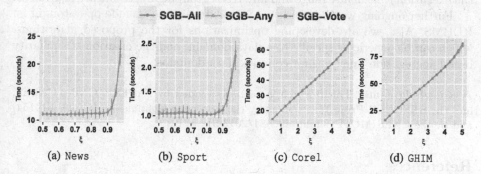

Fig. 7. Query time for textual (News and Sport) and image datasets (Corel and GHIM) using *Range-SGB*, considering varying ξ value.

processing time for *KNN-SGB* operators for these datasets, where the horizontal axis represents the threshold k. Figures 7 and 8 shows that there is no significant differences for the execution time of any operator.

260   N.A. Laverde et al.

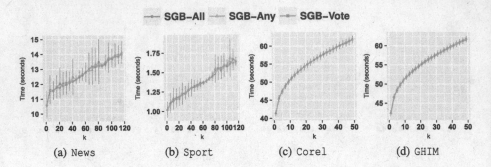

Fig. 8. Query time for textual (News and Sport) and image datasets (Corel and GHIM) using *KNN-SGB*, considering varying *k* value.

6 Conclusions

In this work, we proposed the *SGB-Vote*, a novel way to group metric data by similarity. We also generalized the existing SGB operators to deal also with predicates based on a given amount of neighbors. We provided simple algorithms to execute every operator and employed them to evaluate the proposed operators using both real and synthetic datasets from different domains, varying from low to high dimensional datasets. All the operators were implemented in a framework using the Arboretum library.

The evaluation highlighted that *Range-SGB-Vote* is the best option for the most useful values of radii and that *KNN-SGB-Vote* is the best choice for grouping based on the number of elements in groups. The experiments also showed that the execution time of any range and any KNN grouping operators are equivalent.

Further ongoing works include an SQL extension to include operators in an RDBMS. Also, we are developing optimizations for the proposed algorithms, which can be generic or specific to a data domain, and the creation of similarity based aggregation functions for complex data domains.

Acknowledgments. This research is partially funded by FAPESP, CNPq, CAPES, and the RESCUER Project, as well as by the European Commission (Grant: 614154) and by the CNPq/MCTI (Grant: 490084/2013-3).

References

1. Barioni, M.C.N., Kaster, D.D.S., Razente, H.L., Traina, A.J.M., Traina Jr., C.: Querying Multimedia Data by Similarity in Relational DBMS. In: Yan, L., Ma, Z. (eds.) Advanced Database Query Systems: Techniques, Applications and Technologies, chap. 14, pp. 323–359. IGI Global, Hershey, NY, USA (2010)
2. Barioni, M.C.N., Razente, H.L., Traina, A.J.M., Traina Jr., C.: SIREN: a similarity retrieval engine for complex data. In: VLDB, pp. 1155–1158. ACM (2006)
3. Carvalho, L.O., de Oliveira, W.D., Pola, I.R.V., Traina, A.J.M., Traina, C.: A wider concept for similarity joins. JIDM 5(3), 210–223 (2014)

4. Garcia-Molina, H., Ullman, J.D., Widom, J.: Database Systems: The Complete Book. Prentice Hall Press, Upper Saddle River (2008)
5. Gray, J., Bosworth, A., Layman, A., Pirahesh, H.: Data cube: a relational aggregation operator generalizing group-by, cross-tab, and sub-total. In: ICDE, pp. 152–159. IEEE Computer Society (1996)
6. Han, J., Kamber, M.: Data Mining: Concepts and Techniques. Morgan Kaufmann, Burlington (2000)
7. Jacox, E.H., Samet, H.: Metric space similarity joins. ACM Trans. Database Syst. **33**(2), 7:1–7:38 (2008)
8. Kaster, D.S., Bugatti, P.H., Traina, A.J.M., Traina, C.: FMI-SiR: a flexible and efficient module for similarity searching on oracle database. JIDM **1**(2), 229–244 (2010)
9. Li, C., Wang, M., Lim, L., Wang, H., Chang, K.C.: Supporting ranking and clustering as generalized order-by and group-by. In: SIGMOD Conference, pp. 127–138. ACM (2007)
10. Marri, W.J.A., Malluhi, Q., Ouzzani, M., Tang, M., Aref, W.G.: The similarity-aware relational intersect database operator. In: Traina, A.J.M., Traina, C., Cordeiro, R.L.F. (eds.) SISAP 2014. LNCS, vol. 8821, pp. 164–175. Springer, Cham (2014). doi:10.1007/978-3-319-11988-5_15
11. Oliveira, P.H., Fraideinberze, A.C., Laverde, N.A., Gualdron, H., Gonzaga, A.S., Ferreira, L.D., Oliveira, W.D., Rodrigues Jr., J.F., Cordeiro, R.L.F., Traina Jr., C., Traina, A.J.M., de Sousa, E.P.M.: On the support of a similarity-enabled relational database management system in civilian crisis situations. In: ICEIS (1), pp. 119–126. SciTePress (2016)
12. Pola, I.R.V., Cordeiro, R.L.F., Traina, C., Traina, A.J.M.: A new concept of sets to handle similarity in databases: the SimSets. In: Brisaboa, N., Pedreira, O., Zezula, P. (eds.) SISAP 2013. LNCS, vol. 8199, pp. 30–42. Springer, Heidelberg (2013). doi:10.1007/978-3-642-41062-8_4
13. Schallehn, E., Sattler, K., Saake, G.: Advanced grouping and aggregation for data integration. In: CIKM, pp. 547–549. ACM (2001)
14. Sedgewick, R., Wayne, K.: Algorithms, 4th edn. Addison-Wesley, Boston (2011)
15. Silva, Y.N., Aly, A.M., Aref, W.G., Larson, P.: SimDB: a similarity-aware database system. In: SIGMOD Conference, pp. 1243–1246. ACM (2010)
16. Silva, Y.N., Aref, W.G., Ali, M.H.: Similarity group-by. In: ICDE, pp. 904–915. IEEE Computer Society (2009)
17. Silva, Y.N., Aref, W.G., Ali, M.H.: The similarity join database operator. In: ICDE, pp. 892–903. IEEE Computer Society (2010)
18. Tang, M., Tahboub, R.Y., Aref, W.G., Atallah, M.J., Malluhi, Q.M., Ouzzani, M., Silva, Y.N.: Similarity group-by operators for multi-dimensional relational data. IEEE Trans. Knowl. Data Eng. **28**(2), 510–523 (2016)
19. Zezula, P., Amato, G., Dohnal, V., Batko, M.: Similarity Search - The Metric Space Approach. Advances in Database Systems, vol. 32. Kluwer, Dordrecht (2006)
20. Zhang, C., Huang, Y.: Cluster By: a new sql extension for spatial data aggregation. In: GIS, p. 53. ACM (2007)

Succinct Quadtrees for Road Data

Kazuki Ishiyama[1], Koji Kobayashi[2], and Kunihiko Sadakane[1(✉)]

[1] Graduate School of Information Science and Technology,
The University of Tokyo, Tokyo, Japan
{kazuki_ishiyama,sada}@mist.i.u-tokyo.ac.jp
[2] National Institute of Informatics, Tokyo, Japan
kobaya@nii.ac.jp

Abstract. We propose succinct quadtrees, space-efficient data structures for nearest point and segment queries in 2D space. We can compress both the tree structure and point coordinates and support fast queries. One important application is so called map matching, given GPS location data with errors, to correct errors by finding the nearest road. Experimental results show that our new data structure uses 1/25 working memory of a standard library for nearest point queries.

1 Introduction

In this paper we consider a fundamental problem in computational geometry: data structures for nearest neighbor queries. Though there exist a lot of researches on the topic [3,5,12], our research is different in the sense that we focus on queries on compressed data using compressed data structures. In a standard approach, we are given a set of n points stored in an array, and construct a data structure on the array for efficient queries. Then the space requirement is typically $nw + \mathcal{O}(n \log n)$ bits where w is the number of bits necessary to represent coordinates of a point. The term $\mathcal{O}(n \log n)$ is the size of the data structure because a pointer to the array uses $\mathcal{O}(\log n)$ bits. The value of w is at least 64 for two dimensional points and the hidden constant in the big-O is large in general. The value of n is also huge for real data, for example, road networks in Japan or USA have more than 50 million edges[1]. Therefore a standard approach requires huge memory.

1.1 Related Work

There exist some compressed data structures for point sets. de Bernardo et al. [9] proposed compact representations of raster data, which represent one attribute for each point in 2D space, using an extension of the k^2-tree [7]. Gagie et al. [13] proposed compressed quadtrees for storing coordinates of 2D point sets, which is also a variant of the k^2-tree. Venkat and Mount [21] proposed succinct quadtrees

---

This work was supported by JST CREST Grant Number JPMJCR1402, Japan.

[1] http://www.dis.uniroma1.it/challenge9/download.shtml.

© Springer International Publishing AG 2017
C. Beecks et al. (Eds.): SISAP 2017, LNCS 10609, pp. 262–272, 2017.
DOI: 10.1007/978-3-319-68474-1_18

for proximity queries on 2D point sets. These data structures are for storing not line segments but points. The difference is that we have to keep orders of points to store line segments, which makes compressing coordinates more difficult.

The interleaved k^2-tree [6] is a data structure for indexing 3D points, which can be used for storing line segments by adding an integer ID to each point in line segments. However the data structure cannot use the fact that points with similar ID's are located close and the index size will be large.

There exist data structures for searching for line segments such as R⁺-tree and PMR quadtree [15]. These are disk resident and use 20 bytes for a line segment, which is huge.

1.2   Our Contribution

In our approach, the input data are compressed, and the data structure is also compressed. We propose *succinct quadtrees* which support efficient nearest point or segment queries on compressed data using a small data structure. Though our data structure has many applications, we focus on the problem called *map matching*; given a query point in a two dimensional space and a set of line segments, we find k-nearest line segments to the point in the set. This is a typical query in GIS (geographic information systems). Line segments represent a road network, and a point represents the location of a car or a person obtained by GPS. Because GPS data contain errors, we do map matching to align the location to a road. We consider a system that collects locations of many cars or people to detect events such as accidents or to discover knowledge from the data. Therefore it is important to support such queries efficiently using small resources.

We give experimental results on real road network data in Japan. We compare our succinct quadtrees with one of the most famous libraries for nearest neighbor queries called ANN [3]. Our succinct quadtree uses 1/100 working memory of that of ANN, at the cost of increase in nearest point query time by a factor of 5 to 10.

2   Preliminaries

2.1   Data Structures for Nearest Queries

Quadtrees have been proposed by Finkel and Bentley [12]. A two-dimensional space is subdivided into four rectangles by cutting the space by two orthogonal lines. Each subspace is further divided recursively. Therefore the regions are represented by a tree, each internal node of which has four children, called *Quadtree*. Though it is easy to search the space recursively from the root of the quadtree, high-dimensional variants of the quadtree are not efficient. In such a case, k-d tree [5] is used. A space is divided into two sub regions recursively. In high-dimensional space, exact nearest neighbor search is difficult. Therefore approximate nearest search algorithms have been also proposed [3].

As for compression of data and indexes, there exist several results. Arroyuelo et al. [2] proposed data structures for two-dimensional orthogonal range search.

Indeed their data structure can be made implicit; no extra space is required beyond that for the data points. Brisaboa et al. [8] proposed space-efficient data structures for two-dimensional rectangles. Coordinates of rectangles are sorted in each dimension and stored as increasing integer sequences. Therefore space for coordinates are compressed, however we have to store permutations representing correspondence of coordinates. There are also compressed data structures for integer grids [11,17] and for nearest neighbor searches in high dimensions [20].

2.2 Succinct Data Structures

Succinct data structures are data structures which use asymptotically optimal space for encoding data and support efficient queries. In this paper, we use one for sparse bit-vectors [14,18] and one for ordered trees [1,4].

A bit-vector of length n with m ones and $n - m$ zeros is encoded in $m(2 + \lceil \log \frac{n}{m} \rceil) + o(m)$ bits so that the position of i-th one from the left is computed in constant time. We use it for compressing pointers to data structures.

We also use the DFUDS representation of an ordered tree [4]. An n-node ordered tree is encoded in $2n + o(n)$ bits so that parent, i-th child, subtree size, etc. can be computed in constant time. A node with degree d is represented by d open parentheses, followed by a close parenthesis. However for application to quadtrees, there are only two types of nodes: an internal node with four children and a leaf. Therefore instead of '(((()' and ')', we can encode '1' and '0'. Operations on the tree are also supported in constant time [16].

3 Succinct Quadtrees

Consider to store a point set P on a plane in a quadtree T. Assume that all the points in P are in a universe $R = [0, W) \times [0, H)$. The root node of T corresponds to R. If the number n of points in R is greater than a constant C, R is divided into four sub-regions $R_0 = [W/2, W) \times [H/2, H)$, $R_1 = [0, W/2) \times [H/2, H)$, $R_2 = [0, W/2) \times [0, H/2)$, and $R_3 = [W/2, W) \times [0, H/2)$. The set P is also divided into $P_0, \ldots, P_3$ so that points in P_i are contained in R_i for $i = 0, \ldots, 3$. The root node of T has four children, which are quadtrees storing P_i's. The edge from a node for R to its child for R_i are labeled i. We divide a region into four if it contains more than C points and the depth of the corresponding node in T is at most a constant D. Otherwise the node becomes a leaf. A leaf contains at most D points, and may contain no points.

We can identify nodes of the quadtree in several ways. First, each node has a unique preorder in the quadtree, which can be computed in constant time using the DFUDS representation. We can also give preorders for only leaves. Next, each node can be identified with the label on the edges from the root to that node. This is equivalent to use the coordinate of a point in the region represented by the node and the depth of the node. We assume that the coordinates of the points are distinct. Therefore we can identify a point by its coordinates. If we store auxiliary data to a point, the data are indexed by the coordinates of the point.

We encode the tree T using the DFUDS representation. Because we always divide a region at the middle, it is not necessary to store the coordinates of cut lines. The DFUDS and the values W, H are enough to represents coordinates of each sub-region. Each leaf of T stores a (possibly empty) set of points. The points are stored in different places from that for T. Therefore we need to store pointers to the places. We store the points in preorder of leaves, and the pointers to them are compressed by the data structure for sparse bit-vectors [18].

Query algorithm for the nearest point query is the same as that of the original quadtree. Given a query region (a circle with radius ϵ, a rectangle, etc.), we first find children of the root node of the quadtree which overlaps the query region, then recursively search them. If we arrive at a leaf, we check all the points in it.

3.1   Storing Points

To compress coordinates of points, points in P are classified into two: base points and diff points. We classify the points so that for any diff point q, there exists a base point $b \neq q$ in the same leaf of the quadtree as q and the distance between q and b is at most a constant L. We say that b is the base point of q.

We define a point query to find all points $p \in P$ such that $dist(q, p) \leq \epsilon$. We also define a k-nearest point query to find k-nearest points with distance at most ϵ.

To prove our main result, we give some lemmas. We assume that the distance function is a metric, that is, it is symmetric and satisfies the triangle inequality.

Lemma 1. *Let q be a query point of a query. Assume that there exists a point p such that $dist(p, q) \leq \epsilon$. Then there is a point b such that b is the base point of q and $dist(q, b) \leq \epsilon + L$.*

Proof. It is obvious from the definition of base points and the triangle inequality.   □

Therefore we can perform a point query by first finding the set B of all base points b with $dist(q, b) \leq \epsilon + L$, then checking all diff points p whose base point is in B. The following theorem gives a worst-case time bound.

Theorem 1. *Let $N(P, q, \epsilon)$ be the number of points in a set P within distance ϵ from a point q, and $T_A(P, q, \epsilon)$ be the time to enumerate those points by an algorithm A. Let Q be a set of base points for points in P. Then there exists an algorithm B for k-nearest point query satisfying $T_B(P, q, \epsilon) = T_A(Q, q, \epsilon + L) + \mathcal{O}(N(P, q, \epsilon + 2L)) + \mathcal{O}(k \log k)$.*

Proof. We can create the algorithm B as follows. For each base point p, we store a list of diff points for p so that we can enumerate the diff points in time proportional to the number of those points. Then we create a data structure for a point query from the set Q of base points using the algorithm A. The algorithm B first enumerates all base points within distance $\epsilon + L$ from a query point q. Then the algorithm B enumerates all diff points of the base points by scanning the lists, and checks for each diff point if it is within distance ϵ from q. The algorithm B can find any diff point p within distance from q because if such p exists, there is the base point b of p and distance between p and b is at most L. Therefore b can be found by the point query from Q. Next we estimate the number of diff points which are associated with the base points. Because the distance between q and a base point is at most $\epsilon + L$, and the distance between a base point and its diff point is at most L, the distance between q and any diff point is at most $\epsilon + 2L$. Therefore the number of diff points to check is at most $N(P, q, \epsilon + 2L)$. For a k-nearest query, we use a Fibonacci heap. We insert all the points in the Fibonacci heap in $\mathcal{O}(N(P, q, \epsilon + 2L))$ time. Then we can extract k smallest values in $\mathcal{O}(k \log k)$ time. $\qquad\square$

Theorem 1 says that there is an algorithm using indexes for a sampled point set Q and simple lists for other points without indexes. Therefore we can reduce the index size. The theorem also says that we can use any encoding for the diff points if it supports sequential decoding. Therefore we can compress coordinates of diff points. Note that the theorem holds for any data structure for nearest point queries, not only for the quadtree, and for any distance function which is metric.

We show how to store points compactly. The coordinates of a point p is represented by the label of the leaf containing p and relative coordinates inside the region of the leaf. For base points, we sort them in increasing order of their x-coordinates, then they are encoded as differences from the previous value using the delta-code [10]. Note that because y-coordinates are not sorted, we have to encode negative values.

For diff points whose base point is the same, we sort them by x-coordinates, and store differences of x-coordinates from the previous value using the delta-code. We also store the number of diff points for each base point. Then we can define a unique number to each point (either a base or a diff point). Namely, let $b_1, b_2, \ldots, b_k$ be base points and $d_1, d_2, \ldots, d_k$ be the number of diff points associated with the base points, respectively. Then the number for i-th base point is $1 + d_1 + 1 + d_2 + \cdots + 1 + d_{i-1} + 1 = i + \sum_{j=1}^{i-1} d_j$. The unique number for a base point can be computed in constant time using a sparse bit-vector which represents prefix sums of the numbers of diff points for base points [18]. The x-coordinate of a diff point is the summation of all differences from the corresponding base point.

3.2   Storing Polygonal Chains

We consider to store a set of polygonal chains. A typical application is to store road maps. A polygonal chain is a set of line segments $(v_0, v_1), (v_1, v_2), \ldots, (v_{m-1}, v_m)$ where v_i are points.

The query we consider is nearest segment query. That is, given a query point p, we find line segments in the set of polygonal chains so that the distance between p and a line segment is at most δ. We define the distance $dist(p, \ell)$ between a point p and a line segment $\ell = (u, v)$ as follows. Let w be the foot of a perpendicular from p to ℓ. If w is inside ℓ, we define $dist(p,\ell) \equiv dist(p,w)$, otherwise $dist(p,\ell) \equiv \min\{dist(p,u), dist(p,v)\}$.

We reduce a segment query into point queries. First we show basic facts.

Lemma 2. *Let q be a query point of a segment query. Assume that there exists a line segment $\ell = (u,v)$ such that $dist(q,\ell) \leq \epsilon$. Then it holds that $\min\{dist(q,u), dist(q,v)\} \leq \epsilon + dist(u,v)/2$ and $\max\{dist(q,u), dist(q,v)\} \leq \epsilon + dist(u,v)$.*

Proof. The claim holds if the nearest point in a segment (u,v) from q is an endpoint because $\min\{dist(q,u), dist(q,v)\} = dist(q,\ell) \leq \epsilon$ and $\max\{dist(q,u), dist(q,v)\} \leq \min\{dist(q,u), dist(q,v)\} + dist(u,v) = dist(q,\ell) + dist(u,v) \leq \epsilon + dist(u,v)$. If the nearest point is inside the segment, let w be the foot of a perpendicular from q to ℓ. Then $\min\{dist(q,u), dist(q,v)\} \leq dist(q,w) + \min\{dist(w,u), dist(w,v)\} \leq \epsilon + dist(u,v)/2$. We can also show $\max\{dist(q,u), dist(q,v)\} \leq dist(q,w) + \max\{dist(w,u), dist(w,v)\} \leq \epsilon + dist(u,v)$. □

Let P be all the points in a given set of polygonal chains. We first construct the quadtree T from P. Then for each point $p \in P$, a leaf of the quadtree to which p belongs is defined. If there exists a line segment (v_{i-1}, v_i) in a polygonal chain such that v_{i-1} and v_i belongs to different leaves and v_i is not the tail of the polygonal chain, we cut the polygonal chain at v_i.

We greedily classify the points. Namely, we define v_0 to be a base point and let $q = v_0$. Then, for $i = 1, 2, \ldots, v_m$, if the distance between v_i and q is at most L, we define v_i to be a diff point. Otherwise we define v_i to be another base point and let $q = v_i$.

We reduce a segment query into point queries on base point.

Lemma 3. *Let q be a query point of a segment query. Assume that there exists a line segment $\ell = (u,v)$ such that $dist(q,\ell) \leq \epsilon$. Then there exists a base point b such that $dist(q,b) \leq \epsilon + dist(u,v) + L$.*

Proof. From Lemma 2, both $dist(q,u)$ and $dist(q,v)$ are at most $\epsilon + dist(u,v)$. If u or v is a base point, the claim holds. Otherwise, there is a base point b such that $dist(u,b) \leq L$ or $dist(v,b) \leq L$. Then it holds $dist(q,b) \leq \max\{dist(q,u), dist(q,v)\} + \max\{dist(u,b), dist(v,b)\} \leq \epsilon + dist(u,v) + L$. □

Therefore if we know the maximum length of line segments, we can find a desired line segment by first finding candidate end points of line segments and checking the distances.

Theorem 2. *Let $N(P,q,\epsilon)$ be the number of points in a set P within distance ϵ from a point q, and $T_A(P,q,\epsilon)$ be the time to enumerate those points by an algorithm A. Let P be a set of points in polygonal chains and Q be a set of base points for points in P. Let M be the maximum length of a line segment. Then there exists an algorithm for k-nearest segment query running in time $T_A(Q,q,\epsilon + M + L) + \mathcal{O}(N(P,q,\epsilon + M + 2L)) + \mathcal{O}(k \log k)$.*

K. Ishiyama et al.

Proof. Let (u, v) be a line segment which is of distance at most ϵ from the query point q. From Lemma 3, the distance between q and the base point for u or v is at most $\epsilon + dist(u, v) + L \leq \epsilon + M + L$. Therefore we can find all base points for candidates of desired line segments in time $T_A(Q, q, \epsilon + M + L)$. Then we scan lists of diff points for those base points. Because all diff points are within distance L from their base points, the distance between those diff points and q is at most $\epsilon + M + 2L$. Therefore the number of diff points which are candidates of end points of desired line segments is at most $N(P, q, \epsilon + M + 2L)$ and we can enumerate them in time proportional to the number of the points. By using a Fibonacci heap, we can find k-nearest segments in $\mathcal{O}(k \log k)$ time. □

The encoding or coordinates of the points are almost the same as the case of storing points. However because diff points for a base point have an order, we cannot sort them. We encode them in the same order as in the polygonal chain.

If there exist few number of long segments, we cut them into short segments by adding dummy nodes on them. Then we can lower the value M, the maximum length of line segments. Though the number of points increases, queries will be faster because search radius decreases.

We can improve the compression ratio by giving a reasonable assumption of polygonal chains that the curvature is small. Instead of differences of coordinates, we encode differences of differences (twice differential). Then the absolute values decrease and the compression ratio improves.

4 Digital Road Map

Digital Road Map (DRM for short) is a standard format of road map in Japan[2]. The universe is partitioned into primary meshes, each of which is a rectangle of 1 degree in longitude and 2/3 degree in latitude. Each primary mesh is further partitioned into 8×8 secondary meshes. Then each secondary mesh is roughly a square of 10 Km times 10 km. The labels of primary and secondary meshes are defined by the coordinates. The location of a point is represented by the label of the secondary mesh to which the point belongs, and the normalized coordinates inside the secondary mesh. The normalized coordinates are represented by a pair of numbers from 0 to 10000 (not necessarily integers). If a line segment overlaps two secondary meshes, it is divided into two segments, each of which is contained in a secondary mesh.

Because in DRM the map is completely divided into secondary meshes, it is convenient to align regions in our quadtree with secondary meshes. Consider to represent the region on the earth with longitude 100 to 164 degrees and latitude 20 to 62.666666 degrees that contains the whole Japan by our quadtree. Then a node in the quadtree with depth 6 corresponds to a primary mesh, and that with depth 9 corresponds to a secondary mesh.

Each road (polygonal chain) has a unique ID. Therefore the space for storing IDs should be taken into account. Our data structure can also store IDs compactly. An ID of a polygonal chain consists of the label of the secondary mesh

---

[2] http://www.drm.jp/english/drm/e_index.htm.

containing the polygonal chain and a unique integer inside the secondary mesh. The former is determined by the coordinates, therefore we need not to store it. The latter can be changed inside the secondary mesh. Because we permute data for better compression, we change the IDs so that they coincide with the order in a data structure. Therefore we can implicitly encode the IDs.

5   Experimental Results

We compare our succinct quadtree with a standard library for nearest point searches, ANN: A Library for Approximate Nearest Neighbor Searching[3]. The library supports both exact and nearest point queries, but we used only exact ones. It uses the k-d tree [5]. Unfortunately there exist no publicly available source codes for line segment databases.

Because ANN does not support nearest segment queries, we show experimental results for only nearest point queries. However the performance of our segment queries is similar to that of point queries because the difference is only the search radius and distance calculation. Though ANN supports both bounded and unbounded radius searches, we use only bounded radius searches. For experiments, we used a Linux machine with 64GB memory and Intel Xeon CPU E5-2650 (2.60GHz). The algorithms use single core. We used g++ compiler 4.4.7. To measure the memory usage, we use the getrusage function. The function returns the value Maxrss, the amount of memory used in the algorithm. Therefore we can obtain the precise memory usage. To measure the query time, we use the clock function, which returns the processor time.

For a data set, we create a point set from real road map data in Japan. The number of points is about 54 million (53,933,415). Figure 1 shows the k-d tree for the point set, drawn by ann2fig command of the ANN library. Though we cannot show the quadtree because we do not have a tool, its shape is similar to the k-d tree. A point is represented by two real numbers truncated to 6 decimal places. In succinct quadtrees, the numbers are converted to integers by multiplying 10^6. If we store the coordinates as two 32-bit integers or float numbers, 411 MB are necessary.

The quadtree has 8,948,040 nodes and 838,879 leaves. Therefore each leaf contains 64 points on average. The DFUDS for the tree uses 2.5 MB, and compressed coordinates of the points occupy 174 MB. Therefore the size of the quadtree is 177 MB, which is smaller than a naive encoding of coordinates. We use the original encoding of nodes in DFUDS for simplicity; an internal node uses five bits and a leaf uses one bit. The coordinate of a two-dimensional point is compressed to 27 bits, including pointers for random access. Because the number of nodes of the quadtree is not so small compared with that of input points (the ratio is about 1/6), it is important to efficiently encode coordinates of regions. In succinct quadtrees, they are implicitly represented. This is a reason that succinct quadtrees use less space.

---

[3] http://www.cs.umd.edu/~mount/ANN/.

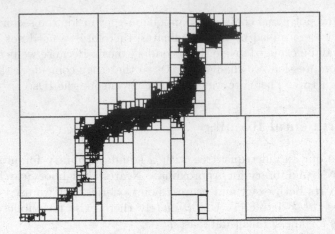

Fig. 1. The k-d tree for road map in Japan created by ann2fig command of the ANN library.

5.1   The Effect of Leaf Cache

We show experimental results on the effect of leaf cache. Let k be a fixed integer. When the query algorithm decodes a leaf of the succinct quadtree, the decoded information is stored in a cache. We keep recent k leaves decoded during the algorithm. We use a simple algorithm to update the cache; leaf cache is stored in a linked list, which is updated by the move-to-front rule. When the query algorithm tries to decode a leaf, it first scans the list to search for the leaf. If it is found, we do not decode the leaf again, and it is moved to the head of the list. If it is not found, the leaf is decoded and stored at the head of the list, and the tail entry is removed if there are more than k entries.

Table 1 shows the experimental results. We performed 1 million queries. We tried cache size (the number of leaves which can be stored in the cache) from 0 to 1000. We measured query time, Maxrss, and the numbers of cache hit and miss. From the experiments, we choose leaf cache size 50 because it achieves good trade-off between memory usage and query time.

5.2   Comparison with ANN

We compare our succinct quadtree with the ANN library. We find 10 nearest points to a query point. The distribution of query points follows that of real GPS location data of cars.

In succinct quadtrees, we set the leaf cache size as 50. In ANN, we set the search radius as 0.001, which is about 100 meters on the earth. For a set of 1 million queries, the ANN took 4.61 seconds using 6,532,780 KB memory. As shown in Table 1, our succinct quadtree with leaf cache size 50 took 21.70 seconds using 52,348 KB memory, and one with no cache took 49.80 seconds using 51,300 KB memory. The working memory of the succinct quadtree is less than 1/100

Table 1. The effect of leaf cache. Maxrss is the amount of memory used.

Cache size	Time (s)	Maxrss (KB)	Cache hit	Cache miss
0	49.80	51300	0	15583041
10	48.15	51812	1945984	13637057
20	36.13	51904	7805431	7777610
30	26.56	51948	11365034	4218007
40	22.52	52164	13620269	1962772
50	21.70	52348	14191071	1391970
60	21.65	52404	14202692	1380349
70	21.44	52592	14206189	1376852
80	21.98	52600	14208726	1374315
90	21.55	52888	14210652	1372389
100	21.72	52972	14212504	1370537
1000	26.70	62808	14281407	1301634

of that of the ANN! Note that the working memory of the succinct quadtree is smaller than the index size, 177 MB. This is because we use the mmap (memory map) function of the operating system. The part of the data structure that is accessed by the query algorithm is loaded into the memory. For a fair comparison, we assume that all the data structure is loaded into the memory. Even in this case, the succinct quadtree uses 1/25 memory of the ANN. Queries are five to ten times slower, depending on the size of leaf cache.

6   Concluding Remarks

We have proposed succinct quadtrees, data structures for two-dimensional nearest point and segment queries. Succinct quadtrees use much less space than an existing library for nearest point queries. Succinct quadtrees can be also used for indexing three or high dimensional space. An application will be to compress coordinates of three dimensional triangle meshes [19].

References

1. Arroyuelo, D., Cánovas, R., Navarro, G., Sadakane, K.: Succinct trees in practice. In: Proceedings 11th Workshop on Algorithm Engineering and Experiments (ALENEX), pp. 84–97. SIAM Press (2010)
2. Arroyuelo, D., et al.: Untangled monotonic chains and adaptive range search. Theor. Comput. Sci. **412**(32), 4200–4211 (2011)
3. Arya, S., Mount, D.M.: Approximate nearest neighbor queries in fixed dimensions. In: Proceedings of the Fourth Annual ACM-SIAM Symposium on Discrete Algorithms, SODA 1993, pp. 271–280, Philadelphia, PA, USA, 1993. Society for Industrial and Applied Mathematics (1993)

4. Benoit, D., Demaine, E.D., Munro, J.I., Raman, R., Raman, V., Rao, S.S.: Representing trees of higher degree. Algorithmica **43**(4), 275–292 (2005)
5. Bentley, J.L.: Multidimensional binary search trees used for associative searching. Commun. ACM **18**(9), 509–517 (1975)
6. Brisaboa, N.R., de Bernardo, G., Konow, R., Navarro, G.: K^2-Treaps: Range Top-k Queries in Compact Space. In: Moura, E., Crochemore, M. (eds.) SPIRE 2014. LNCS, vol. 8799, pp. 215–226. Springer, Cham (2014). doi:10.1007/978-3-319-11918-2_21 K ^{2}-Treaps: Range Top-k Queries in Compact Space. In: Moura, E., Crochemore, M. (eds.) SPIRE 2014. LNCS, vol. 8799, pp. 215–226. Springer, Cham (2014). doi:10.1007/978-3-319-11918-2_21
7. Brisaboa, N., Ladra, S., Navarro, G.: Compact representation of web graphs with extended functionality. Inf. Syst. **39**(1), 152–174 (2014)
8. Brisaboa, N., Luaces, M., Navarro, G., Seco, D.: Space-efficient representations of rectangle datasets supporting orthogonal range querying. Inf. Syst. **35**(5), 635–655 (2013)
9. Bernardo, G., Álvarez-García, S., Brisaboa, N.R., Navarro, G., Pedreira, O.: Compact Querieable Representations of Raster Data. In: Kurland, O., Lewenstein, M., Porat, E. (eds.) SPIRE 2013. LNCS, vol. 8214, pp. 96–108. Springer, Cham (2013). doi:10.1007/978-3-319-02432-5_14
10. Elias, P., Interval, R.R., Coding, S.: Two on-line adaptive variable-length schemes. IEEE Trans. Inf. Theor. **33**(1), 3–10 (1987)
11. Farzan, A., Gagie, T., Navarro, G.: Entropy-bounded representation of point grids. Comput. Geom. Theory Appl. **47**(1), 1–14 (2014)
12. Finkel, R.A., Bentley, J.L.: Quad trees a data structure for retrieval on composite keys. Acta Informatica **4**(1), 1–9 (1974)
13. Gagie, T., González-Nova, J.., Ladra, S., Navarro, G., Seco, D.: Faster compressed quadtrees. In: Proceedings 25th Data Compression Conference (DCC), pp. 93–102 (2015)
14. Grossi, R., Vitter, J.S.: Compressed suffix arrays and suffix trees with applications to text indexing and string matching. SIAM J. Comput. **35**(2), 378–407 (2005)
15. Hoel, E.G., Samet, H.: A qualitative comparison study of data structures for large line segment databases. In: Proceedings of the 1992 ACM SIGMOD international conference on Management of data - SIGMOD 1992, vol. 21, pp. 205–214, New York, USA, 1992. ACM Press (1992)
16. Jansson, J., Sadakane, K., Sung, W.-K.: Ultra-succinct representation of ordered trees with applications. J. Comput. Syst. Sci. **78**(2), 619–631 (2012)
17. Navarro, G., Nekrich, Y., Russo, L.: Space-efficient data-analysis queries on grids. Theoret. Comput. Sci. **482**, 60–72 (2013)
18. Okanohara, D., Sadakane, K.: Practical entropy-compressed rank/ select dictionary. In: Proceedings of Workshop on Algorithm Engineering and Experiments (ALENEX) (2007)
19. Rossignac, J.: Edgebreaker: connectivity compression for triangle meshes. IEEE Trans. Visual Comput. Graphics **5**(1), 47–61 (1999)
20. Tellez, E.S., Chavez, E., Navarro, G.: Succinct nearest neighbor search. Inf. Syst. **38**(7), 1019–1030 (2013)
21. Venkat, P., Mount, D.M.: A succinct, dynamic data structure for proximity queries on point sets. In: Proceedings of the 26th Canadian Conference on Computational Geometry, CCCG 2014, Halifax, Nova Scotia, Canada, 2014 (2014)

Applications and Specific Domains

On Competitiveness of Nearest-Neighbor-Based Music Classification: A Methodological Critique

Haukur Pálmason[1], Björn Þór Jónsson[1,2(✉)], Laurent Amsaleg[3],
Markus Schedl[4], and Peter Knees[5]

[1] Reykjavík University, Reykjavik, Iceland
bjorn@ru.is
[2] IT University of Copenhagen, Copenhagen, Denmark
[3] CNRS–IRISA, Rennes, France
[4] Johannes Kepler University Linz, Linz, Austria
[5] TU Wien, Vienna, Austria

Abstract. The traditional role of nearest-neighbor classification in music classification research is that of a straw man opponent for the learning approach of the hour. Recent work in high-dimensional indexing has shown that approximate nearest-neighbor algorithms are extremely scalable, yielding results of reasonable quality from billions of high-dimensional features. With such efficient large-scale classifiers, the traditional music classification methodology of aggregating and compressing the audio features is incorrect; instead the approximate nearest-neighbor classifier should be given an extensive data collection to work with. We present a case study, using a well-known MIR classification benchmark with well-known music features, which shows that a simple nearest-neighbor classifier performs very competitively when given ample data. In this position paper, we therefore argue that nearest-neighbor classification has been treated unfairly in the literature and may be much more competitive than previously thought.

Keywords: Music classification · Approximate nearest-neighbor classifiers · Research methodology

1 Introduction

The traditional role of nearest neighbor classification in music information retrieval research is that of a straw man opponent: after the music features have been fine-tuned to optimize the results of the learning approach of the day, which typically means aggregating and compressing the audio features, the k-NN classifier is applied to those features directly, inevitably losing. We argue, however, that since the strength of k-NN classification lies precisely in the ability to handle a large quantity of data efficiently, this methodology is grossly misleading.

© Springer International Publishing AG 2017
C. Beecks et al. (Eds.): SISAP 2017, LNCS 10609, pp. 275–283, 2017.
DOI: 10.1007/978-3-319-68474-1_19

1.1   Trends in High-Dimensional Indexing

In the music domain, Schnitzer [22] developed a content-based retrieval system that operates on a collection of 2.3 million songs and can answer audio queries in a fraction of a second by using a filter-and-refine strategy. As state-of-the-art content description models build upon high-dimensional Gaussian distributions with costly similarity calculations, an approximate projection into a vector space is used to perform high-speed nearest neighbor candidate search, before ranking candidates using the expensive model. Using the efficiency of k-NN, the combined approach speeds up queries by a factor of 10–40 compared to a linear scan.

In the image retrieval domain, recent work in high-dimensional indexing has shown that approximate nearest-neighbor algorithms are extremely scalable, as several approaches have considered feature collections with up to several billions of features. Jégou et al. [9] proposed an indexing scheme based on the notion of product quantization and evaluated its performance by indexing 2 billion vectors. Lejsek et al. [12] described the NV-tree, a tree index based on projections and partitions, with experiments using 2.5 billion SIFT feature vectors. Babenko and Lempitsky used the inverted multi-index to index 1 billion SIFT features [2] and deep learning features [3] on a single server. Finally, the distributed computing frameworks Hadoop and Spark have been used with collections of up to 43 billion feature vectors [8,14]. Evaluation of the quality of these approaches shows that approximate nearest neighbor retrieval yields results of reasonable quality. In particular, many studies have shown that while results from individual queries may lack quality, applications with redundancy in the feature generation usually return results of excellent quality (e.g., see [12,14]). These results indicate that in order to fairly evaluate k-NN classification as an independent methodology, we must supply the classifiers with massive collections of highly redundant features, and then aggregate the results to classify individual items [5].

1.2   Critique of Methodology

In this position paper, we present a case study of k-NN classification using a well-known MIR genre classification benchmark with well-known music features from the literature. Unlike previous studies, however, we generate an extensive collection of music features to play to the strength of k-NN classifiers. We first use an exhaustive (exact) k-NN classifier, which scans the entire feature collection sequentially, to tune the parameters of the feature generation. Then we use a very simple approximate k-NN classifier to show that equivalent results can be obtained in a fraction of the time required for the exhaustive search.

The results of this case study show that the approximate k-NN classifier strongly outperforms any k-NN classification results reported in the literature. Furthermore, the approximate k-NN classifier actually outperforms learning-based results from the literature obtained using this particular music collection and these particular music features. We therefore argue that nearest-neighbor-based classification is much more competitive that previously thought. Furthermore, we argue that when evaluating the quality of k-NN classification, the evaluators must work with the strengths of k-NN classification, namely scalability and efficiency, for a fair comparison. Anything else is much like inviting a fish to a tree-climbing competition with a monkey!

2   Music Classification: A (Very Brief) Literature Review

Automatic genre classification of music has received considerable attention in the MIR community in the last two decades, albeit partly (and recently) in the form of criticism [11, 18, 26, 27]. Due to space limitations, we only survey a few key works of content-based methods that are particularly relevant to our case study. For a more detailed discussion on the general subject we refer the reader to [27], and for a review of context/web-based methods to [10].

It is notable that genre classification results reported in the literature are realized on very different datasets with different numbers of classes/genres, as there is (for valid reasons) no commonly agreed upon standard dataset for this task (cf. [1, 6, 17]). For instance, the ISMIR2004 dataset partitions its 1,458 full song examples into 6 genre classes, while the the USPOP collection used in the MIREX 2005 Genre Classification set comprises 8,746 tracks by 400 artists classified into 10 genres (of which 293 artists belong to "Rock") and 251 styles. The seminal work of Tzanetakis and Cook [28] uses a dataset with 10 musical genres, and proposes to classify 1,000 song excerpts of 30 seconds based on timbral features, rhythmic content, and pitch content. While their collection (referred to as GTZAN) exhibits inconsistencies, repetitions, and mislabeling, as confirmed by Tzanetakis and investigated in detail by Sturm [25], it has since been used by several researchers, and is indeed the collection used in this study.

Tzanetakis and Cook [28] used a 30-dimensional feature vector extracted from each of the song excerpts in their collection. An accuracy of 61% is reported using a Gaussian Mixture Model classifier and 60% using k-NN. Li et al. [13] used the same features as Tzanetakis and Cook, with the addition of Daubechies Wavelet Coefficient Histograms (DWHC). Wavelets decompose the audio signal based on octaves with each octave having different characteristics. Support Vector Machine (SVM) classification yielded 78.5% classification accuracy on GTZAN while only 62.1% was achieved using k-NN. One feature vector per song was used for the experiments. Instead of using just one vector for each song, Bergstra et al. [4] extract features in frames of 1,024 samples and aggregate frames into segments before using AdaBoost for classification, yielding 82.5% accuracy on GTZAN. They state that the best results are achieved by aggregating between 50 and 100 frames for a segment. In the MIREX 2005 Audio Genre Classification competition, however, this setting could not be pursued as calculation for all songs would have taken longer than the allowed 24 h. Panagakis et al. [19] achieved 78.2% accuracy using "multiscale spectro-temporal modulation features," where each audio file is converted to a third-order feature tensor generated by a model of auditory cortical processing. Non-negative matrix factorization is used for dimensionality reduction and SVM for classification. The highest accuracy results for GTZAN are achieved by Panagakis et al. [20] with 91% accuracy. The authors extract auditory temporal modulation representation of each song and examine several classifiers. Their best results are achieved using sparse representation-based classification (SRC), while their results using k-NN are below 70% for GTZAN. Seyerlehner et al. [24] propose to compute several spectrum- and rhythm-based features from cent-scaled audio spectrograms on the level of (overlapping) blocks. Training an SVM gave an accuracy of 78%, but with a Random Forest classifier the accuracy was 87% [23].

Recent approaches for genre recognition, music similarity, and related retrieval tasks, aim at modeling finer grained musical concepts, also as an intermediate step for genre classification [21], or try to optimize audio features using user preference patterns by means of deep learning [15]. While these approaches are specifically devised with scalability of indexing in mind, they cannot be compared to existing work due to the proprietary industrial data used. In an effort to make large-scale learning approaches comparable, van den Oord et al. [16] apply deep learning models, trained using the Million Song Dataset, to classification of GTZAN (transfer learning), achieving 88.2% accuracy.

3 Case Study: Genre Classification

In this section, we report on our classification experiments using two k-NN classifiers, one exact and one approximate. We start by describing the experimental setup. Then we detail experiments using a full (exact) sequential scan to analyze various aspects of the music feature configuration. We then use the best feature configuration to study the impact of the approximate k-NN classifier.

3.1 Experimental Setup

For our experiments, we used the MARSYAS framework for feature extraction [28] that accompanies the GTZAN collection. As the purpose of this case study is not to create the best classifier, or generate the best classification results, but rather to point out methodological issues in previous studies, we chose to use a collection that is (a) extensively studied in the literature, and (b) comes with easily available and flexible feature extraction software.

The GTZAN collection consists of 1,000 song excerpts, where each excerpt is 30 seconds long. Each song is sampled at 22,050 KHz in mono. The songs are evenly distributed into 10 genres: Blues, Classical, Country, Disco, HipHop, Jazz, Metal, Pop, Reggae and Rock. We used randomized and stratified 10-fold cross-validation, by (a) shuffling songs within each genre, and then concatenating the genre files, and (b) by ignoring, when computing distances, all feature vectors from songs that are located within the same 10% of the collection as the query. We ignore the first feature vectors of each song, due to overlap between songs. All experiments were performed on an Apple MacBook Pro with 2.66 GHz Duo Core Intel processors, 4 GB RAM and a 300 GB 5.4Krpm Serial ATA hard disk.

3.2 Exact Classification: Impact of Feature Parameters

The first k-NN classifier performs a sequential scan of all database descriptors and calculates the Manhattan distance between each query vector and each database vector. Once the scan has computed the nearest neighbors for all query feature vectors, the total score for the query song is aggregated, by counting how many neighbors "vote" for each genre and returning the ranked list of genres.

Table 1. Effect of feature selection on accuracy

Features	Dim.	Accuracy(%)
TF	34	75.4
TF + SFM	82	80.4
TF + SFM + SCF	130	80.0
TF + SFM + LSP	118	80.0
TF + SFM + LPCC	106	79.8
TF + SFM + CHROMA	106	79.4

Table 2. Effect of varying k

Neighbors (k)	Accuracy (%)	Time (min)
1	80.4	207.4
2	80.7	208.3
3	80.8	209.3
4	80.6	210.1
5	80.5	212.3

For feature extraction, we used window and hop sizes of 512 and a memory size of 40, as they performed best in experiments. For feature selection, we started from timbral features (TF), namely MFCCs, and added additional features; the results are summarized in Table 1. The table shows that adding spectral flatness measure (SFM) increases the accuracy for this test set from 75.4% to 80.4%. Adding further information to the feature vector, however, does not improve results, and in fact actually hurts accuracy slightly. In the remainder of this study, we hence use the TF+SFM features, with a dimensionality of 82.

Finally, we experiment with the k parameter: how many neighbors from the collection are considered for each query feature. We ran the sequential scan with k ranging from 1 to 10. Table 2 shows the results (no further change is observed beyond $k = 5$). As the table shows, varying k does not affect the classification accuracy much, nor does it have significant impact on the classification time.

3.3 Approximate Classification: Impact of eCP Parameters

Having reached a respectable 80.8% accuracy in 3.5 h using a sequential scan, we now turn our attention to reducing the classification time by employing the eCP high-dimensional indexing method [7]. Very briefly, the algorithm works as follows. For index construction, C points are randomly sampled from the feature collection as cluster representatives. Each feature vector is then assigned to the closest representative, and the resulting clusters are stored on disk. In order to facilitate efficient assignment, as well as efficient search, a cluster index is created by recursively sampling and clustering the representatives. At search time, the b clusters that are closest to each query feature are found via the cluster index, and then scanned to find the closest k neighbors. The approximation comes from

Table 3. Effect of varying number of clusters C

Clusters	Features	Clustering	Classification	Accuracy
(C)	(average per cluster)	(sec)	(sec)	(%)
50	25,867	7	784	80.0
100	12,933	7	470	80.6
500	2,587	7	114	80.2
1,000	1,293	7	55	78.4
5,000	259	9	15	79.7

the fact that only b clusters are scanned (retrieving a fraction b/C of the feature collection, on average), and some neighbors may be missed, because they are close to a boundary or due to the approximate nature of the index tree traversal.

Table 3 shows the impact of the number of clusters C on clustering time, classification time and classification accuracy. In this experiment $b = k = 1$. We observe that even with very small clusters (larger C), where only about 260 features are read on average, the classification accuracy is nearly the same as with the sequential scan. The classification time, however, is in the order of a few minutes or less, compared to over 200 min for the sequential scan.

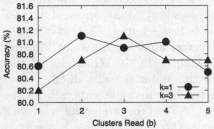

Fig. 1. Effect of b on accuracy   **Fig. 2.** Effect of b on classification time

Figure 1 shows the impact of b on accuracy (note the narrow y-axis), for $C = 100$ and $k = \{1, 3\}$. In other configurations, not reported here due to space constraints, similar effects were observed. When b is increased, the eCP algorithm retrieves and analyses more feature vectors. When the first clusters are added, results are improved. As more clusters are added, however, near vectors belonging to other genres are also found, but no change is observed beyond $b = 5$. These fluctuations are due to the fact that eCP is an approximate technique and since songs quite often receive a similar number of votes from two genres, limiting the number of audio features retrieved can have this effect. But with several configurations we actually achieve higher accuracy than with the sequential scan.

Figure 2, on the other hand, shows the time required to retrieve the clusters. As the figure shows, there is some initial time needed to rank the cluster representatives, but after that the cost of reading clusters grows linearly. The cost, however, is not affected significantly by the k parameter.

3.4 Discussion

The key result is that using approximate k-NN classification we can very efficiently obtain accuracies in the range 80–81%. Table 4 summarizes classification results from the literature for the used GTZAN collection, as well as k-NN classification results where reported. As the table shows, the results from our case study rank fourth and far outperform any reported k-NN classification results from the literature. While previous results have denounced k-NN classification, this is presumably because those studies focused on using a single feature, or very few features, for each song. The fact that we can obtain these results within minutes, on very limited hardware, leads us to believe that k-NN classification is a viable option at a large scale.

Table 4. Comparison to some genre classification results for the GTZAN collection

Reference	Accuracy	k-NN	Comments
Panagakis et al., 2009 [20]	91.0%	$\simeq$68%	Sparse repr. + dim. reduction
v.d. Oord et al., 2014 [16]	88.2%	–	Transfer learning from MSD tags
Seyerlehner et al., 2011 [23]	87.0%	–	Random forest, opt. feat.
This work	**80–81%**		**k-NN + MFCC + SFM**
Li et al., 2003 [13]	78.5%	62.1%	SVM
Panagakis et al., 2008 [19]	78.2%	–	Multilinear approach
Tzanetakis & Cook, 2002 [28]	61.0%	60.0%	Gaussian classifier + MFCC

4 Conclusions

Very efficient and scalable approximate algorithms for high-dimensional k-NN retrieval have recently been developed in the computer vision and databases communities. Using one such approximate algorithm, we have shown that k-NN classification may be used with good results for music genre classification, if provided the appropriate feature collection to work with. Indeed, the reported results are the best realized with a k-NN classifier for this particular music collection (GTZAN). Observing the difference between our results and previous results reported for k-NN classification points to a methodological problem with previous studies: the k-NN classifiers were compared unfairly to the competing approaches. We thus argue that when evaluating the quality of k-NN classification, the evaluators must work with the strengths of k-NN classification, namely scalability and efficiency, for a fair comparison.

Acknowledgements. This work was partially supported by Icelandic Student Research Fund grant 100390001, Austrian Science Fund (FWF) grant P25655 and Austrian FFG grant 858514 (*SmarterJam*).

References

1. Aucouturier, J.J., Pachet, F.: Representing musical genre: a state of the art. J. New Music Res. **32**(1), 83–93 (2003)
2. Babenko, A., Lempitsky, V.S.: The inverted multi-index. TPAMI **37**(6), 1247–1260 (2015)
3. Babenko, A., Lempitsky, V.S.: Efficient indexing of billion-scale datasets of deep descriptors. In: Proceedings of CVPR, Las Vegas, NV, USA (2016)
4. Bergstra, J., Casagrande, N., Erhan, D., Eck, D., Kégl, B.: Aggregate features and ADABOOST for music classification. Mach. Learn. **65**(2–3), 473–484 (2006)
5. Boiman, O., Shechtman, E., Irani, M.: In defense of nearest-neighbor based image classification. In: Proceedings of CVPR (2008)
6. Fabbri, F.: A theory of musical genres: two applications. Popular Music Perspect. **1**, 52–81 (1981)
7. Guðmundsson, G., Amsaleg, L., Jónsson, B.: Distributed high-dimensional index creation using Hadoop, HDFS and C++. In: Proceedings of CBMI (2012)
8. Guðmundsson, G., Amsaleg, L., Jónsson, B., Franklin, M.J.: Towards engineering a web-scale multimedia service: a case study using Spark. In: Proceedings of MMSys (2017)
9. Jégou, H., Douze, M., Schmid, C.: Product quantization for nearest neighbor search. TPAMI **33**(1), 117–128 (2011)
10. Knees, P., Schedl, M.: A survey of music similarity and recommendation from music context data. ACM TOMCCAP **10**(1), 1–21 (2013)
11. Knees, P., Schedl, M.: Music Similarity and Retrieval - An Introduction to Audio- and Web-based Strategies. Springer, Heidelberg (2016)
12. Lejsek, H., Jónsson, B., Amsaleg, L.: NV-Tree: nearest neighbours at the billion scale. In: Proceedings of ICMR (2011)
13. Li, T., Ogihara, M., Li, Q.: A comparative study on content-based music genre classification. In: Proceedings of SIGIR, Toronto, Canada (2003)
14. Moise, D., Shestakov, D., Guðmundsson, G., Amsaleg, L.: Indexing and searching 100M images with map-reduce. In: Proceedings of ICMR (2013)
15. van den Oord, A., Dieleman, S., Schrauwen, B.: Deep content-based music recommendation. In: Proceedings of NIPS (2013)
16. van den Oord, A., Dieleman, S., Schrauwen, B.: Transfer learning by supervised pre-training for audio-based music classification. In: Proceedings of ISMIR (2014)
17. Pachet, F., Cazaly, D.: A taxonomy of musical genre. In: Proceedings of RIAO (2000)
18. Pálmason, H., Jónsson, B., Schedl, M., Knees, P.: Music genre classification revisited: An in-depth examination guided by music experts. In: Proceedings of CMMR (2017)
19. Panagakis, I., Benetos, E., Kotropoulos, C.: Music genre classification: a multilinear approach. In: Proceedings of ISMIR (2008)
20. Panagakis, Y., Kotropoulos, C., Arce, G.R.: Music genre classification via sparse representations of auditory temporal modulations. In: Proceedings of EUSIPCO (2009)

21. Prockup, M., Ehmann, A.F., Gouyon, F., Schmidt, E.M., Celma, Ò., Kim, Y.E.: Modeling genre with the music genome project: comparing human-labeled attributes and audio features. In: Proceedings of ISMIR (2015)
22. Schnitzer, D.: Indexing content-based music similarity models for fast retrieval in massive databases. Dissertation, Johannes Kepler University, Austria (2012)
23. Seyerlehner, K., Schedl, M., Knees, P., Sonnleitner, R.: A refined block-level feature set for classification, similarity and tag prediction. In: Proceedings of MIREX (2011)
24. Seyerlehner, K., Widmer, G., Pohle, T.: Fusing block-level features for music similarity estimation. In: Proceedings of Digital Audio Effects (DAFx) (2010)
25. Sturm, B.L.: An analysis of the GTZAN music genre dataset. In: Proceedings of MIRUM (2012)
26. Sturm, B.L.: Classification accuracy is not enough. JIIS **41**, 371–406 (2013)
27. Sturm, B.L.: A survey of evaluation in music genre recognition. In: Nürnberger, A., Stober, S., Larsen, B., Detyniecki, M. (eds.) AMR 2012. LNCS, vol. 8382, pp. 29–66. Springer, Cham (2014). doi:10.1007/978-3-319-12093-5_2
28. Tzanetakis, G., Cook, P.: Musical genre classification of audio signals. IEEE Trans. Speech Audio Process. **10**(5), 293–302 (2002)

DS-Prox: Dataset Proximity Mining
for Governing the Data Lake

Ayman Alserafi[1,2]([✉]), Toon Calders[2,3], Alberto Abelló[1], and Oscar Romero[1]

[1] Universitat Politècnica de Catalunya - BarcelonaTech, Barcelona, Catalunya, Spain
{alserafi,aabello,oromero}@essi.upc.edu
[2] Université Libre de Bruxelles (ULB), Brussels, Belgium
{aalseraf,toon.calders}@ulb.ac.be
[3] Universiteit Antwerpen (UAntwerp), Antwerp, Belgium
toon.calders@uantwerp.be

Abstract. With the arrival of Data Lakes (DL) there is an increasing
need for efficient dataset classification to support data analysis and information retrieval. Our goal is to use meta-features describing datasets to
detect whether they are similar. We utilise a novel proximity mining
approach to assess the similarity of datasets. The proximity scores are
used as an efficient first step, where pairs of datasets with high proximity
are selected for further time-consuming schema matching and deduplication. The proposed approach helps in early-pruning unnecessary computations, thus improving the efficiency of similar-schema search. We
evaluate our approach in experiments using the OpenML online DL,
which shows significant efficiency gains above 25% compared to matching without early-pruning, and recall rates reaching higher than 90%
under certain scenarios.

1 Introduction

Data Lakes (DL) [1] are huge data repositories covering a wide range of heterogeneous topics and business domains. Such repositories need to be effectively
governed to gain value from them; they require the application of data governance techniques for extracting information and knowledge to support data
analysis and to prevent them from becoming an unusable *data swamp* [1]. This
involves the organised and automated extraction of metadata describing the
structure of information stored [15], which is the main focus of this paper.

The main challenge for data governance posed by DLs is related to information retrieval: identify related datasets to be analysed together as well as duplicated information to avoid repeating analysis efforts. To handle this challenge
it was previously proposed in [2] to utilise schema matching techniques which
can identify similarities between attributes of different datasets. Most techniques
proposed by the research community [4] are designed for 1-to-1 schema matching
applications that do not scale up to large-scale applications like DLs prone to
gather thousands of datasets.

© Springer International Publishing AG 2017
C. Beecks et al. (Eds.): SISAP 2017, LNCS 10609, pp. 284–299, 2017.
DOI: 10.1007/978-3-319-68474-1_20

To facilitate such holistic schema matching and to deal with the sheer size of the DL, [4] proposed to utilise the strategy of early pruning which limits the number of comparisons of pairs of datasets. We apply this approach in this paper by proposing a technique which approximates the proximities of pairs of datasets using similarity-comparisons of their meta-features. More specifically, we use a supervised machine learning approach to model topic-wise related classification of datasets. We then utilise this model in assigning proximities between new datasets and those already in the DL, and then predicting whether those pairs should be compared using schema matching (i.e., have related information) or not. We implement this technique in the datasets-proximity (DS-Prox) approach presented in this paper. Our focus is on early-pruning of unnecessary dataset comparisons prior to applying state-of-the-art schema matching and deduplication (the interested reader is referred to [4,13] for more details on such techniques).

Our contributions include the following: 1. a novel proximity mining approach for calculating the similarity of datasets (Sect. 4), 2. applying our new technique to the problem of early-pruning in holistic schema matching and deduplication within different scenarios for maintaining the DL (Sects. 2, 3), and finally, 3. testing the proposed proximity mining approach on a real-world DL to demonstrate its effectiveness and efficiency in early-pruning (Sect. 5).

2   Problem Statement

Our goal is to automate information profiling, defined in [2], which aims at efficiently finding relationships between datasets in large heterogeneous repositories of *flat semi-structured data* (i.e., tabular data like CSV, web tables, spreadsheets, etc.). Those repositories usually include datasets uploaded multiple times with the same data but with different transformed attributes. Such datasets are structured as groups of *instances* describing real-world entities, where each instance is expressed as a set of *attributes* describing the properties of the entity. We formally define a dataset D as a set of instances $D = \{I_1, I_2, ...I_n\}$. The dataset has a schema of attributes $S = \{A_1, A_2, ...A_m\}$, where each attribute A_i has a fixed type, and every instance has a value of the right type for each attribute. We focus on two types of attributes: continuous numeric attributes and categorical nominal attributes, and two types of relationships for pairs of datasets $[D_1, D_2]$:

- $Rel(D_1, D_2)$: Related pairs of datasets describe similar real-world objects or concepts from the same domain of interest. These datasets store similar information in (some of) their attributes. Typically, the information contained in such attributes partially overlap. An example would be a pair of datasets describing different human diseases, like one for diabetes patients and another for hypertension patients. The datasets will have similar attributes (partially) overlapping their information like the patient's age, gender, and some common lab tests like blood samples.

– $Dup(D_1, D_2)$: Duplicate pairs of datasets describe the same concepts. They convey the same information in *most* of their attributes, but such information can be stored using differences in data. For example, two attributes can describe the weight of an object but one is normalised between 0 and 1 and the other holds the raw data in kilograms. Both attributes are identified to be representing similar information although their data are not identical.

Examples. We scrutinise the relationship between two pairs of datasets in Fig. 1. Each dataset has a set of attributes. An arrow links similar attributes between two datasets. For example, attributes 'A1' from D_2 and D_3 are nominal attributes with two unique values, making them similar. A numeric attribute like 'A2' in D_2 holds similar data as attributes 'A3' and 'A4' from D_3, as expressed by the intersecting numeric ranges. In our approach we extract meta-features from the datasets (for this example, the number of distinct values and means respectively) to assess the similarity between attributes of a given pair of datasets. The *Rel* and *Dup* properties are then used to express datasets similarities. For example, $Dup(D_1, D_2)$ returns '1' because they have similar information in most attributes (even though 'A5' and 'A3' do not match). Based on these two properties, our proposed approach will indicate whether two datasets are possibly related (e.g., $Rel(D_2, D_3) = $'1') and should be considered for further scrutinising by schema matching, or if they are possibly duplicated (e.g., $Dup(D_1, D_2) = $'1') and should be considered for deduplication efforts.

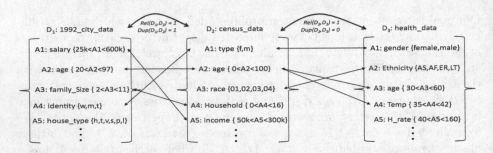

Fig. 1. Similarity relationships between two pairs of datasets

Scenarios. We aim at governing the DL by maintaining the *Rel* and *Dup* relationships between the datasets it contains. We consider two typical scenarios. In **scenario (a)**, we want to dredge a data swamp which we don't know any relationships for, thus, for all pairs in the DL we need to find if they are related or duplicated. In **scenario (b)**, we have an existing DL for which we know all relationships between the datasets. However, given the dynamic nature of DLs new datasets are frequently ingested. Thus, we need to compare this dataset against the datasets already in the DL to find its relationships with them.

3   Related Work

As described in [15], metadata describing the information stored in datasets need to be collected to effectively govern big data repositories. Such metadata are usually automatically collected across multiple datasets using data profiling techniques like schema matching [11], which seeks to identify schematic overlaps between datasets. This involves detecting related objects (instances or attributes) and matching instances between two different schemata [4]. The main line of research in this field is focused around improving the efficiency of matching techniques for two very large schemata. In our research, however, we focus on matching attributes between *multiple large amounts of schemata*, closely related to the field of holistic schema matching [4,13]. A more restrictive case of schema matching involves *deduplication* [13]; finding highly overlapping instances [6]. Similar to our special requirements for schema matching, we also seek to detect *duplicated schemata* instead of instances. This is when schemata have similar overlapping attributes, not necessarily the same instances.

As described in [4], it is recommended to utilise early-pruning mechanisms for holistic schema matching, which filters out unnecessary matching efforts using less complex techniques. This is commonly done using similarity search techniques which seek to eliminate unnecessary comparisons of datasets [12]. Several techniques for *instance-based matching* were proposed including techniques like clustering [3,6,8], hashing [12], and indexing [10,12]. Alternatively, we propose to focus on *attribute-based* matching across multiple-schemata for governing the DL which needs new and efficient techniques. This field was not sufficiently studied before, with only preliminary results in [14]. We propose a new approach utilising a novel technique of computationally cheaper meta-features proximity comparisons. We seek to prevent unnecessary and expensive schema matching computations in further steps. We propose a machine learning approach for early-pruning that is based on metadata collected from datasets. Such learning techniques were proposed for future research in similarity search [5] where they use a supervised machine learning model based on SVM to find similar strings for deduplication. [5] shows that using machine learning leads to more accurate similarity search from different domains of knowledge.

4   The DS-Prox Approach

We propose a proximity computation based on overall meta-features extracted from the datasets, which we call DS-Prox. We are seeking to have approximate similarity comparisons of pairs of datasets for the early-pruning task. Here we apply cheap computation steps for the overall similarity search, to prevent further expensive detailed analysis of the content of datasets which are estimated to be dissimilar. Similar to our previous work in [2], we seek to profile the datasets ingested in the DL by extracting some *meta-features* describing the overall content and attributes in the datasets. We compute distances between each of the meta-features as proximity metrics. We take a sample of pairs of datasets which are analysed by a data analyst and annotated whether they hold *related* or *duplicate* data by means of the *Rel* and *Dup* properties. *Rel* and *Dup* are boolean functions retrieving either 1 (similar/duplicate respectively) or 0 (dissimilar/not duplicate). We then use machine learning techniques over the proximity metrics to create two independent models which can classify pairs of datasets according to $Rel(D_1, D_2)$ and $Dup(D_1, D_2)$ respectively. The classification models are used to *score* pairs of datasets with a similarity measure $Sim(D_1, D_2)$. The similarity score '*Sim*' is defined independently for each of the relationships $Rel(D_1, D_2)$ and $Dup(D_1, D_2)$ as a number between 0 and 1, where 0 means dissimilar and 1 means most similar: $Sim(D_1, D_2) \in [0, 1]$.

4.1 The Meta-Features Distance Measures

For each dataset in the DL, we extract meta-features using data profiling techniques. This includes general statistics about the dataset and its attributes as described in Table 1. Our purpose for those meta-features is to describe the general structure and content of the datasets for an approximate comparison using our proximity metric and classification models. We compute distances for each meta-feature m_i from Table 1 between each pair of datasets $[D_1, D_2]$ using Eq. 1 which gives the relative difference as a number between 0 and 1. Those distances we feed to the supervised machine learning algorithm in our approach.

$$dist_{m_i}(D_1, D_2) = \frac{\max\{m_i(D_1), m_i(D_2)\} - \min\{m_i(D_1), m_i(D_2)\}}{\max\{m_i(D_1), m_i(D_2)\}} \quad (1)$$

4.2 The Approach

The approach proposed for early-pruning depends on classical machine learning which is divided into two phases: *Supervised Learning* Phase and *Scoring and Classification* Phase. In the first phase, which can be seen in Fig. 2, we build a classification model for each of the properties *Rel* and *Dup* using supervised learning techniques. First, for each dataset we extract its meta-features from Table 1 which returns its data profile (In Fig. 2 we see a sample of two meta-features: number of attributes 'nAttr', and number of instances 'nIns'). Then, for each dataset, we generate all pairs with each of the other datasets and compute the distances between their meta-features using Eq. 1. We also present the pairs of datasets to a human-annotator who manually decides whether they satisfy (assign '1') or not satisfy (assign '0') $Rel(D_1, D_2)$ and $Dup(D_1, D_2)$. Any pair annotated as a match for $Dup(D_1, D_2)$ must also be annotated as a match for $Rel(D_1, D_2)$ (i.e., all duplicate pairs of datasets are also related). We feed both the annotated pairs of datasets with their distances as training examples to a learner which creates two classifiers: M_{rel} and M_{dup}.

In the second phase, we apply the classifiers to the scenarios discussed in Sect. 2, to score each new pair of previously unseen datasets. In *scenario (a)*, we have a setting where there are two DLs. DL_1 has a group of datasets which have previously known annotations of all their $Rel(D_1, D_2)$ and $Dup(D_1, D_2)$

Table 1. DS-Prox meta-features

Type	Meta-feature	Description
General	Number of instances	The number of instances in the dataset
	Number of attributes	The number of attributes in the dataset
	Dimensionality	The ratio of number of attributes to number of instances
Attributes by type	Number per type	The number of attributes per type (Nominal or Numerical)
	Percentage per type	The percentage of attributes per type (Nominal or Numerical)
Nominal attributes	Average number of values	The average number of distinct values per nominal attribute
	Standard deviation of number of values	The standard deviation in the number of distinct values per nominal attribute
	Minimum/maximum number of values	The minimum and maximum number of distinct values per nominal attribute
Numeric attributes	Average numeric mean	The average of the means of all numeric attributes
	Standard deviation of the numeric mean	The standard deviation of the means of the numeric attributes
	Minimum/maximum numeric mean	The minimum and maximum mean of numeric attributes
Missing values	Missing attribute count	The number of attributes with missing values
	Missing attribute percentage	The percentage of attributes with missing values
	Minimum/maximum number of missing values	The minimum and maximum number of instances with missing values per attribute
	Minimum/maximum missing values percentage	The minimum and maximum percentage of instances with missing values per attribute
	Mean number of missing values	The mean number of missing values from each attribute
	Mean percentage of missing values	The mean percentage of missing values from each attribute

relationships between all pairs of datasets. On the other hand, DL_2 is without any annotations of such relationships and is therefore a data swamp we would like to dredge. Therefore, we need to learn the models for $Rel(D_1, D_2)$ and

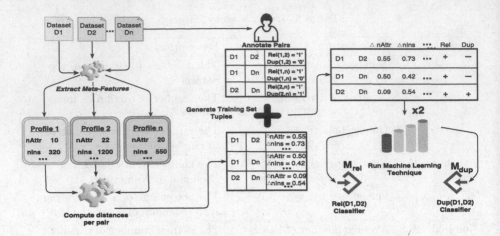

Fig. 2. DS-Prox: supervised machine learning

$Dup(D_1, D_2)$ from DL_1 and apply them to DL_2 which has different datasets. In *scenario (b)*, we have an existing DL for which we know all relationships between the datasets. We need to deal with a new dataset as it arrives in this DL. We learn the models from the DL, and we apply them to each new dataset D_i ingested within the same DL. The models should identify all datasets in the DL which are related or duplicate of D_i.

When applying the classifiers, we compute for each pair of datasets the similarity score of $Sim_{rel}(D_1, D_2)$ and $Sim_{dup}(D_1, D_2)$ using the classifiers extracted in the previous phase. The *Sim* score is the positive-class distribution value generated by each classifier. The predicted distribution-value achieved for the 'true' class from each classifier is checked against a minimum threshold to indicate whether the pair of datasets are overall related or duplicates. In our approach, pairs of datasets are evaluated first if they match the $Dup(D_1, D_2)$ relationship (indicating that it also matches $Rel(D_1, D_2)$. If it fails this duplicate test, then we evaluate if the pair still satisfies $Rel(D_1, D_2)$. The output classifiers can classify in the future any new pairs of datasets as either related or duplicate according to two matching approaches: *1-to-1 matching* or *cluster matching*.

1-to-1 matching: all pairs satisfying $Rel(D_1, D_2)$ and $Dup(D_1, D_2)$ need to be selected for further schema matching and deduplication. The calculations are performed under the assumption that each and every pair of matching datasets should be correctly identified using our models.

Cluster-based matching: It is common to use clustering based approaches for the matching process [3,4,8]. Groups of datasets with close proximity are segmented into clusters. In our case, the relationship *Rel* can be used to cluster the datasets in the DL, after all relationships are discovered. We therefore relax our requirements for the second phase so that a new dataset should match with *any single* dataset in the same cluster in order to consider it a positive match. Therefore, if a dataset matches one or more dataset(s) from a cluster, we consider all pairs of datasets in this cluster as positively matching pairs (even if the classifier did not indicate a positive match for some of those pairs separately). The rationale behind this approach is that in a real holistic schema matching setting, a new dataset ingested should be compared to *all* the datasets in a cluster it matches to. Clustering can take place after schema matching identifies the relationships between datasets (which is outside the scope of this paper, but the reader can refer to [3,8] for such clustering in instance-based matching).

We illustrate our general approach with a toy example in Fig. 3. Suppose we have two meta-features $nIns$ and $nAttr$ for each dataset. To classify a pair $[(nIns_1, nAttr_1), (nIns_2, nAttr_2)]$ we compute the relative differences. In Fig. 3(a) we have plotted $(\Delta nIns, \Delta nAttr)$ for all pairs in the training data. '+' indicates a matching pair, '−' a non-matching pair. Based on this data we learn a classifier, for instance a separating hyperplane as shown in Fig. 3(a) by the red line. Here, for simplification, we show pairs of datasets plotted based on the distances of only two meta-features ($nIns$ and $nAttr$). The actual approach would consider all meta-features in Table 1.

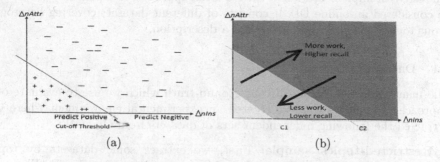

Fig. 3. DS-Prox cut-off thresholds tuning (Color figure online)

Most classification models produce a score instead of a binary output. In the example of the separating hyperplane the obtained distance to the hyperplane can be used as a score. This score can be compared against different cut-off thresholds to decide on the final classification '+' or '−'. The threshold can be chosen to lead to different results, as seen in Fig. 3(b). If we choose the cut-off threshold 'C1' we restrict the classifier to return less pairs of high proximity (i.e., low distance), leading to lower recall but less work. Alternatively, if we alter the cut-off threshold to 'C2', we relax the classifier to return pairs of lower proximity. This leads to more pairs (i.e., more work) returned by the classifier as positive matches and higher recall of positive cases, but, with more pairs marked incorrectly as matching. Therefore, the cut-off threshold can be tweaked by the data scientist according to practical requirements in order to increase recall at the expense of more work or vice versa. This is the trade-off which we seek to optimise in our experiments when selecting different thresholds. We can use different thresholds 'c_{rel}' and 'c_{dup}' for each of the classifiers evaluated.

This means that we consider a positive match if the classifier scores a new pair of datasets with a score greater than the threshold as in Eqs. (2) and (3).

$$Rel(D_1, D_2) = \begin{cases} 1, & Sim_{rel}(D_1, D_2) > c_{rel} \\ 0, & \text{otherwise} \end{cases} \quad (2)$$

$$Dup(D_1, D_2) = \begin{cases} 1, & Sim_{dup}(D1, D_2) > c_{dup} \\ 0, & \text{otherwise} \end{cases} \quad (3)$$

The complexity of our approach is quadratic in the number of datasets, however, it applies the cheapest computational steps for early-pruning (just computing distances in Eq. 1 and the classifier scoring model on each pair). This way, we save unnecessary expensive schema matching processing in later steps.

5   Experimental Evaluation

We tested an implementation of the DS-Prox approach on OpenML[1], which can be considered an online DL. It consists of different datasets covering heterogeneous topics, each having a name and a description.

5.1   Datasets

The main challenge is to create the ground-truth which we use to evaluate our approach. To achieve this, we created an experimental environment where we extracted the following independent sets of datasets from OpenML:

- **Restricted-topics sample:** First, we extract some datasets by topic using 11 keywords-search over OpenML, e.g., "Disease", "Cars", "Flights", "Sports", etc. This restricted sample consists of 130 datasets and we consider them to be similar if they belong to the same topic.
- **All-topics sample:** This is an independent set of other datasets collected from OpenML. To collect this sample, we scraped the OpenML repository to extract all datasets not included in the restricted-topics sample and having a description of more than 500 characters. Out of the 514 datasets retrieved we selected 213 with descriptive descriptions (i.e., excluding datasets whose descriptions do not allow to interpret its content and to assign a topic).

Therefore, we created two new groups of datasets from OpenML for our experiments, each having its own independent set of datasets without any overlap. Having two independent sets strengthens our results and allows us to generalise our conclusions. A domain expert and one of the authors collaborated to manually label the pairs of datasets with the same topic as duplicated and/or related. The interested reader can download the two annotated datasets from GitHub[2]. The details of each sample is summarised in Table 2, which lists the

---

[1] http://www.openml.org.

[2] https://github.com/AymanUPC/datasets_proximity_openml.

Table 2. A description of the OpenML samples collected

Sample	Datasets	Topics	Top topics	$Rel(D_1, D_2)$	$Dup(D_1, D_2)$
Restricted-Topics	130	29	Diseases (45), Health (31), Cars (13), Academic Courses (6), Sports (5)	1205	72
All-Topics	213	79	Computer software defects (17), citizens census data (12), digit handwriting recognition (12), Diseases (11)	570	128

number of datasets, the number of topics, top topics by the number of datasets, and the number of related and duplicated pairs per sample.

Some of the pairs from the all-topics sample can be seen in Table 3. Dataset with ID 23 should match all datasets falling under the topic of 'census data' like dataset 179. Both datasets have data about citizens from a population census. In rows 4 and 5 we can see examples of duplicated datasets, which have highly intersecting data in their attributes. Duplicate pairs in row 4 have the same number of instances, but described with different number of attributes, which are overlapping. The duplicate pairs in row 5 have identical number of attributes, yet, the attributes are transformed using pre-processing techniques and there are different number of instances between both datasets, so in essence the second dataset is a transformed and cleaned version of the first. We aim to detect such kind of scenarios using our DS-Prox approach.

Table 3. An example of pairs of datasets from the all-topics sample from OpenML

No.	DID 1	Dataset 1	DID 2	Dataset 2	Topic	Relationship
1	23	cmc	179	adult	Census data	related
2	14	mfeat-fourier	1038	gina_ agnostic	Digit handwriting recognition	related
3	55	hepatitis	171	primary-tumor	Disease	related
4	189	kin8nm	308	puma32H	Robot motion sensing	duplicate
5	1514	micro-mass	1515	micro-mass	Mass spectrometry data	duplicate

294   A. Alserafi et al.

5.2 Experimental Setup

In order to evaluate our approach, we create an experimental setup where we have two sets of datasets for each experiment: 1. Training set and 2. Test set. The training set is used in the supervised learning phase to create the classification models. The classification models are then evaluated using the test set. We use the restricted-topics sample as a training set, and we use both the restricted-topics and the all-topics samples in the scoring phase as test sets to evaluate our approach. We describe how we used those samples to create the training and test sets within our experiments for the two scenarios from Sect. 2:

- **Scenario (a)** from Sect. 2: We evaluate our approach by using the restricted-topics sample as the training set and the all-topics sample as the test set. In this case the testing set is an independent collection of datasets. We evaluate both of the 1-to-1 matching and cluster matching approaches for $Rel(D_1, D_2)$. We also evaluate the 1-to-1 matching with $Dup(D_1, D_2)$.
- **Scenario (b)** from Sect. 2: We evaluate our approach using a leave-one-out (LOO) variant evaluation method and the restricted-topics sample. Here we remove a dataset and all its pairs from the original training set and we use those pairs for evaluation of the output classifiers as a separate test set. We also remove all duplicate pairs of this dataset from the training set to guarantee independence between the training and evaluation environments. We repeat this for every dataset in the input training set. We use the 1-to-1 matching approach in our evaluation.

To execute our experiments, we profile the datasets to extract their meta-features. We use the training set of annotated datasets with the WEKA[3] tool to create the classification models using different supervised techniques: **Bayesian** (Bayesian Network with K2 search, Naïve Bayes), **Regression** (LogitBoost), **Support Vector Machines** (Sequential Minimal Optimization), and **Decision Trees** (Random Forest). We also use **Ensemble Learners** [7]: AdaBoost (with Decision Stump classifier), Classification Via Regression (with M5 Tree classifier), and Random Subspace (with Regression Tree classifier). We tested different techniques because it was suggested by [7] that some individual techniques can outperform the ensemble learners in classification problems. We evaluate the classifiers with 10 different cut-off thresholds for 'c_{rel}' and 'c_{dup}' from Eqs. (2) and (3), in order to cover a wide range of values. We benchmark the techniques against the decision table technique [9] which simply assigns the majority class based on matching the features to a table of learned examples.

5.3 Results

We evaluate the effectiveness of our approach using the recall, precision, and efficiency-gain measurements, as described in Eqs. (4), (5) and (6) respectively. Here, TP means true-positives which are the pairs of datasets correctly classified

---
[3] https://weka.wikispaces.com/Use+WEKA+in+your+Java+code.

by the classifier. FN are false negatives, FP are false-positives, TN are true-negatives, and N indicates the total number of possible pairs of datasets. The efficiency gain measures the amount of reduction in work required, in terms of number of pairs of datasets eliminated by the classifier.

$$recall = \frac{TP}{TP + FN} \tag{4}$$

$$precision = \frac{TP}{TP + FP} \tag{5}$$

$$efficiency - gain = \frac{TN + FN}{N} \tag{6}$$

We change the cut-off thresholds, and we aim to maximize this as much as possible while maintaining the highest recall possible. The effectiveness of our approach is evaluated by recall and precision. By applying our approach with the different scenarios and relationships, we conduct 4 sets of experiments as in Table 4. The results are depicted in the graphs in Fig. 4.

Table 4. A description of the experiments conducted

Experiment	Graphs in Fig. 4	Matching approach	Scenario	Relationship
1	row 1: (a) and (b)	1-To-1	(b)	$Rel(D_1, D_2)$
2	row 2: (c) and (d)	1-To-1	(a)	$Rel(D_1, D_2)$
3	row 3: (e) and (f)	Cluster-based	(a)	$Rel(D_1, D_2)$
4	row 4: (g) and (h)	1-To-1	(a)	$Dup(D_1, D_2)$

The measures shown in the graphs are all averages from all datasets involved in the test sets for a specific data mining technique and a certain cut-off threshold for the proximity score (darker points have higher cut-off values). The common measure for all graphs, which is the recall plotted on the y-axis, is highlighted by having some of its main values labelled on each graph. Graphs (a) and (b) are the same graphs as (c) and (d) respectively but for the 1-to-1 matching approach applied with the different scenarios. We select for the experiments certain target results, which are minimum expected values for each measure. All area above those values are shaded as follows: **Min. recall**: 0.75 for 1-to-1 matching & 0.9 for cluster matching, **Min. efficiency gain**: 0.33 for 1-to-1 matching & 0.25 for cluster matching, **Min. precision**: 0.25 for all approaches. This means that we were targeting at least 75% recall rate for 1-to-1 matching and 90% recall rate for the cluster based matching (which improves our previous results in [2]). We aimed for at least 25% efficiency gains with the cluster matching approach, which exceeds those achieved in [3]. However, we acknowledge that their approach applied to instances-matching within the same dataset, not cross-schema attribute-matching as in our case. For experiment 4 for $Dup(D_1, D_2)$, we aim for a min. recall of 0.9, min. efficiency gain of 0.75, and min. precision of

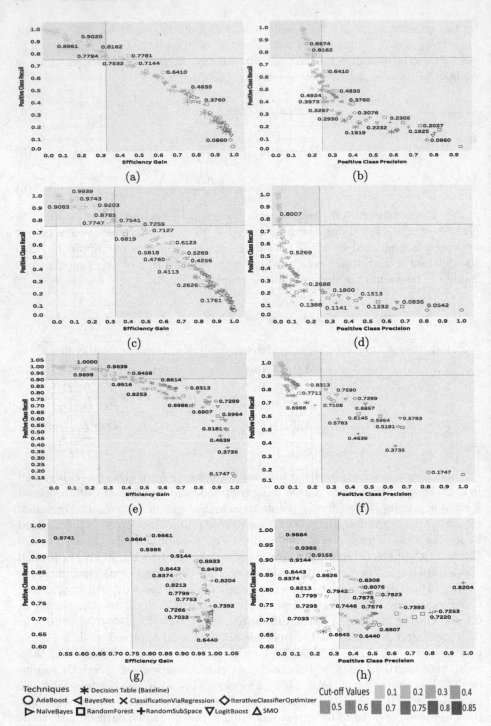

Fig. 4. Recall-efficiency plots (left column) and recall-precision plots (right column) for experiments 1,2,3 and 4 in each row respectively

0.33. In real-world applications, the data scientist can choose different minimum thresholds for each measure according to practical requirements.

5.4   Discussion

General trend. From the results depicted in Fig. 4, the optimum technique and cut-off threshold is the one in the top-right quadrant of each graph, optimising both measures plotted. The recall-precision and recall-efficiency plots follow the general trend expected which indicate the trade-off between both measures in each plot, yet, more optimised solutions are possible for balancing recall-efficiency, as seen by the classifiers performing in the top-right quadrant. As the cut-off thresholds increase, there is a drop in recall against an increasing efficiency gain. Still, the top mining techniques and thresholds can be used to achieve high efficiency gain and recall. This is discussed for each property below. As the precision rates are generally low, we conclude that our approach can only be used as an early-pruning step, and should be followed by other more expensive and more detailed matching steps. Yet, a good compromise can still achieve high recall and efficiency gains; efficiency gains up to 0.5 for *Rel* and 0.8 for *Dup*. Such efficiency gains can make an important difference for computationally-expensive applications of holistic schema matching in the DL environment.

Rel evaluation. The recall-efficiency plots indicated that it was possible to achieve an optimum technique and threshold in the top-right quadrant, which represent the compromise of not sharply losing recall with higher efficiency gain. For example, from Fig. 4(e) for experiment 3, using the AdaBoost technique at a threshold of 0.5 can lead to 0.42 efficiency gains while still maintaining 0.95 recall. If a recall of 1.0 is required, then this can be achieved by the cut-off threshold of 0.3 for the same technique, but only 0.13 efficiency gain is achieved. The data scientist will have to decide if this efficiency gain is sufficient and whether a recall rate of 100% is critical in their application, else, a 0.05 drop in recall should be allowed to achieve much higher efficiency gain using the techniques and thresholds in the top quadrant. For the 1-to-1 matching in Fig. 4(c), we can achieve 0.75 recall and 0.35 efficiency gain. There is a drop in recall, as would be expected, because the classifier has more challenges in matching all possible 'related' datasets, while in the cluster matching approach, a single match to a dataset in a cluster acts like a pivot which results in matching all the required related datasets in the same cluster. The cluster-matching approach shows an improved performance over the 1-to-1 matching approach, therefore it is recommended to use DS-Prox with the clustering-based approach.

Dup evaluation. For the results in Fig. 4(g) and (h), the top performing techniques were Random subspace and Random Forest at 0.2 cut-off thresholds. This achieved about 0.97 recall and 0.76-0.8 efficiency gain. The baseline method was not able to differentiate at different cut-off thresholds, and had best recall of 0.65, except for the lowest cut-off of 0.1 where it achieved a jump to 0.94 recall. Since the recall was very high for our target efficiency gain using the 1-to-1 approach, the cluster-based approach did not yield any better results.

Baseline comparisons. Different techniques can yield better results than the baseline for several of our experiments. There is not one single technique which is best, yet, ensemble learners tend to perform better than their counterparts. However, simple techniques like logistic regression and Naïve Bayes can still have good performance as seen in the graph (e) top-right quadrant. The baseline technique was never in the top-quadrant of graph (e) and many techniques outperformed it. In the 1-to-1 matching in graphs (a) and (c), the baseline classifier was comparable with the other techniques. The top techniques include the iterative optimiser in graph (c) with 0.75 recall and 0.35 efficiency gain. Nearly 1.0 recall was possible using the same classifier at a lower threshold, yet with only 0.09 efficiency gain. For experiment 1, Random Forest and Random Subspace outperformed the baseline with 0.3 cut-off thresholds.

Generalizability. Although our approach is generic and does not apply to a specific domain only, we note that we do not claim that the classifiers for one type of data or of a certain domain will have the same guaranteed effectiveness when applied in another setting. The approach might need to be adjusted and retrained within other settings. Albeit, our results from experiments 2 and 3 show a positive indicator of the possibility to train the model on specific domains, independent of those used in the test set (or real-world setting), and still be effective. We think that this needs further experimentation in the future.

6   Conclusion and Future Work

This paper presented a novel approach of similarity search within a DL based on a proximity mining technique for early-pruning in holistic dataset schema matching and deduplication applications. The approach uses supervised machine learning techniques based on meta-features describing semi-structured datasets. Experiments on a real-life DL demonstrate the effectiveness in achieving high recall rates and efficiency gains. Proposed techniques support data governance in the DL by identifying relationships between datasets. The drawback of our approach, however, is that it needs some manual effort to annotate training examples for the classifiers. In the future, we will test the generalizability of applying the same classifier to different data sources. We plan to experiment with more detailed meta-features which might lead to improved results. We will also test our approach on other kinds of semi-structured data (like RDF or XML).

Acknowledgement. This research has been funded by the European Commission through the Erasmus Mundus Joint Doctorate (IT4BI-DC).

References

1. Abelló, A.: Big data design. In: Proceedings of ACM DOLAP, pp. 35–38 (2015). doi:10.1145/2811222.2811235

2. Alserafi, A., Abelló, A., Romero, O., Calders, T.: Towards information profiling: data lake content metadata management. In: DINA Workshop, ICDM (2016). doi:10.1109/ICDMW.2016.0033
3. Ares, L.G., Brisaboa, N.R., Ordóñez Pereira, A., Pedreira, O.: Efficient similarity search in metric spaces with cluster reduction. In: Navarro, G., Pestov, V. (eds.) SISAP 2012. LNCS, vol. 7404, pp. 70–84. Springer, Heidelberg (2012). doi:10.1007/978-3-642-32153-5_6
4. Bernstein, P.A., Madhavan, J., Rahm, E.: Generic schema matching, ten years later. Proc. VLDB Endowment 4(11), 695–701 (2011)
5. Bilenko, M., Mooney, R.J.: Adaptive duplicate detection using learnable string similarity measures. In: ACM SIGKDD, pp. 39–48 (2003)
6. Cordero Cruz, J.A., Garza, S.E., Schaeffer, S.E.: Entity recognition for duplicate filtering. In: Traina, A.J.M., Traina, C., Cordeiro, R.L.F. (eds.) SISAP 2014. LNCS, vol. 8821, pp. 253–264. Springer, Cham (2014). doi:10.1007/978-3-319-11988-5_24
7. Džeroski, S., Ženko, B.: Is combining classifiers with stacking better than selecting the best one? Mach. Learn. 54(3), 255–273 (2004)
8. Figueroa, K., Paredes, R.: List of clustered permutations for proximity searching. In: Brisaboa, N., Pedreira, O., Zezula, P. (eds.) SISAP 2013. LNCS, vol. 8199, pp. 50–58. Springer, Heidelberg (2013). doi:10.1007/978-3-642-41062-8_6
9. Kohavi, R.: The power of decision tables. In: Lavrac, N., Wrobel, S. (eds.) ECML 1995. LNCS, vol. 912, pp. 174–189. Springer, Heidelberg (1995). doi:10.1007/3-540-59286-5_57
10. Lokoč, J., Čech, P., Novák, J., Skopal, T.: Cut-Region: a compact building block for hierarchical metric indexing. In: Navarro, G., Pestov, V. (eds.) SISAP 2012. LNCS, vol. 7404, pp. 85–100. Springer, Heidelberg (2012). doi:10.1007/978-3-642-32153-5_7
11. Naumann, F.: Data profiling revisited. ACM SIGMOD Rec. 42(4), 40–49 (2014)
12. Patella, M., Ciaccia, P.: Approximate similarity search: a multi-faceted problem. J. Discrete Algorithms 7(1), 36–48 (2009)
13. Rahm, E.: The case for holistic data integration. In: Pokorný, J., Ivanović, M., Thalheim, B., Šaloun, P. (eds.) ADBIS 2016. LNCS, vol. 9809, pp. 11–27. Springer, Cham (2016). doi:10.1007/978-3-319-44039-2_2
14. Stonebraker, M., et al.: Data curation at scale: the data tamer system. In: 6th Biennial Conference on Innovative Data Systems Research (CIDR) (2013)
15. Varga, J., Romero, O., Pedersen, T.B., Thomsen, C.: Towards next generation BI systems: the analytical metadata challenge. In: Bellatreche, L., Mohania, M.K. (eds.) DaWaK 2014. LNCS, vol. 8646, pp. 89–101. Springer, Cham (2014). doi:10.1007/978-3-319-10160-6_9

DeepBrowse: Similarity-Based Browsing Through Large Lists (Extended Abstract)

Haochen Chen, Arvind Ram Anantharam, and Steven Skiena[✉]

Stony Brook University, Stony Brook, NY 11790, USA
{haocchen,aranantharam,skiena}@cs.stonybrook.edu

Abstract. We propose a new approach for browsing through large lists in the absence of a predefined hierarchy. DeepBrowse is defined by the interaction of two fixed, globally-defined permutations on the space of objects: one ordering the items by similarity, the second based on magnitude or importance. We demonstrate this paradigm through our *WikiBrowse* app for discovering interesting Wikipedia pages, which enables the user to scan similar related entities and then increase depth once a region of interest has been found.

Constructing good similarity orders of large collections of complex objects is a challenging task. Graph embeddings are assignments of vertices to points in space that reflect the structure of any underlying similarity or relatedness network. We propose the use of graph embeddings (DeepWalk) to provide the features to order items by similarity.

The problem of ordering items in a list by similarity is naturally modeled by the Traveling Salesman Problem (TSP), which seeks the minimum-cost tour visiting the complete set of items. We introduce a new variant of TSP designed to more effectively order vertices so as to reflect longer-range similarity. We present interesting combinatorial and algorithmic properties of this formulation, and demonstrate that it works effectively to organize large product universes.

1 Introduction

Browsing is a form of information retrieval, where one does not know exactly what they want but hope to recognize it when they see it. Browsing through menus or lists of items is a very common component of user interface design for web and mobile applications. Menus are effective for presenting small sets of possible selections to the user, but rapidly become unwieldy and tedious to use beyond a dozen or so possibilities. Hierarchical systems, like faceted search or DAG-like structures help to efficiently navigate through large sets of possibilities, but constructing such taxonomies generally requires considerable effort and domain expertise.

We propose a new approach to list navigation, permitting serendipitous discovery over lists of hundreds of thousands of items without the need for a predefined hierarchy. Our approach *DeepBrowse* is based on two basic concepts:

© Springer International Publishing AG 2017
C. Beecks et al. (Eds.): SISAP 2017, LNCS 10609, pp. 300–314, 2017.
DOI: 10.1007/978-3-319-68474-1_21

1. We organize the universe of items in a fixed order, according to *similarity*. Thus local regions in the full list will be *coherent*, because each item should be similar to its neighbors.

2. To provide the diversity necessary for serendipitous discovery, we modulate the set of items visible from our current position by *significance*. At the highest level of navigation, we select from a broad array of the most popular or important items, but once we identify an item of interest we want to see more choices like that. Our interface enables the user to move between smaller and larger item universes at will.

These concepts and their presentation are extremely simple, but realizing them with minimum domain knowledge requires considerable technology under the hood. Consider the problem of constructing effective similarity orderings. How can we computationally measure the pairwise similarity between members of item-universes like books, movies, and music? We propose a general approach based on deep learning, namely the use of graph embeddings to construct similarity orderings.

How can we construct the most effective similarity order for browsing? We define an appropriate and novel optimization criteria for this task. Although it is NP-complete to construct the optimal order, we provide an approximation algorithm and heuristics which construct excellent orderings in practice. And finally, how can we order arbitrary items (e.g. books, movies, and music) by relative significance? We propose series of generally-available proxies to capture this notion.

This paper is organized as follows. We first introduce the *DeepBrowse* interface paradigm, and discuss its implementation through Android apps over three distinct item universes: Wikipedia pages, movies, and dictionary words. We then delve into the technical details of constructing similarity orderings through graph embeddings and combinatorial optimization. Finally, we report the results of a user study gauging the effectiveness of our interface, and review the research literature in several topics relevant to our work.

Specifically, our work makes the following contributions:

- **A Paradigm for List-Oriented Browsing.** We abstract the notion of browsing to the basic operations of *scanning* and *deepening*: scanning along a fixed similarity order, and deepening to expose more specialized items once a region of interest has been identified. We show that these operations permit us to access any list item of known position in $O(\log n)$ operations, assuming the two orderings are independent.

  We note that our approach can naturally be integrated with faceted search interfaces, by using the facet selection values as conditionals to block undesired items from appearing in the display window.

- **Implementation in Three Domains.** We have created three Android apps implementing the *DeepBrowse* paradigm in three distinct domains: Wikipedia pages (*WikiBrowse*), movies (*MovieBrowse*), and vocabulary words (*WordBrowse*). We have released these apps in the Android app store, and encourage the reader to play with it to get a feel for the interface in action.

– **Deep Learning/Embedding Approach for Measuring Item Similarity.** The notion of graph embeddings (*DeepWalk*) serves as a unifying approach to measure pairwise similarity between classes of items as diverse as people, movies, and vocabulary words. Starting from a partial similarity network, *DeepWalk* learns a high-dimensional embedding for each vertex in the graph, generalizing its structure and reducing the problem of computing pairwise distance to a vector difference and dot product.

We demonstrate that this approach scales to entity universes with hundreds of thousands of items, and provides a general and convenient abstraction for quantifying item similarity.

– **Similarity Orders for Effective Browsing.** The famous traveling salesman problem (TSP) seeks the minimum cost (maximum similarity) tour over a set of n items. However, the objective function in browsing is different than in transportation problems: we seek to maximize the similarity among items in each visible window, not just adjacent pairs.

To improve browsing orders, we develop the notion of k-*robust* TSP tours, generalizing the traditional (or 1-robust) TSP problem. To the best of our knowledge, this variant of TSP has never been previously studied in the literature.

We give efficient and effective heuristics to construct k-robust tours, and present experimental results that they achieve their objective of increasing categorical coherence with the parameter k.

– **User Study.** Browsing implies that the user does not have a well-defined task in mind, complicating the question of how to evaluate the success of their venture. Still, we perform a modest user study, demonstrating that our similarity order improved both performance and user experience on serendipitous discovery tasks over alphabetical order. We also demonstrate that user performance increases rapidly with exposure to the interface.

2   Related Work

Serendipitous Browsing. Serendipity is defined as the occurrence of something unexpected in a happy or beneficial way. André et al. [2,3] summarizes serendipity related research along two dimensions: the activity engaged in when encountering serendipitous information (directed browsing/nondirected browsing/none), and what type of information was found (relevant/not relevant to the goal). *DeepBrowse* falls into the category of nondirected browsing [13,25] where users do not have a pre-defined goal while using it. [17,21] show that organizing images according to similarity is useful for serendipitous browsing on images. StumbleUpon [14] allows users to stumble through the Web one (semi-random) page at a time. Bordino et al. [8] investigates the potential of entities in promoting serendipitous search from user-generated content (UGC). Clarke et al. [10] proposes a framework that systematically rewards novelty and diversity in information retrieval evaluation.

Graph Embeddings. Extensive literature has discussed different methods for graph embeddings. Multidimensional scaling [11], Laplacian eigenmaps [6] and IsoMap [24] all have good performance on small graphs, but the time complexity of these algorithms is too high to fit for a large-scale graph. Thus, these methods are not applicable to the network of Wikipedia people, which consists of about 500,000 vertices. Recently, methods are developed for building graph embeddings for large-scale graph. Deepwalk [20] presents an efficient online algorithm for learning representation of vertices in a network. By performing truncated random walk in the graph, Deepwalk treats the walks as sentences in a language model, and utilizes the Skip-gram model [18] on the random walks to train the vertex embeddings.

Traveling Salesman Problem. Traveling salesman problem (TSP) is a widely studied algorithmic problem [15]. Given a weighted graph, the TSP problem seeks a minimum weighted Hamiltonian cycle. The Euclidean TSP is proved to be NP-complete [19], so the main interest is in developing approximation algorithms [22]. Heuristics like 2-opt [12], Lin-Kerninghan [16], all considerably improve the solution quality.

Variants of the standard TSP problem have also drawn researchers attention. The maximum-scatter TSP [4] is perhaps the most relevant work to our definition of k-robust TSP; it maximizes the minimum distance between each vertex and all of its neighbors which are at most m points away in the tour. Another related TSP variant is discussed by [7], which requires constructing a tour that minimizes $\sum_{i=1}^{n} l(i)$. Here, $l(i)$ is the distance traveled before the i-th vertex in the TSP tour. These two problems both take the distance between each vertex and its close neighbors into consideration, which is similar to our notion of k-robust TSP. Although approximation methods provide strict theoretical bound for these problems, their time complexities are at least quadratic in n, which is not feasible for the large-scale graphs we consider here.

3   The DeepBrowse Paradigm

3.1   Formulation

The *DeepBrowse* search paradigm is defined over any universe U of n items by the interaction among two permutations (similarity and significance) by two operations on these permutations (scanning and deepening). *DeepBrowse* is designed to facilitate efficient browsing through very large item universes on small displays capable of representing only a few items simultaneously.

The similarity and significance permutations are defined as follows:

- *Similarity* – This permutation P_1 over U orders items by semantic similarity, so items $x = P_1(i)$ and $y = P_1(i+j)$ should be similar or related, for $1 \le i \le n-1$ and $1 \le j \le w$, where w reflects the size of the display window. Similarity permutations can naturally be constructed given a pairwise-distance function $d(x,y)$, although other approaches can be built on clustering (pairs in the same cluster have smaller distance than intra-cluster pairs), text description similarity, customer co-purchase data, etc. In this paper, we will generally employ a method using L_2 distance on graph embeddings of the underlying universe. We preserve a single precomputed similarity order of items for all users.

- *Significance* – The permutation P_2 over U orders items by popularity, importance, or merit, so for items $x = P_2(i)$ and $y = P_2(i+1)$ then x is deemed more significant than y, for $1 \leq i \leq n-1$. Thus the most significance item is $P_2(1)$ and the least significance $P_2(n)$.

The significance order of items is of great importance for efficient browsing. Item universes often exhibit power-law behavior, where a small fraction of the items command a large fraction of user interest and attention. The significance permutation explicitly encodes this into the search process. Natural measures of significance include sales, views, downloads, likes, frequency of use, critical reviews, and ranking functions built by a combination of such variables. Network centrality algorithms like PageRank provide potential ranking functions on similarity networks even in the absence of such metadata.

We note that these permutations are very small indexing structures, each requiring $n \lg n$ bits for an n-item universe. Each such permutation takes only 200 KB for $n = 100,000$, and 2.5 MB for $n = 1,000,000$, although more space may be necessary to turn this into an efficient index. These permutations are all precomputed for the given universe U.

The *scanning* and *deepening* operations rely on state variables m and p, both of which are bounded between 1 and n:

- The *significance horizon* m defines the size of the currently active universe, namely the set of the m items highest ranked by significance. The deepening operation increases or decreases this value m, further enlarging or restricting the size of this active universe.
- The *position* p defines the point currently of central interest in similarity permutation P_1. The scanning operation increases or decreases p, moving us forward or backward in this similarity permutation.

We presume that the display is capable of displaying a sequence of w elements at any given time. The central item displayed items is $P_1(p)$. The $w/2$ items above (below) it are those items from $P_1(p)$ to $P_1(p+x)$ such that $P_1(p+i)$ is displayed iff the rank of $P_1(p+i)$ in $P_2 \leq m$, and x is as small as possible. The $w/2$ items below $P_1(p)$ are defined analogously.

When m is large relative to n, a large fraction of the items are suitable for display, and it is efficient to simply walk past the items in P_1 of insufficient magnitude. However, for $m \ll n$, it may require a prohibitive amount of skipping to identify the nearest items for display. We recommend keeping two separate data structures for P_1: the first over the full universe and the second over only the top $\sqrt{n}$ items from P_2, and toggle between them for different values for m.

3.2   Implementation

We have implemented *DeepBrowse* as an Android application, so far instantiated over three separate search domains:

- *WikiBrowse* – Here we seek to identify interesting people to read about in Wikipedia. Our dataset here consists of the pages of $n = 496,614$ historical figures in Wikipedia, with a natural network defined by the links between their Wikipedia pages.
- *MovieBrowse* – Here we seek to identify interesting movies to watch from the Internet Movie Database (IMDB). Our dataset here consists of $n = 73,232$ films defining the vertex of a network, with an undirected edge (v_i, v_j) when IMDB recommends movie i (j) to viewers of movie j (i).
- *WordBrowse* – Here we seek to identify interesting words, worth checking the definition of in an on-line dictionary. Our dataset here consists of $n = 100,232$ English words, each associated with its vector representation from the Polyglot multilingual word embeddings [1].

In this section, we will introduce our basic user interface design, instantiated as *WikiBrowse* for a motivating example. Figure 1 presents several screenshots, which we use to illustrate the major components of our design: the magnitude slider and the content scroll.

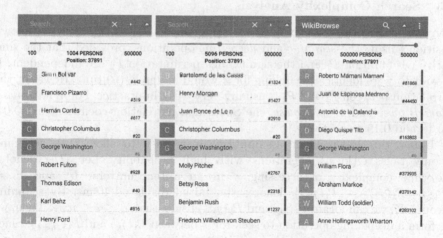

Fig. 1. Screenshots of the *WikiBrowse* interface, supporting searching through all historical figures appearing in Wikipedia. In this series, with *George Washington* selected as the central item, the universe size is expanded from the top 1000 historical figures to 5000, and finally 500,000 items, as we move from left to right. Each progressive shift successively exposes items of more specialized interest in the similarity order, those with closer connections to the central entity.

The *magnitude slider*, on top of our user interface, modulates the effective size of the entity universe, here restricting focus to the top 1,004 people in significance order. The current selection, *George Washington* appears in 37,891th position in the full TSP tour over almost 500,000 historical figures, although all but the top 1,004 are currently hidden from view. The size of the displayed universe can be modulated by sliding left or right.

These figures are accessible by scrolling up and down through the names, ordered by similarity. Thus, Spanish explorers Hernando Cortez and Francisco Pizarro appear as neighbors, as do inventors Robert Fulton and Thomas Edison, and automotive pioneers Karl Benz and Henry Ford. The names, here centered around George Washington, all have significance rank better than 1,004, as reflected by the number to the right of their names and color encoded in the box to the left of their name. Clicking on this box brings up the relevant Wikipedia page. The color strip on right reflects the category/cluster associated with each entity: for George Washington this is *Politicians – U.S.*

These screenshots also illustrate how the neighbors of a given item change as we slide right to increase the allowable significance rank. The neighborhood around George Washington gets progressively filled with lower-wattage figures of closer association, such fellow patriots of the American revolution Molly Pitcher and Betsy Ross. At its most expansive setting (right) Washington's neighbors are very obscure but relevant figures, e.g. William Flora, Abraham Markoe, and William Todd all served as soldiers under Washington during the revolution.

3.3   Search Complexity Analysis

Here we analyze the complexity of accessing a specific item $x = P_1(i)$ of known position i using the *scanning* and *deepening* operations, provided that the similarity permutation P_1 and the significance permutation P_2 are independent. If P_1 and P_2 were identical, searching for x would require $O(n)$ time. However, the correlation between P_1 and P_2 is usually very weak in practice. For *WikiBrowse*, *MovieBrowse* and *WordBrowse*, the Spearman correlation coefficient are 0.04, −0.18 and 0.19 respectively.

Under this independent assumption, it is easy to show that accessing x only takes $O(\log n)$ operations with high probability. Let $f(n)$ denotes the time complexity of accessing x in an n-item universe U, and $Y = \{P_1(j_1), P_1(j_2), \cdots, P_1(j_w)\}$ denotes the initial displayed items. By scanning through Y, we can locate $P_1(j_k)$ and $P_1(j_{k+1})$ such that $j_k \leq i \leq j_{k+1}$. Then, we perform a deepening operation to seek for x between $P_1(j_k)$ and $P_1(j_{k+1})$. Since P_1 and P_2 are independent, there are approximately $\frac{n}{w}$ items between $P_1(j_k)$ and $P_1(j_{k+1})$. This gives the following recursion:

$$f(n) = f(\frac{n}{w}) + O(w) \tag{1}$$

By applying the master theorem we have $f(n) = O(\log n)$ since w is a constant.

4 Constructing the Permutations

Here we describe a general pipeline for generating the significance and similarity permutations for n items, which works well for the three domains we describe: Wikipedia pages, vocabulary words, and movies.

4.1 Ranking Construction

The ranking permutation P_2 is best constructed from domain-specific metadata for the given entity universe. For our three initial search domains:

- *WikiBrowse* – To measure the significance of people in Wikipedia, we use the historical rankings published in [23]. It uses Wikipedia as its main data source, performing a statistical factor analysis of criteria such as PageRank, article readership (hits), length, and editing history – although each of these component variables defines an independent ranking. Details of this analysis, including corrections to reflect historical aging, appear in [23].
- *MovieBrowse* – To measure the significance of movies, we use the number of votes the film received in IMDB to indicate its importance. Alternate permutations might be defined by box-office gross, critical scores, quality ratings, or some combination of these variables.
- *WordBrowse* – To measure the significance of words, we sorted them according to their frequency of appearance in the English edition of Wikipedia. Weighting these by TF-IDF score or Google search frequency would give other reasonable criteria.

4.2 Similarity Permutation Construction

As a general approach to measuring pairwise similarity between arbitrary pairs of items, we start with a partial domain-similarity network of items:

- *WikiBrowse* – Here we start with the network where the vertices are Wikipedia pages, and the edges are links between pages: (a, b) implies that page a refers to page b.
- *MovieBrowse* – Here we start with the network where the vertices are movies from IMDB, and the edges are recommendation links: (a, b) implies that movie b is recommended to people who liked movie a.
- *WordBrowse* – Here we use the Polyglot word embeddings space directly: (a, b) implies that word a has word b as one of its k-nearest neighbors.

To generalize from these partial graphs, we employ the *DeepWalk* [20] technique to construct a high-dimensional vector representation for each item. *DeepWalk* performs random walks over this graph to generate sequences of vertices, which can be interpreted "sentences" over the "vocabulary" of vertices. Using this formalism, Skip-gram embeddings can be constructed to build a vector representation for each vertex.

This high-dimensional representation allows for the fast computing of item similarity. We used the Euclidean distance between the high-dimensional representations of the two vertices, but the cosine distance/vector dot product could alternately be used. Specifically, more details about the similarity measurement between Wikipedia figures can be found in [9].

4.3 k-robust TSPs

The problem of optimizing similarity order has a natural connection to the famous traveling salesman problem (TSP). In particular, if we define the similar order over the vertices of a graph/network, the natural optimization goal is to minimize the total distance between adjacent vertices in the tour, i.e. find the TSP.

However, since phone screen is capable of displaying multiple items at the same time, it is desirable that in our interface mutually similar items appear throughout the same screen, not merely as neighboring elements. Thus the TSP objective function does not result in a browsing order which is optimally visually appealing.

To address this issue, we propose (to our knowledge) the novel combinatorial notion of a k-robust TSP tour. Given a graph $G = (V, E)$, the k-robust cost $C(k, T)$ of tour $T = \{t_1, t_2, \ldots t_{|V|}, t_1\}$ is defined as:

$$C(k, t) = \sum_{i=1}^{|V|} \sum_{j=1}^{k} d(t_i, t_{i+j}) \tag{2}$$

This cost is the weight of the kth power of the tour in a graph theoretic sense. Thus we take into account the cost between a vertex and all the vertices within a window size of k.

Fig. 2. The optimal k-robust TSP tour over all the U.S. state capital cities, for $k = 1$ (left), $k = 2$ (center), and $k = 3$ (right). The tours show greater regional coherence and more zigzags with increasing k.

For $k = 1$, this objective function is the same as a standard TSP tour, but this is not the case for larger values of k. Figure 2 illustrates this by showing the optimal k-robust tours of U.S. state capital cities for $1 \leq k \leq 3$. These tours clearly differ. Further they show greater geographic coherence as k increases.

This is more easily appreciated by considering the points on a $\sqrt{n} \times \sqrt{n}$ unit grid graph, as shown in Fig. 3. The crinkly space-filling Hilbert curve has a bend at every possible location, as opposed the straight-edged snake tour, which

Fig. 3. All Hamiltonian cycles of the unit grid graph are minimum cost TSP tours for the grid. However, the space-filling Hilbert curve defines a better 2-robust tour than the conventional snake order, by a factor of $\sim\sqrt{2}$.

minimizes bends. The cost $d(t_i, t_{i+2})$ is thus $\sqrt{2} = 1.414$ for the Hilbert curve vs. $d(t_i, t_{i+2}) = 2$ for the snake tour.

Approximating Optimal k-robust Tours. That the complexity of finding the optimal k-robust tour is NP-complete follows directly from the hardness of TSP for $k = 1$. This motivates the question of finding provable approximations to the optimal k-robust tour.

In fact, the optimal TSP tour approximates the optimal k-robust tour to within a factor of k:

Theorem 1. *The optimal TSP tour $T = (t_1, \ldots, t_n)$ serves as a $\Theta(k)$ approximation to the optimal k-robust tour, on metric graphs where n is relatively prime to $k!$.*

Proof. First, observe that the k-robust distance function on a metric graph satisfies the triangle inequality. This gives a trivial bound on $d_{i,i+k}$, namely

$$d_{i,i+k} \leq \sum_{j=1}^{k} d(l_i, t_{i+j}). \tag{3}$$

Let OPT be the TSP cost of optimal tour T. Thus $OPT = \sum_{i=1}^{n} d(i, i+1)$, giving a simple upper bound on the k-robust cost of T is $C(k, 2) \times OPT$. This suggests an $O(k^2)$ approximation ratio.

But we can tighten this bound by observing that the edges of the form $(i, i+j)$ for all $1 \leq i \leq n$ form a closed tour visiting all n points, if j is relatively prime to n. The cost of this tour must be $\geq OPT$, or else it would have defined the optimal tour. Thus the sum of the edges in the kth power of T must be at most $k \times OPT$, yielding the result.

4.4 Heuristic Optimization for k-robust TSP

We use a nearest neighbor based heuristic for building the initial TSP tour. We start from a random vertex in the graph, and repeatedly prepend the nearest neighbor of the current TSP tour to it. Formally, if the current partial TSP tour is $T = (v_1, v_2, \cdots, v_n)$, the nearest neighbor u to this TSP tour is defined as:

$$\arg\min_{u} \sum_{i=1}^{k} d(u, v_i), u \in V - T \tag{4}$$

The algorithm runs in $O(D|V|^2)$ (D is the dimensionality of vertex representation), but we can speed it up to $O(mD|V|)$ by sampling only m candidate vertices for consideration for insertion, at some cost in quality. Here, we sample the m nearest neighbors of each vertex as candidates by constructing a ball tree on the graph. The algorithm for querying the m nearest neighbors in a ball tree runs in $O(D|V|log|V|)$, thus the whole algorithm runs in $O(mD|V| + D|V|log|V|)$. In the experiments below, we choose $m = 500$, $D = 64$.

Then, we adapt the widely used 2-opt heuristic to improve the resulting greedy tour. The 2-opt approach repeatedly swaps a pair of vertices (v_i, v_j) in the TSP tour, thus reversing the tour between v_i and v_j. We accept a swap if it reduces the k-robust cost. This process is repeated until the tour is locally optimal. For large-scale graphs, we only perform 2-opt until the tour cost reaches a predefined threshold.

We use two metrics to assess the k-robustness of our tours:

- *Label consistency rate* (LC) – Items in a real-world universe tend to occur into natural domain-specific clusters. A good browsing order should respect these clusters, positioning items in the same cluster close to each other in the tour. The *label consistency rate* measures the ratio of neighbors of v which share the same cluster label. Each vertex v_i in $G = (V, E)$ is assigned a cluster label c_i. The label consistency rate l of tour $T = (v_1, v_2, \cdots, v_{|V|}, v_1)$ is defined as:

$$l = \frac{\sum l_{i,j}}{2k|V|}, 1 \leq i \leq |V|, i - k \leq j \leq i + k \tag{5}$$

where:

$$l_{i,j} = \begin{cases} 0, c_i = c_j \\ 1, c_i \neq c_j \end{cases} \tag{6}$$

- Average k-neighbor distance (D) – Here d_k measures the average distance between each vertex and all its neighbors in a window of size k, so

$$d_k = \frac{1}{|V|} \sum_{i=1}^{|V|} \sum_{j=1}^{k} d(t_i, t_{i+j}) \tag{7}$$

4.5 k-robust TSP: Experimental Results

We evaluated our k-robust TSP heuristic on each of the three datasets associated with our WikiBrowse, MovieBrowse, and WordBrowse apps. The cluster label for each dataset was generated by the K-means++ algorithm [5], with the number of clusters is set to 8 for each dataset.

We employed our heuristic to built k-robust TSP tours for each $1 \leq k \leq 4$, and measure the cost of each of these tours under the distance and label consistency metrics for various k' values from 1 to 16. The heuristic explicitly seeks to minimize the distance, but implicitly seeks to maximize label consistency.

Our results are shown in Table 1. The dominant results for each column are shown in bold, and highlight along the main diagonal for both evaluation metrics.

Table 1. Experimental result for k-robust TSP tour optimization for the Wikipedia dataset. D@k denotes the average k-neighbor distance for the given tour, while LC@k denotes the k-neighbor label consistency rate. Tours designed to optimize robustness for a given $1 \leq k \leq 4$ dominate the performance for the criteria they were designed for under both measures of quality. Smaller values of $D@k$ and larger values of $LC@k$ indicate better performance.

WikiBrowse	D@1	D@2	D@3	D@4	D@8	D@16	LC@1	LC@2	LC@3	LC@4	LC@8	LC@16
1-Robust Tour	**1.62**	1.73	1.81	1.87	2.04	2.25	**93.0**	91.9	91.0	90.2	87.8	84.2
2-Robust Tour	1.67	**1.71**	1.77	1.82	1.96	2.13	92.7	**92.3**	91.7	91.1	89.3	86.6
3-Robust Tour	1.69	1.73	**1.76**	1.81	1.93	2.09	92.5	92.1	**91.8**	91.3	89.9	87.6
4-Robust Tour	1.71	1.75	1.78	**1.80**	**1.91**	**2.05**	92.4	92.0	91.7	**91.4**	**90.2**	**88.2**

In particular, tours optimized for k-robustness tend to perform best for the k they were optimized for, which confirms the soundness of our optimization heuristic. Further, designs for $k = 4$ outperform tours optimized for smaller k when tested for higher levels of robustness, namely $k' = 8$ and $k' = 16$.

5   User Study

Evaluating a user interface for serendipitous browsing is complicated by the nature (or lack) of the task. Success is achieved when the user finds something interesting to them, not a particular item proposed by the investigator.

To provide a baseline for comparison and assess the value of the similarity order, we constructed two versions of the *WikiBrowse* app for both user studies. The standard version uses our 4-robust TSP tour to provide conceptual orders. The alternate version uses alphabetical order instead. we asked our subjects to browse on the app for 5 min, and mark all the items they find interesting enough to read its Wikipedia page. Since by default the app displays the 100 most significant people who are already well-known to most subjects, we asked the participants to mark only those people who are not within the top 100. 16 different subjects (9 males and 7 females) were recruited to participate in this study. All the subjects were students at a local university, all of whom have normal or corrected-to-normal eye vision. Each participant was given a brief questionnaire about their experience at the conclusion of their task. To rule out order effects, half of the participants tried the similarity order version first, while the other half used the alphabetical order version first.

For the serendipity discovery task, the average number of interesting people found within five minutes of browsing is 17.7 ($SD = 7.95$)·in the similarity order app, versus 11.6 in the alphabetical order app ($SD = 6.20$), which means the similarity order app yields a much higher chance of encountering interesting people. Also, subjects who used the alphabetical order app first made significantly more serendipitous discoveries when they switched to the similarity order app (from 11.5 to 20.6 in the mean). In contrast, subjects who used the similarity order app first showed degraded performance with alphabetical order

Table 2. Average scores for the user interface satisfaction questionnaire, using a 7-point Likert scale, with 1 = strongly disagree, 7 = strongly agree. Alpha denotes the average score and standard deviation for the alphabetical order app, while Sim stands for that of the similarity order app.

Question	Alpha	Sim
I notice the similarities between the type of people appearing near each other	1.8 (0.66)	**5.9** (1.17)
The interface is good for discovering interesting Wikipedia people	4.1 (1.27)	**5.4** (1.00)
The interface is good for finding a specific person in Wikipedia	**4.9** (1.39)	4.0 (1.06)
It was easy to learn how to use this interface	5.9 (1.03)	**6.1** (0.70)

(from 14.75 to 11.75 in the mean). Thus, the similarity order app proves superior to the alphabetical order app in encouraging serendipitous discoveries.

Table 2 reports the results of our user satisfaction survey, graded on a 7-point Likert scale. Subjects clearly noticed that the neighboring items were more similar with the k-robust tour than alphabetical order. By contrast, the alphabetical order was deemed better for locating specific figures, exactly the expected result since there is no way to locate specific items by name in the similarity order short of exhaustive search, versus binary search for alphabetical order. Questions addressing user satisfaction generally yielded approval. Subjects generally felt the interface as easy to use and good for discovering interesting people.

6   Conclusion

We have proposed a new design for browsing-oriented user interfaces. Implementing a domain-specific DeepBrowse interface can be reduced to the problem of constructing two permutations, one measuring the similarity between items, and the other the relative significance of each item over the universe. We show how to construct these permutations in a systematic way using graph embeddings and combinatorial optimization. Finally, we report a user study which demonstrates that our interface meets its basic design objectives.

References

1. Al-Rfou, R., Perozzi, B., Skiena, S.: Polyglot: distributed word representations for multilingual NLP. In: CoNLL 2013, p. 183 (2013)
2. André, P., Teevan, J., Dumais, S.T.: From x-rays to silly putty via Uranus: serendipity and its role in web search. In: Proceedings of the SIGCHI Conference on Human Factors in Computing Systems, pp. 2033–2036. ACM (2009)
3. André, P., Teevan, J., Dumais, S.T., et al.: Discovery is never by chance: designing for (un) serendipity. In: Proceedings of the Seventh ACM Conference on Creativity and Cognition, pp. 305–314. ACM (2009)

4. Arkin, E.M., Chiang, Y.J., Mitchell, J.S.B., Skiena, S.S., Yang, T.: On the maximum scatter TSP. SIAM J. Comput. **29**(2), 515–544 (2000)
5. Arthur, D., Vassilvitskii, S.: k-means++: the advantages of careful seeding. In: Proceedings of the Eighteenth Annual ACM-SIAM Symposium on Discrete Algorithms, pp. 1027–1035. Society for Industrial and Applied Mathematics (2007)
6. Belkin, M., Niyogi, P.: Laplacian Eigenmaps and spectral techniques for embedding and clustering. In: NIPS, vol. 14, pp. 585–591 (2001)
7. Blum, A., Chalasani, P., Coppersmith, D., Pulleyblank, B., Raghavan, P., Sudan, M.: The minimum latency problem. In: Proceedings of the Twenty-sixth Annual ACM Symposium on Theory of Computing, pp. 163–171. ACM (1994)
8. Bordino, I., Mejova, Y., Lalmas, M.: Penguins in sweaters, or serendipitous entity search on user-generated content. In: Proceedings of the 22nd ACM International Conference on Information and Knowledge Management, pp. 109–118. ACM (2013)
9. Chen, Y., Perozzi, B., Skiena, S.: Vector-based similarity measurements for historical figures. In: Amato, G., Connor, R., Falchi, F., Gennaro, C. (eds.) SISAP 2015. LNCS, vol. 9371, pp. 179–190. Springer, Cham (2015). doi:10.1007/978-3-319-25087-8_17
10. Clarke, C.L., Kolla, M., Cormack, G.V., Vechtomova, O., Ashkan, A., Büttcher, S., MacKinnon, I.: Novelty and diversity in information retrieval evaluation. In: Proceedings of the 31st Annual International ACM SIGIR Conference on Research and Development in Information Retrieval, pp. 659–666. ACM (2008)
11. Cox, T.F., Cox, M.A.: Multidimensional Scaling. CRC Press, Boca Raton (2000)
12. Croes, G.A.: A method for solving traveling-salesman problems. Oper. Res. **6**(6), 791–812 (1958)
13. De Bruijn, O., Spence, R.: A new framework for theory-based interaction design applied to serendipitous information retrieval. ACM Trans. Comput. Hum. Interact. (TOCHI) **15**(1), 5 (2008)
14. Hauff, C., Houben, G.J.: Serendipitous browsing: stumbling through wikipedia. In: Searching4Fun! Workshop (2012)
15. Hoffman, K.L., Padberg, M., Rinaldi, G.: Traveling salesman problem. In: Encyclopedia of Operations Research and Management Science, pp. 1573–1578. Springer (2013)
16. Lin, S., Kernighan, B.W.: An effective heuristic algorithm for the traveling-salesman problem. Oper. Res. **21**(2), 498–516 (1973)
17. Liu, H., Xie, X., Tang, X., Li, Z.W., Ma, W.Y.: Effective browsing of web image search results. In: Proceedings of the 6th ACM SIGMM International Workshop on Multimedia Information Retrieval, pp. 84–90. ACM (2004)
18. Mikolov, T., Sutskever, I., Chen, K., Corrado, G.S., Dean, J.: Distributed representations of words and phrases and their compositionality. In: Advances in Neural Information Processing Systems, pp. 3111–3119 (2013)
19. Papadimitriou, C.H.: The Euclidean travelling salesman problem is NP-complete. Theoret. Comput. Sci. **4**(3), 237–244 (1977)
20. Perozzi, B., Al-Rfou, R., Skiena, S.: DeepWalk: online learning of social representations. In: Proceedings of the 20th ACM SIGKDD International Conference on Knowledge Discovery and Data Mining, pp. 701–710. ACM (2014)
21. Rodden, K., Basalaj, W., Sinclair, D., Wood, K.: Evaluating a visualisation of image similarity as a tool for image browsing. In: IEEE Symposium on Information Visualization, pp. 36–43. IEEE (1999)
22. Rosenkrantz, D.J., Stearns, R.E., Lewis, P.M.: An analysis of several heuristics for the traveling salesman problem. SIAM J. Comput. **6**(3), 563–581 (1977)

23. Skiena, S.S., Ward, C.B.: Who's Bigger? Where Historical Figures Really Rank. Cambridge University Press, Cambridge (2013)
24. Tenenbaum, J.B., De Silva, V., Langford, J.C.: A global geometric framework for nonlinear dimensionality reduction. Science **290**(5500), 2319–2323 (2000)
25. Toms, E.G.: Serendipitous information retrieval. In: DELOS Workshop: Information Seeking, Searching and Querying in Digital Libraries, Zurich (2000)

Malware Discovery Using Behaviour-Based Exploration of Network Traffic

Jakub Lokoč[1], Tomáš Grošup[1], Přemysl Čech[1], Tomáš Pevný[2,3],
and Tomáš Skopal[1(✉)]

[1] SIRET Research Group, Department of Software Engineering,
Faculty of Mathematics and Physics, Charles University, Prague, Czech Republic
skopal@ksi.mff.cuni.cz
[2] Faculty of Electrical Engineering, Czech Technical University,
Prague, Czech Republic
[3] Cisco Systems, Inc., Cognitive Research Center in Prague,
Prague, Czech Republic
http://www.siret.cz

Abstract. We present a demo of behaviour-based similarity retrieval in network traffic data. The underlying framework is intended to support domain experts searching for network nodes (computers) infected by malicious software, especially in cases when single client-server communication does not have to be sufficient to reliably identify the infection. The focus is on interactive browsing enabling dynamic changes of the retrieval model, which is based on a recently proposed statistical description (fingerprint) of a communication between two network hosts and the bag of features approach. The demo/framework provides unique insight into the data and enables annotation of the data and model modifications during the search for more effective identification of infected hosts.

1 Introduction

Due to the continuous sophistication of computer viruses (malware), the prevalent approach to detect them relying on identification of a specific sequence of bytes in stored files, mail attachments, or packet content became insufficient [12]. Consequently, researchers and vendors of intrusion detection/prevention systems (IDS/IPS) are searching for alternatives, one of them being behaviour-based approach, in which the IDS describes the behaviour of legitimate users/programs and tries to detect anomalies or behaviours specific for infected computers [2,5,10]. For example, infected computers are frequently used to send spam, hence, a computer suddenly sending large number of e-mails is suspicious and should be verified if it is not infected. Similarly, upload of large volume of data might be suspicious for ex-filtering company sensitive data. The behaviour can be described on the level of system calls, TCP/IP traffic, or HTTP(s) network traffic. The last two are particularly interesting, as they enable to monitor large number of computers just from observing their network traffic, without actually deploying any tool on computers themselves or otherwise annoying the user.

© Springer International Publishing AG 2017
C. Beecks et al. (Eds.): SISAP 2017, LNCS 10609, pp. 315–323, 2017.
DOI: 10.1007/978-3-319-68474-1_22

Behaviors can be modeled on different levels of communication, e.g., above we have exemplified high-level behavior requiring the knowledge of the "purpose" of the communication (e.g., send mail). Examples of lower level behaviors specific for particular types of malware are frequencies and sizes of messages or even packets used in communication with its controlling server. It has been shown [7,8] that same malware have a specific footprint that can be detected.

Nevertheless, the behaviour-based detection of infected computers has its weaknesses, which is the high false positive rate. Therefore, any framework that would simplify the verification of infections, finds computers behaving similarly to already known infections, or improves the understanding of how the infected computer behaves is highly needed. The framework proposed here aims to group computers with a similar behaviour and highlight reasons why the computers inside the group behave similarly. By doing so, it increases the understanding of malware's behaviour and helps to tackle the above problems.

At this moment, we would like to clarify the terminology because we consider similarity search associated with content-based retrieval. In intrusion detection community the *content* means usually only the payload of the packet, but not the metadata[1] available without inspecting packet's payload. However, in this paper the *content* refers only to the metadata, i.e., to the set of connection properties that can be aggregated in a d-dimensional Euclidean space, as will be clear from the text. We adopted this terminology to be more aligned with the terminology used in image retrieval by which our methods are inspired.

Although the presented framework is general, we aimed it to use persistent HTTP(s) connections as introduced in [7,8] as the most elementary items in the model of computer's communication. The reason for choosing general features is the availability of the data to us and the fact that they also model HTTPs connection, which are more and more frequently adopted for legitimate but also malicious purposes (using HTTPs protocol effectively rules out the signature-based detection methods). Methods presented in [7,8] are designed to detect/classify single persistent communication by means of so-called *communication fingerprints* aggregating persistent communication properties. We believe that identifying infected computers on the basis of a set of their persistent connections should improve the accuracy, as some malware have persistent connections that are similar to that of harmless traffic individually, but quite unique together.

For the prototype (demo) of our framework described in Sect. 3, we have used data from HTTP traffic which has been partially labelled by Cisco's CTA engine [1]. The labelling of network data is very difficult, as we are never certain, that the computer is actually infected.[2] Thus, the labelling is not perfect, some unlabelled (background) computers can be actually infected, and the computers infected by one type of malware can be infected by other malware simultaneously.

---

[1] E.g., number of sent bytes, length of the connections, IP addresses, port used, etc.
[2] No one wants to work on infected computers and simulated infections work poorly.

2   Retrieval Model

The retrieval model used by our framework employs the bag-of-features (BoF) representation [14], where the descriptors (feature vectors) aggregate fingerprints of communication between clients and servers [7,8]. The descriptors and a parameterizable distance are then used to evaluate k-nearest neighbor queries necessary for our interactive browsing interface. In the following text, the fingerprints and descriptors derived from the fingerprints are described in more detail.

2.1   Communication Fingerprints

For client-server communication, multiple requests are observed; each modeled by a vector $r = (r_{up}, r_{down}, r_{td}, r_{ti})$, where r_{up} is bytes sent, r_{down} is bytes received, r_{td} is duration of the request, and r_{ti} is inter-arrival time between this and previous request from the same client. The distribution of multiple requests r_i between the client and the server can be then aggregated into a four dimensional joint soft histogram h that serves as a fingerprint. In [7,8] authors demonstrated that fingerprints alone can be used to distinguish various application servers. However, unlike traditional web applications and services, the identification of fingerprints representing malicious software is more challenging, because only a group of fingerprints can be discriminative for some malicious softwares.

2.2   Descriptors on Fingerprints

We assume malicious software installed on one client that sends requests to various servers. Thus, the descriptors used by our framework are based on an aggregation of fingerprints for each client. So, each descriptor corresponds to a client that is initially represented as a set of fingerprints. Since each fingerprint can be interpreted as a high-dimensional vector in R^n, where n is the number of bins in the fingerprint (i.e., in the joint histogram), the client can be interpreted as a set of points in R^n. In order to aggregate points for each client, the retrieval model employs the BoF representation that is based on a so-called *codebook* evaluated in a preprocessing phase. The codebook consists of *codewords* that correspond to the centroids obtained by a clustering of a sampled set of fingerprints. In this case, we used k-means clustering to create codebook with $k = 10000$ codewords. Given a representative codebook, the utilized descriptor representing a client is a k-dimensional (sparse) vector, where each bin corresponds to a normalized frequency of client fingerprints assigned to a particular codeword, or, to a set of codewords in the soft-assignment variant of the descriptor. The BoF compacts the descriptors into histograms which simplifies also the group of fingerprints identification problem. Instead of identifying various groups of malicious fingerprints that are common for an infected set of clients, the model restricts the problem to identification of a group of histogram bins, corresponding to malicious codewords. The simplification helps with the efficiency of the retrieval, although, it brings also some ambiguity to the client representation.

2.3   Distance on Descriptors

Given the descriptors and a distance, the framework can evaluate k-nearest neighbour queries [4,15] needed for browsing operations and similarity graph construction (see next section). We consider the weighted cosine measure and weighted euclidean distance that allow the user to specifically control the impact of individual codewords on the entire retrieval process. For example, when users browse and locate a suspicious communication from some set of clients, they can decide to focus just on a specific set of codewords (weights of other codewords are set to zero) to find more clients with this type of communication. Note that we assume potentially inaccurate annotation of the data and so the system heavily relies on the interaction with the users. Such use cases require specific user interfaces and supportive visualizations.

3   Exploration Framework (demo)

The (demo of our) exploration framework[3] consists of two main parts – server side component and user interface. The server side is responsible for data preprocessing (communication fingerprint extraction, codebook generation and client representation) and also retrieval model evaluation (k-nearest neighbor query evaluation, distance matrix for set of clients construction). The user interface consists of a browsing interface and a similarity analysis part. The user interface uses graph-based visualization of data with additional information enabling faster orientation in the explored data and manual distance modifications. Since the server side operations are straightforward routines, in the following sections we focus on the description of the user interface in more detail.

3.1   Browsing Interface

The browsing interface presents clients as graph nodes and their similarity relations as edges, based on the employed BoF representation and a selected distance measure. In order to graphically display similarity relations[4] between actually displayed clients, the *force directed placement* [3] is utilized, where more/less similar clients are attracted/repulsed (see Fig. 1). Since not all clients can be displayed at once, only a selected subset of k clients is initially presented (page zero). The subset can be selected randomly, or, by a structured query over metadata related to clients (e.g., IP mask of involved clients/servers or extreme values in histogram bins). Given a page zero view, users can select a specific client, executing thus a k-nearest neighbor query. The result of the query forms a new view (i.e., graph-based layout), where users can select another client. In such way, users can browse the set of clients in order to find specific clusters of clients. Note that the placement of clients as well as the similarity distance is unsupervised and the annotation of classes is displayed as the color of the nodes.

---

[3] The demo is available as web app at herkules.ms.mff.cuni.cz/NetworkData.

[4] The similarity relations are represented by distance matrix evaluated by the server.

Figure 1 is an example of what users see in the browser view. One can see how clients infected by the CAMZ01 or CADP01[5] (light blue) malwares formed a tight cluster, being very different from the rest. Similarly, clients infected by CADP01 (dark green) are very different from the healthy clients (dark blue, clients not infected by the malware). Notice a healthy client in the middle of clients infected by CADP01 malware, which is with a high probability actually a client infected by a malware, and its blue color is due to the imperfect labelling. Another nice cluster is the dark violet with COPR01 and CAMZ01 malwares.

The browsing interface is inspired by multimedia exploration systems where users can browse image collections using a visual similarity model [6,9,11,13]. However, unlike image exploration systems where the nodes are represented as images with understandable contents for users, the clients represented by frequency histograms of fingerprints are less intuitive (even for experienced users). Therefore, the displayed graph requires additional information that can help users with fluent browsing, without the need to frequently inspect property files of each displayed node. For example, the graph can contain additional nodes and edges representing selected servers connected to the actually displayed clients. Such augmented graph can help users to distinguish similar clients connected to the same servers from similar clients connected to different servers.

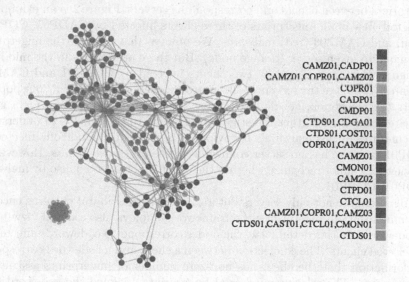

Fig. 1. An example of the browsing interface, where coloured circles represent clients, edges correspond to client similarities and colors represent annotations. (Color figure online)

---

[5] The name of malware families are as reported by the Cisco CTA engine.

3.2 Similarity Analysis

Once users locate a suspicious group of clients using the browsing interface, further investigation is needed to confirm/reject the suspicion. For this purposes, the similarity analysis part focuses on a selected set of clients (by right click) and shows more detailed information about the clients. More specifically, the actual version of the framework supports two different detail views based on either servers or more general codewords corresponding to similar behavioural patterns. In the detail view, all involved servers/codewords and all their corresponding connections to the investigated clients are displayed to uncover all the relations hidden by the retrieval model used by the browsing interface. The detailed view enables users to investigate, analyze, better understand and even select significant servers/codewords and directly modify the retrieval model used by the browsing interface. The modified retrieval model can be used in the browsing interface to locate clients with similar selected properties. The modification of the retrieval model is intended as just temporal or user profile-based because permanent modifications (e.g., permanent zero weights for some codewords) can reduce the ability of the model to identify novel malwares.

The detail view based on clients and servers contains fixed nodes as selected clients and additional nodes as servers. The connections depict communication fingerprints between clients and corresponding servers. Figure 2 is an example of the detail view of all fingerprints of three clients infected by CADP01, COPR01 (green) and CAMZ03 (red) malwares. We observe that most of the fingerprints are unique to given server (orange node). But there are servers in the middle of the figure which are visited by two clients infected by COPR01 and CAMZ03 malwares. These are the servers of the interest, because after filtering out popular servers (e.g. facebook, google, etc.) those are likely to be specific for a given malware family and further investigation of traffic to them helps to understand particularities of this malware's behaviour. The client in the bottom infected by CADP01 malware has no server connected to the other two clients. However, in this view we cannot recognize whether the client is not infected also by malwares CADP01, COPR01.

To investigate not only servers but also similar behavioural patterns encoded by communication fingerprints, the framework employs also a detail view based on codewords. In this detail view, the nodes correspond to codewords affected by the selected clients. The connection between a client and a codeword corresponds to information that the client has non-zero number of fingerprints assigned to the codeword. The relative number of fingerprints assigned the codeword from all selected clients is visualized as an intensity of the codeword nodes. Figure 3 is an example of the detail view created for the same clients as used in Fig. 2. We observe that most of the fingerprints are unique to a given codeword. However, now the client in the bottom infected by CADP01 malware has several codewords connected to the other two clients (see three clusters marked by blue circle). This means that the client has similar behavioural patterns as two infected clients visiting different servers. Such codewords require further investigation in order to identify potential additional infection not recognized by Cisco's CTA engine.

Fig. 2. An example of the client-server detail visualization. (Color figure online)

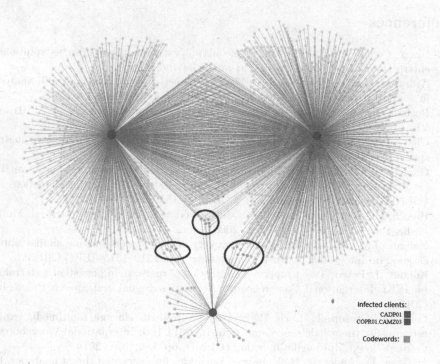

Fig. 3. An example of the client-codeword detail visualization. (Color figure online)

After inspecting the fingerprints and their relations, the users can annotate codewords, or even set a codeword weight (i.e., priority). Using weights, users can focus only on codewords corresponding to malicious fingerprints. This affects the retrieval model used in the browsing interface, so once users switch back to the browsing interface, new infected clients can be detected for another inspection.

4   Conclusion and Future Work

We presented (a demo of) a framework focusing on behaviour-based similarity retrieval in network traffic data. Such framework can be used as a support tool by domain experts trying to identify network nodes with malicious software. We presented the basic interfaces and utilized a retrieval model based on a fingerprint representation of a client-server communication and the bag of features approach. In the future, we plan to investigate options for automatic training of the retrieval model using available annotations, automatic analysis of missing annotations and visualization of such annotations in the browsing interface. We also plan to use the framework to identify malicious edges (i.e., fingerprints).

Acknowledgements. This research has been supported by Czech Science Foundation (GAČR) project 15-08916S and Charles University grant (GAUK) 201515.

References

1. Cisco Cognitive Threat Analytics, http://www.cisco.com/c/en/us/solutions/enterprise-networks/cognitive-threat-analytics/index.html
2. Arora, A., Garg, S., Peddoju, S.K.: Malware detection using network traffic analysis in android based mobile devices. In: NGMAST, pp. 66–71. IEEE (2014)
3. Bostock, M., Ogievetsky, V., Heer, J.: D3 data-driven documents. IEEE Trans. Vis. Comput. Graphics **17**(12), 2301–2309 (2011)
4. Chávez, E., Navarro, G., Baeza-Yates, R., Marroquín, J.L.: Searching in metric spaces. ACM Comput. Surv. **33**(3), 273–321 (2001)
5. Guofei, G., Perdisci, R., Zhang, J., Lee, W., et al.: Botmfiner: clustering analysis of network traffic for protocol-and structure-independent botnet detection. In: USENIX Security Symposium, vol. 5, pp. 139–154 (2008)
6. Heesch, D.: A survey of browsing models for content based image retrieval. Multimedia Tools Appl. **40**(2), 261–284 (2008)
7. Kohout, J., Pevny, T.: Automatic discovery of web servers hosting similar applications. In: Integrated Network Management, pp. 1310–1315. IEEE (2015)
8. Kohout, J., Pevny, T.: Unsupervised detection of malware in persistent web traffic. In: IEEE International Conference on Accoustics, Signal and Speech Processing (2015)
9. Lokoč, J., Grošup, T., Čech, P., Skopal, T.: Towards efficient multimedia exploration using the metric space approach. In: 2014 12th International Workshop on Content-Based Multimedia Indexing (CBMI), pp. 1–4, June 2014
10. McGrew, D., Anderson, B.: Enhanced telemetry for encrypted threat analytics. In: ICNP, pp. 1–6, November 2016

11. Nguyen, G.P., Worring, M.: Interactive access to large image collections using similarity-based visualization. Visual Lang. Comput. **19**(2), 203–224 (2008)
12. Roesch, M.: Snort - lightweight intrusion detection for networks. In: USENIX Conference on System Administration, LISA 1999, pp. 229–238 (1999)
13. Schaefer, G.: A next generation browsing environment for large image repositories. Multimedia Tools Appl. **47**, 105–120 (2010)
14. Sivic, J., Zisserman, A.: Video google: a text retrieval approach to object matching in videos. In: IEEE International Conference on Computer Vision, vol. 2 (2003)
15. Zezula, P., Amato, G., Dohnal, V., Batko, M.: Similarity Search: The Metric Space Approach. Springer, US (2005)

Visual Analytics and Similarity Search: Concepts and Challenges for Effective Retrieval Considering Users, Tasks, and Data

Daniel Seebacher[1]([✉]), Johannes Häußler[1], Manuel Stein[1], Halldor Janetzko[2], Tobias Schreck[3], and Daniel A. Keim[1]

[1] University of Konstanz, Konstanz, Germany
daniel.seebacher@uni-konstanz.de,
{seebacher,haubler,stein,keim}@dbvis.inf.uni-konstanz.de
[2] University of Zurich, Zürich, Switzerland
halldor.janetzko@geo.uzh.ch
[3] TU Graz, Graz, Austria
tobias.schreck@cgv.tugraz.at

Abstract. A major challenge of the contemporary information age is the overwhelming and increasing data amount, especially when looking for specific information. Searching for relevant information is no longer manually possible, but has to rely on automatic methods, specifically, similarity search. From a formal perspective, similarity search can be seen as the problem of finding entities, which are considered to be similar to a query with respect to certain describing features. The question which features or which weighted combination of features to use for a given query creates a need for semi-automatic methods to address the needs of diverse users. Furthermore, the quality of the results of a similarity search is more than effectiveness, measured by precision and recall. The user ideally needs to trust the results and understand how they were computed. We propose to apply Visual Analytics methodologies, for synergistic cooperation of user and algorithms, to integrate three key dimensions of similarity search: users, tasks, and data for effective search. However, there exists a gap in knowledge how user, task as well as the available data influence each other and the similarity search. In this concept paper, we envision how Visual Analytics can be used to tackle current challenges of similarity search.

Keywords: Similarity search · Recommender systems · Visual analytics

1 Introduction

Humans assess two objects as being similar if they are considered to be comparable with respect to certain properties. These properties can be either physical properties (e.g., dimensions, light reflectance, material, etc.) or semantic meta information (e.g., armchairs and chairs are functionally similar). For example, two books can be judged similar if they share similar content, or two movies

C. Beecks et al. (Eds.): SISAP 2017, LNCS 10609, pp. 324–332, 2017.
DOI: 10.1007/978-3-319-68474-1_23

if they have the same combination of genres. At the same time, two books can be similar because of equally colored covers and movies might be considered similar because of common actors. The notion of similarity is compound of different factors, including users, preferences, different options to define and measure properties, and also uncertainty. Besides the goal of searching for similar items, there are several other tasks that a user might want to accomplish. According to the *exploration-search axis*, introduced by Zahálka and Worring [28] in the field of Multimedia Analytics, there are two extreme values, namely Exploration and Search. In between those tasks, there are a variety of other tasks such as *Browsing, Summarization, and Ranking, etc.* which have to be considered as well when it comes to effective retrieval since an analytical work-flow may not only consist of similarity-search.

Digital data storage and processing enabled the research of automated similarity queries and founded the scientific area of information retrieval. A manual search for similar objects might be appropriate for small collections. However, with the advent of computers, the size of collections typically found is increasing rapidly. Prominent examples are *Spotify* with over 30 million songs and *Amazon* with over 200 million products. These volumes of information clarify the need for (semi-)automatic methods to retrieve and rank data items. A first mentioning of automatic retrieval of similar objects by Holmstrom [10] dates back as far as 1948. However, due to increasingly more complex objects, larger collections, and new user demands, automated similarity assessment is still an active research field. The existence of challenges, such as the Netflix Prize and conferences, such as the ACM RecSys, illustrate the practical importance and relevance of working on data- and user-adaptive similarity search. Among other things, the interaction with Recommender Systems (RSs) and helping users understand how their actions influence the recommendations are open challenges in the field of RSs [20]. The effectiveness of a RSs is dependent on more factors than just the quality of the similarity assessment method alone [26]. Similarity search should create trust, should be comprehensible, and transparent. In this paper, we identify interdepending factors influencing similarity search. We highlight arising research aspects and envision a Visual Analytics approach solving the introduced challenges.

2   Foundations

Many influencing factors need to be considered when engaging with the subject of similarity search. We categorize the influencing factors as *building blocks* of the respective *pillars* of similarity search. An overview about the identified pillars can be seen in Fig. 1. In the following paragraphs, we describe and explain the three pillars *data*, *task*, and *user* in detail.

Data. Users need data to perform their retrieval tasks. Therefore, it is essential to pay particular attention when working with data, since errors made in early steps, for example during preprocessing, persist within the system and will negatively impact the quality of the results. In the case of IR or RSs, data might

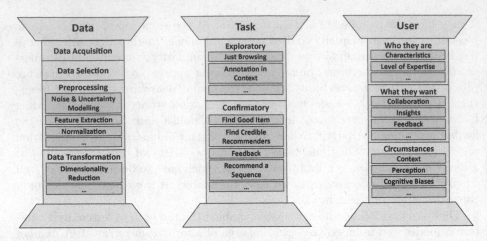

Fig. 1. An overview of the three pillars of similarity, *Data*, *Task*, and *User*. Each pillar consists of multiple building blocks, which in turn can have more building blocks.

already be available beforehand, e.g., provided by a database with records of some kind of media (music, videos, products, images, etc.). Metadata describing the raw data, such as annotations, tags or derived data is usually available as well. Finally, there is also user-generated data. Bobadilla et al. [5] describe the two ways of user data acquisition, in the case of IR and RS. Data can either be acquired explicitly, e.g., through user ratings, comments, etc., or implicitly, e.g., by the number of times a song was played. However, it is crucial to consider the noise or uncertainty in the data, especially for RSs, since there is not only natural, but also malicious noise [17]. Since RSs use real-world data, often provided by users, preprocessing is vital to enable similarity search providing relevant results. In data preprocessing, relevant descriptive features are derived and computed. These features should describe the represented objects very accurately and simultaneously enable similarity assessments. Amatriain et al. [3] give an overview of preprocessing methods in the context of RSs. The choice of the right similarity measure, for example, should be appropriate for the underlying data, even when already dealing with abstract representation of objects, such as feature vectors. The computation of feature vectors and the selection of similarity (distance) measures is highly domain dependent. A good example for the domain dependency are *tf-idf* vectors for the retrieval of text documents, where cosine similarity is the appropriate choice, since it ignores the length of the text documents and finds items of similar content. However, the similarity measure of genetic code – represented by letters and being textual data from an abstract point of view – employs other algorithms such as, Levenshtein, Needleman-Wunsch [16] or Smith-Waterman [25]. This holds also true for other types of data. Lew et al. [14] provide an overview of such data types.

Task. Tasks in similarity search have different backgrounds and goals, e.g., to explore data and formulate a hypothesis or to confirm/reject an existing hypothesis. Herlocker et al. [9] define eleven common tasks in which RSs are beneficial and helpful for users. We use this tasks exemplary to illustrate how the user's task influences the similarity search. The core recommendation task is to *Find Good Items* with respect to a specific information need. Early RSs [24] implemented this task by providing the user with a sorted list of results. For this kind of task, a range or k-nearest neighbor (k-NN) query using a classical similarity method is sufficient. However, depending on the definition of *good items*, adjustments of the similarity method are needed. For instance, the task *find good hairdressers* should not only consider the rating, but also the location, price, user preference, etc. Another important task is to *Find All Good Items*. In this case, neither the range nor k is known, hence a simple range or k-NN query is not sufficient for this kind of task. A simplified assumption would be, that all good items belong to the same cluster. Then instead of searching for the items themselves, one could search for the nearest cluster prototype. This task is especially important for lawyers or patent examiners, where missing one item can have a great impact. A third task is *Just Browsing*. Here, the user wants to explore the item or data space without a clear objective or information need. The similarity search should provide users with new items that might be of interest.

User. The user, applying similarity search to fulfill tasks on data, is judging the success or failure of the similarity-based application. User requirements are often complex and not always free of ambiguity. Users need to be considered not only by their ways of interaction but also by their characteristics and the search context. Users can have different levels of expertise in one or another field [18]. Behavioral scientists, for example, search for movement patterns differently than sport scientists might. Humans are intrigued by their own perspectives and insights. People are, consequently, often working collaboratively to satisfy their information need. Many more important characteristics for users exist and influence the perceived similarity such as a user's current location or time of day [1]. Additionally, not only the context, but also the perception and cognitive biases of the user have an influence during and after searching [13]. Currently, users are integrated into the process of similarity search by giving explicit feedback, for instance, by rating an item, or implicitly by analyzing the items a user previously viewed or for how long she or he viewed these items. Also, in E-commerce, metadata available on the users are exploited to learn and predict user preferences. Therefore, for a successful similarity search, it is key to understand who the users are, what they want to achieve, and under which circumstances they work with the similarity search.

3 Research Aspects

Although similarity search is a well researched and discussed area, there are many open challenges to tackle. Research aspects are categorized with respect to the previously introduced facets of similarity search. This Section is not intended to be a complete and exhaustive survey of the state-of-the-art, since this would exceed the scope of this paper. We rather envision and describe areas in which future research has to cope with open questions.

Data Accessibility and Usability. One constantly increasing main challenge for nowadays similarity search is that the employed data are often not accessible and usable enough. The *curse of dimensionality* [4], for example, falsifies the assumption that as more describing features are used, the similarity assessment will improve. Instead, severe effects on the similarity search have to be expected with an increasing number of dimensions. A dataset containing 15 dimensions, for example, can have a distance between the nearest neighbor close to the distance of the farthest neighbor. Although state-of-the-art similarity methods [11,15] have shown that similarity search in high-dimensional data is possible to a certain extent, the selection of proper discriminative features and a semantic meaningful combination is crucial and complicated. Another challenge dealing with data is the preservation of privacy as stated in [8,20]. Besides ethical and legal issues it is important to ensure that the intersection of query results of different data sources does not reveal more information than intended.

Models for Data and Context. On top of the data accessibility exists a noticeable lack when it comes to appropriate data and context models. This lack of data and context models is immediately affecting all of the introduced pillars in Sect. 2. For example, automatic methods cannot detect, handle, and remove all noise and uncertainty in the available data of RSs [17]. This can, for example, be illustrated by restaurant recommendations, assuming we have restaurants with noisy data of natural or malicious origin. Should a restaurant with a noisy rating still be considered as a *good item*, if it has otherwise positive attributes such as price and location? Furthermore, offering context-depending results of a similarity search helps in recommending *good items* [1,8].

Visualization and Interaction. Eventually, the easiest way to provide a user with relevant items is to purely rely on the data and a static similarity measure. However, incorporating users by capturing their feedback and allowing them to modify the query and/or the similarity measure already improves the performance [23]. Nevertheless, visualizing an abstract similarity space and explain why results were found or not found is highly application and user dependent. Additionally, a lack of traceability combined with missing transparency [20] may lead to situations in which users are unaware where their insights came from and how the interactions with the system generated the results. As a consequence, the task might change during the process of analysis. *YouTube*, to name one famous example from one of the largest RSs in the world, uses Deep Neural Networks [6] for its recommendations. The shift towards a deep learning approach comes at the loss of transparency. For a given recommendation it is vague, how the data was weighted and which factors influenced the result. However, there are initial works proposing visualizations for neural networks that might help to overcome this problem. For instance, from Rauber et al. [19], which enables the inspection of relationships between neurons and classes.

4   Methodology

In Sect. 3 we described how the multitude facets of similarity search are influenced and influence each other. Understanding how these facets interdepend is crucial in order to improve the design of IR and RSs. In the following, we envision how such a system could be designed to support similarity search in the best possible way.

As we need various opportunities to reflect expert knowledge in the analysis process, we propose to follow a Visual Analytics process, as described by Keim et al. [12]. In Visual Analytics, heterogeneous data sources are processed and used to generate visualizations and models, thus enabling users to apply visual as well as automatic analysis methods. By interacting with the visualizations users are able to share background knowledge and context information via interactions. This information is then used to update the underlying model, which creates or updates models and visualizations. Following such a tight coupling of user and system will result in a continuous and mutual discourse, which will lead to higher confidence and better results.

A high-level description of the human and computer processes in Visual Analytics is given by Sacha et al. [22]. It helps to facilitate an understanding of the individual components and concepts of the Visual Analytics process and their interactions. Their *Knowledge Generation Model for Visual Analytics* can serve as a guideline on how to design new Visual Analytics systems or how to evaluate existing ones. One recent example where this is illustrated is the *Note Taking Environment* of Sacha et al. [21], which design is based on the knowledge generation model. Additionally, they show how Visual Analytics systems can be evaluated by measuring and investigating the trust of the user in the system.

In order to show how applying the Visual Analytics process can help tackle the open research aspects presented in Sect. 3, we incorporated them at the corresponding component in the Visual Analytics process, as illustrated in Fig. 2. With the iterative and interconnected model for Visual Analytics, we are able to reflect the interdependent nature. This enables us to develop an understanding of the interdependencies of the different facets of similarity search and how Visual Analytics can help to tackle the open research aspects. The rationales behind this integration are outlined as follows.

Both the Visual Analytics model as well as our proposed pillars of similarity search have a data component, which serves as a base for the automatic analysis via data mining or similarity methods. With respect to the previously stated research aspects, data accessibility and usability questions are faced here. The transformation of the original raw data into meaningful and descriptive features is key for a successful similarity search. This transformation step is often also iterative and influenced by the curse of dimensionality, especially in the design phase of a similarity search Visual Analytics system.

Models for data and context influencing the similarity search as described as the second research aspect are key to understand how users employ RSs. Another important aspect which still needs more attention in the field of RSs are visualizations of both, the results and the underlying model [8]. It is not

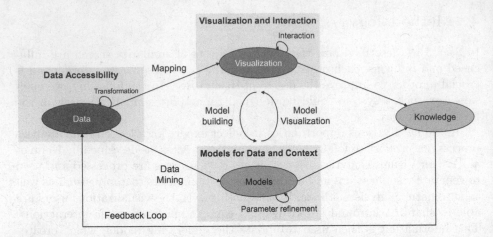

Fig. 2. Main research questions in similarity search integrated in the Visual Analytics model of Keim et al. [12]. The iterative and interconnected model of Visual Analytics reflects the interdependencies of our described research challenges.

only important to identify, how a user can interact with these visualization [2], but also what the rationales behind these interactions are [27]. By capturing these interactions, as well as contextual information, conclusions about the goals of the users can be drawn. This enables us to train the underlying model of the similarity search according to their expectations. Consequently, by Visual Analytics we are able to enrich the similarity search by "Insight Provenance" and traceability of the results.

As the key ingredients of Visual Analytics are visualization and interaction, the overlap to the third research aspect is granted per se. Visualization and user interaction can be used to utilize the user's domain knowledge [7]. In a two-dimensional spatial visualization of documents, documents are distributed by their similarity to each other. By spatially rearranging documents, for example by drag-and-drop, users can communicate to the system, which documents they find similar, which in turn trains the similarity model according to their feedback. As a consequence, the user's domain knowledge is captured, interpreted, and applied to the whole dataset.

5   Conclusion

We believe following a Visual Analytics approach will improve similarity search applications, in particular IR and RSs. With the user-centered focus of Visual Analytics combined with data analytics, information visualization, and interaction, query results can be made transparent and interpretable. Finally, transparent query results will increase users' trust in the similarity search results. However, as a direct consequence of applying the Visual Analytics process on similarity search, new challenges are emerging. There is a need for an increased understanding of the relationship of the components in the process and the influences of the various parameters. This can lead to new insights which help to identify errors, improve robustness, and increase quality of, as well as trust in similarity search.

References

1. Adomavicius, G., Tuzhilin, A.: Context-aware recommender systems. In: Ricci, F., Rokach, L., Shapira, B. (eds.) Recommender Systems Handbook, pp. 191–226. Springer, Boston, MA (2015). doi:10.1007/978-1-4899-7637-6_6
2. Amar, R., Eagan, J., Stasko, J.: Low-level components of analytic activity in information visualization. In: 2005 IEEE Symposium on Information Visualization, INFOVIS 2005, pp. 111–117. IEEE (2005)
3. Amatriain, X., Jaimes, A., Oliver, N., and Pujol, J. M. Data mining methods for recommender systems. In: Ricci, F., Rokach, L., Shapira, B., Kantor, P. (eds.) Recommender Systems Handbook, pp. 39–71. Springer, Boston (2011). doi:10.1007/978-0-387-85820-3_2
4. Beyer, K., Goldstein, J., Ramakrishnan, R., Shaft, U.: When is "Nearest Neighbor" meaningful? In: Beeri, C., Buneman, P. (eds.) ICDT 1999. LNCS, vol. 1540, pp. 217–235. Springer, Heidelberg (1999). doi:10.1007/3-540-49257-7_15
5. Bobadilla, J., Ortega, F., Hernando, A., Gutiérrez, A.: Recommender systems survey. Knowl.-Based Syst. **46**, 109–132 (2013)
6. Covington, P., Adams, J., Sargin, E.: Deep neural networks for YouTube recommendations. In: Proceedings of the 10th ACM Conference on Recommender Systems, pp. 191–198. ACM (2016)
7. Endert, A., Fox, S., Maiti, D., North, C.: The semantics of clustering: analysis of user-generated spatializations of text documents. In: Proceedings of the International Working Conference on Advanced Visual Interfaces, pp. 555–562. ACM (2012)
8. He, C., Parra, D., Verbert, K.: Interactive recommender systems: a survey of the state of the art and future research challenges and opportunities. Expert Syst. Appl. **56**, 9–27 (2016)
9. Herlocker, J.L., Konstan, J.A., Terveen, L.G., Riedl, J.T.: Evaluating collaborative filtering recommender systems. ACM Trans. Inf. Syst. (TOIS) **22**(1), 5–53 (2004)
10. Holmstrom, J.E.: Section III. Opening plenary session. In: The Royal Society Scientific Information Conference. Royal Society (1948)
11. Houle, M.E., Sakuma, J.: Fast approximate similarity search in extremely high-dimensional data sets. In: 2005 Proceedings of the 21st International Conference on Data Engineering, ICDE 2005, pp. 619–630. IEEE (2005)
12. Keim, D., Kohlhammer, J., Ellis, G., Mansmann, F.: Mastering the information age solving problems with visual analytics. Eurographics Association (2010)
13. Lau, A.Y., Coiera, E.W.: Do people experience cognitive biases while searching for information? J. Am. Med. Inform. Assoc. **14**(5), 599–608 (2007)
14. Lew, M.S., Sebe, N., Djeraba, C., Jain, R.: Content-based multimedia information retrieval: state of the art and challenges. ACM Trans. Multimed. Comput. Commun. Appl. (TOMM) **2**(1), 1–19 (2006)
15. Liu, K., Bellet, A., Sha, F.: Similarity learning for high-dimensional sparse data. In: AISTATS (2015)

16. Needleman, S.B., Wunsch, C.D.: A general method applicable to the search for similarities in the amino acid sequence of two proteins. J. Mol. Biol. **48**(3), 443–453 (1970)
17. O'Mahony, M.P., Hurley, N.J., Silvestre, G.: Detecting noise in recommender system databases. In: Proceedings of the 11th International Conference on Intelligent User Interfaces, pp. 109–115. ACM (2006)
18. Picault, J., Ribiere, M., Bonnefoy, D., Mercer, K.: How to get the recommender out of the lab? In: Ricci, F., Rokach, L., Shapira, B., Kantor, P. (eds.) Recommender Systems Handbook, pp. 333–365. Springer, Boston (2011). doi:10.1007/978-0-387-85820-3_10
19. Rauber, P.E., Fadel, S.G., Falcao, A.X., Telea, A.C.: Visualizing the hidden activity of artificial neural networks. IEEE Trans. Visual Comput. Graphics **23**(1), 101–110 (2017)
20. Ricci, F., Rokach, L., Shapira, B.: Recommender systems: introduction and challenges. In: Ricci, F., Rokach, L., Shapira, B. (eds.) Recommender Systems Handbook, pp. 1–34. Springer, Boston, MA (2015). doi:10.1007/978-1-4899-7637-6_1
21. Sacha, D., Boesecke, I., Fuchs, J., Keim, D.A.: Analytic behavior and trust building in visual analytics. In: Proceedings of the Eurographics/IEEE VGTC Conference on Visualization: Short Papers, pp. 143–147. Eurographics Association (2016)
22. Sacha, D., Stoffel, A., Stoffel, F., Kwon, B.C., Ellis, G., Keim, D.A.: Knowledge generation model for visual analytics. IEEE Trans. Visual Comput. Graphics **20**(12), 1604–1613 (2014)
23. Seebacher, D., Stein, M., Janetzko, H., Keim, D.A., Retrieval, P.: A multi-modal visual analytics approach. In: Andrienko, N., Sedlmair, M., (eds.) EuroVis Workshop on Visual Analytics (EuroVA), pp. 013–017. The Eurographics Association (2016)
24. Shardanand, U., Maes, P.: Social information filtering: algorithms for automating word of mouth. In: Proceedings of the SIGCHI Conference on Human Factors in Computing Systems, pp. 210–217. ACM Press/Addison-Wesley Publishing Co. (1995)
25. Smith, T.F., Waterman, M.S.: Identification of common molecular subsequences. J. Mol. Biol. **147**(1), 195–197 (1981)
26. Swearingen, K., Sinha, R.: Beyond algorithms: an HCI perspective on recommender systems. In: ACM SIGIR 2001 Workshop on Recommender Systems, vol. 13, pp. 1–11. Citeseer (2001)
27. Yi, J.S., Ah Kang, Y., Stasko, J.: Toward a deeper understanding of the role of interaction in information visualization. IEEE Trans. Visual Comput. Graphics **13**(6), 1224–1231 (2007)
28. Zahálka, J., Worring, M.: Towards interactive, intelligent, and integrated multimedia analytics. In: 2014 IEEE Conference on Visual Analytics Science and Technology (VAST), pp. 3–12. IEEE (2014)

Author Index

Printed in the United States
By Bookmasters